Principle and Practice in Enhancing Nutrient Use Efficiency

Zhou Jianmin, Du Changwen, Wang Huoyan

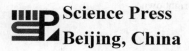
Science Press
Beijing, China

Zhou Jianmin, Du Changwen, Wang Huoyan
Institute of Soil Science, Chinese Academy of Sciences
Nanjing, Jiangsu, China
chwdu@issas.ac.cn
Responsible Editors: Wei Changlong, Zheng Xin

Copyright © 2015 by Science Press
Published by Science Press
16 Donghuangchenggen North Street
Beijing 100717, P. R. China

Printed in Beijing

All rights reserved. No part of this publication may be reproduced, stored in a retrieval system, or transmitted in any form or by any means, electronic, mechanical, photocopying, recording or otherwise, without the prior written permission of the copyright owner.

ISBN 978-7-03-045589-5

Zhou Jianmin

He is a professor at Institute of Soil Science, Chinese Academy of Sciences. He earned his Ph. D degree from the University of Saskatchewan. He is the director of Academic Committee of the State Key Laboratory of Soil and Sustainable Agriculture. His research interests are development of functional fertilizers and the interaction of nutrients in soil. He is the recipient of the award of Outstanding Scientist from Soil Society of South-East Asia. He currently serves as chairman emeritus of Soil Society of China. He has published more than 200 research articles, patents, proceedings and book chapters.

Du Changwen

He is a professor at Institute of Soil Science, Chinese Academy of Sciences. He earned his Ph. D degree from the University of Chinese Academy of Sciences. He is the associate director of the State Engineering Laboratory of Soil Nutrient Management. His research interests are proximal soil sensing and diagnosis of nutrient status in plant and soil. He is the recipient of the award of Scientific Contribution from Chinese Academy of Sciences. He currently serves as associate chairman of New Type Fertilizer Committee of Plant Nutrition and Fertilizer Society of China. He has published more than 180 research articles, patents, proceedings and book chapters.

Wang Huoyan

He is a professor at Institute of Soil Science, Chinese Academy of Sciences. He earned his Ph. D degree from the Huazhong Agricultural University. He is the dean of Plant Nutrition and Fertilizer Science Division in Institute of Soil Science, Chinese Academy of Sciences. His research interests are soil fertility evaluation, nutrients transformation and movement, and fertilizer application techniques. He is the recipient of the award of Potassium Application from International Potassium Institute. He currently serves as council member of Plant Nutrition and Fertilizer Society of China. He has published more than 200 research articles.

For the 20th Anniversary of
Group of Plant Nutrition and Quality Control
Institute of Soil Science, Chinese Academy of Sciences

Preface

Due to the hard work and the exploration innovation of members from the group headed by Professor Zhou Jianmin, rich fruits in research were harvested in recent decade, and these fruits fused the love, feeling, intelligence and diligence of the young scientists in the group during this period. As an old man over eighty of age, when thinking high of the excellent research work with great delight, many thoughts in the past years involuntarily emerged in my mind.

The studies about soil potassium including potassium content, potassium form and potassium availability started to be conducted since 1950s, and then the potassium availability and crop requirement were uncovered and confirmed through a large number of field experiments, technology demonstration and extension as well as long term cooperation with numerous international organizations, which strongly promoted the research and application of potassium in China, especially since 1970s, which then played a vital role in agricultural production in China. However, more topics should be explored in the new era featured with precision agriculture.

Currently many excellent young scientists emerge, and time makes it inevitable that the new indeed will replace the old. Professor Zhou Jianm returned China from Canada with the deep love of our country after 8 years' hard study and work, and after me he led the research group of plant nutrition and quality control as principal investigator; thereafter, he obtained a number of administrative posts and profession affiliations due to his excellent academic performance and terrific leadership as well as estimable catholicity and love. Though he was busy with a myriad of administrative and social affairs, more than 30 MSc or Ph.D students were trained and graduated under his intelligent supervision, and most of them became the young talents in the field of research and education. Centered on the study of efficient use of nutrient, Professor Zhou always holds the keen interests, deeply grasps the discipline frontier, and continuously forges ahead, a large amount of valuable work has been conducted, and this book is the connection and summary of the main results and findings.

Knowledge has no limit, and the study about efficient use of nutrients is a long-term and arduous task. I sincerely hope that our young scientists can carry on the traditions in the group, keep pace with the times, and open up a new way in future.

Professor Xie Jianchang
July 1, 2015, Nanjing

Introduction

China leads the world in the production and consumption of fertilizers, with an annual total production capacity of around sixty million tons. The production and application of fertilizers tremendously propel the food production of our country, and provide safeguard for the national food supply. However, excessive application of fertilizers is being accompanied by increasingly serious issues, including nutrient loss and low use efficiency, waste of resources and energy, and ecological and environmental deterioration, all of which have been plaguing our country in the agricultural and industrial development, as well as the well-being of the public. To address these issues, a goal of zero growth in fertilizer application and increasing efficiency of fertilizer utilization has been put forward. The core of the goal is the utilization efficiency. Therefore, highly efficient utilization of nutrients is not only an important scientific and technological challenge but also a significant national demand by intensive and high-efficiency agricultural systems.

High-efficiency utilization of nutrients is a highly demanding and complicated project, involving numerous factors and thus posing many practical challenges. Overall, four major factors, involving fertilizer, soil, crop and environment, deserve our serious attention. The four factors are further reduced to the type, nutrient ratio and amount of fertilizers, the nutrient preservation and supply capacity of soils, the nutrient absorption and utilization capacity of crops (cultivation measures, although important and effective, are not covered in this book due to beyond the scope of plant nutrient subject), and nutrient environmental circulation processes. In practice, the four factors are usually interwoven together and play a synthetic role in nutrient utilization. Hence, the interaction of these factors should also be considered for the agricultural production in a specific region.

With the goal of high-efficiency utilization of nutrients, and the purpose of serving agricultural production, our group of plant nutrition and regulation at the Institute of Soil Science, Chinese Academy of Sciences, has conducted extensive work about fundamental and applied research as well as related popularization and demonstration, and achieved important progress. This book has summarized the work of our group over the last two decade, and we hope that it will provide reference for highly efficient utilization of nutrients.

This book is divided into four chapters, i.e. functional fertilizers and development, soil fertility evolution and evaluation, soil nutrient cycling, and plant nutrient and quality, to formulate principle and practice of high-efficiency nutrient utilization from different angles of view. This book provides a comprehensive but not exhaustive summery of the main and representative work and advances that our group has accomplished in the past two decade. Please do not hesitate to contact the author if you have comments and suggestions on any aspect of the book.

We sincerely appreciate the support from the National Natural Science Foundation of

China and Special Fund for Agro-scientific Research in the Public Interest. We are sincerely grateful to all the research staff, post-doctors and graduate students in our group for their diligent and consistent work, whose name are as following:

Zhou Jianmin, Wang Huoyan, Dong Caixia, Du Changwen, Sun Lei, Nie Jun, Chen Xiaoqin, Hua Quanxian, Du Zhenyu, Gao Li, Xie Wenjun, Huan Hengfu, Gan Fangqun, Li Tingxuan, Yuan Huimin, Hang Xiaoshuai, Wang Tao, Zhang Wenzhao, Chen Yudong, Zhao Cong, Shen Yazhen, Xing Lu, Shao Yanqiu, Wang Jidong, Zhao Hongtao, Lu Yuzhen, Zhou Zijun.

We are also greatly thankful to the Science Press for their kind assistance in publishing the book.

Zhou Jiamnin, Du Changwen, Wang Huoyan
January, 2015, Nanjing

Contents

Preface

Introduction

1 Theoretical basis and production technology of functional fertilizer ········· 1

    1.1 Characterization of nutrient release from polymer coated controlled release fertilizer ··· 1

    1.2 Enhancement of phosphorus solubility by humic substances in ferrosols ············ 11

    1.3 Evaluation of water-borne coating for controlled release fertilizer using Wurster fluidized bed ········· 16

    1.4 Aqueous polyacrylate/poly (silicone-co-acrylate) emulsions-coated fertilizers for slow nutrient release application ········· 23

    1.5 Biodegradation of a biochar-modified water-borne polyacrylate membrane coating for controlled release fertilizer and its effects on soil bacterial community profiles ···· 34

2 Evolution pathway and evaluation methods of soil fertility ········· 49

    2.1 Plants use alternative strategies to utilize nonexchangeable potassium in minerals ··· 49

    2.2 Nutrient budget and soil nutrient status in greenhouse system ········· 63

    2.3 Potassium movement and transformation in an acid soil as affected by phosphorus ··· 72

    2.4 Potash application patterns and soil potash fertility change ········· 85

    2.5 Identification of reaction products of phosphate fertilizers with soil using chemical and FTIR-PAS methods ········· 91

    2.6 Influence of humic acid on interaction of ammonium and potassium ions on clay minerals ········· 100

    2.7 Minimum data set for assessing soil quality in farmland of Northeast China ······ 113

    2.8 Contributions of greenhouse soil nutrients accumulation to the formation of the secondary salinization ········· 130

3 Nutrient migration and cycling in environment ········· 140

    3.1 Phosphorus fractions in sediment profiles and their potential contributions to eutrophication in Dianchi Lake ········· 140

    3.2 Removal of phosphate from aqueous solution by thermally activated palygorskite ··· 153

    3.3 Short-term effects of copper, cadmium and cypermethrin on dehydrogenase activity and microbial functional diversity in soils after long-term fertilization ··· 165

    3.4 Index models to evaluate the potential metal pollution contribution from washoff of road-deposited sediment ········· 177

    3.5 Phosphorus mobility in soil column experiment with manure application ·········· 192

    3.6 Rapid determination of isotope labeled nitrate using Fourier transform infrared attenuated total reflection spectroscopy ········· 203

4 Nutrition diagnosis and quality control of crops ································· **212**

4.1 Effects of different nitrogen forms on the growth and cytokinin content in xylem sap of tomato (*Lycopersicon esculentum* Mill.) seedlings ···················· 212

4.2 Effects of N fertilizer application time on dry matter accumulation and yield of Chinese potato ·· 225

4.3 Fertilization and catch crop strategies for improving tomato production in North China ·· 236

4.4 Weed community composition after 26 years of fertilization of late rice ········· 246

4.5 Risk assessment of potentially toxic element pollution in soils and rice (Oryza sativa) in a typical area of the Yangtze River Delta ·· 255

4.6 Intraspecific variation in potassium uptake and utilization among sweet potato (*Ipomoea batatas* L.) genotypes ··· 269

4.7 Use of FTIR-PAS combined with chemometrics to quantify nutritional information in rapeseeds (*Brassica napus*) ·· 282

1 Theoretical basis and production technology of functional fertilizer

1.1 Characterization of nutrient release from polymer coated controlled release fertilizer

1.1.1 Introduction

Polymers are versatile materials in most promising and comprehensive fields, and one useful application is in controlled release fertilizers (CRFs), which are called polymer coated fertilizers. Polymer coated fertilizers are the most promising section in controlled release fertilizers, and have the potential of optimal supply of nutrients during growth period of crops, and their application should benefit the environment and economic aspect (Hauk,1985; Shaviv,1999,2001; Peoples et al., 1995). Significant progress in explaining the mechanisms and quantifying the release from urea based (or other single fertilizers) polymer coated CRFs were made in the last years (Shaviv, 2000; Shaviv et al., 2003a, 2003b). The release course consists of three distinct stages (Shaviv et al., 2003a): ①the initial stage during which almost no release is observed (lag period), ②a constant-release stage, and ③a stage of gradual decay of release rate. It is assumed that the duration of the lag period is linked to the time needed for the internal voids of a coated granule to fill with a critical amount of water and thus induce good contact of the solution with the inner side of coating, after which a steady state between water penetrating into the granule and nutrients leaving it is attained. The stage of linear (constant rate) release lasts as long as there is solid fertilizer in the granule and thus a constant gradient between the granule and medium solutions is, practically, maintained. Mathematical mechanistic models were proposed for urea (and possibly for similar single fertilizer CRFs) enabling prediction of the release by utilizing chemo-physical parameters (Lu and Lee, 1992; Shaviv et al., 2003a). Less attention was devoted to the release of nutrients from compound N-P-K coated CRFs, in which case the processes are expected to be more complex than with a single fertilizer/nutrient. The release of each nutrient/ion in such case is expected to depend on its own solubility in solution, diffusivity/permeability through the polymer coating, interactions between ions as well as temperature, water content and medium type effects.

Wilson and Chem (1988) in their critical review, related to release characteristics of slow release fertilizers (SRFs), drew attention to compound N-P-K fertilizers, from which the fractional rate of release of N is greater than that of K, and even more so than the release rate of P. Shoji and Gandeza (1992) demonstrated this phenomenon with polyolefin-coated CRFs. Huett and Gogel (2000) tested release from 17 coated CRFs and reported a consistent trend in nutrient release periods across all CRFs with P>K>N and with differences of around 10% (only) in duration between nutrients. Shaviv (1999, 2000) mentioned such a common trend for polymer-coated CRFs stressing the significantly lower rates of release of P as compared to K and N. Several other recent reports examined the release from compound CRFs but did not emphasize the unique characteristics of the differential release of the different nutrients and their dependence on environmental conditions (e.g., Engelsjord et al., 1997; Smith et al., 1998; Tzika et al., 2003). Gandeza et al. (1991) made an effort to model the temperature effect on nitrogen release from a polyolefin coated CRF but used an empirical (polynomial) expression for this purpose. Jarrel and Boersma (1980) suggested an Arrhenius type expression for modeling the effect of temperature on nitrogen release from sulfur coated urea (SCU). Since their model was specifically developed for the release from SCU it was considered questionable for predicting release from polymer coated CRFs (Shaviv, 2000; Shaviv et al., 2003a).

This paper examines the differential release rates and patterns of nitrate, ammonium, potassium and phosphate from two polymer coated compound CRFs and the effects of temperature and water content (or type of release medium) on it. Special emphasis is put on the changes in the linear release rates and the lag periods for each nutrient/ion as affected by the above mentioned factors. Differences in the rates of linear release into free water at 3 different temperatures were used to estimate the activation energy of the release EA_{rel}.

1.1.2 The Oretical Consideration

Shaviv et al. (2003a) proposed the following, rather simplified, expression for the duration of the lag period, t_{lag}, for a single granule of a polymer coated CRF:

$$t_{lag} = \frac{\gamma r l}{3 P_h \Delta P} \quad (1)$$

Where, P_h is water permeability, $cm^2/(d \cdot Pa)$; l is coating thickness, cm; r is granule radius, cm; ΔP is the difference between vapor pressure of water and saturated urea solution, Pa; γ is the critical volume fraction of voids filled with water. The water potential gradient, ΔP, is not expected to change greatly when water content changes from field capacity to saturation or free water in a vessel and thus P_h remains the main temperature dependent factor, which affects the lag.

The lag period was significantly different for different nutrients in the same polymer coated granule, and this difference couldn't be reflected in above model. Based on the Shaviv's model, Du et al. (2006) deduced a little more complex one considering the dissolving time of nutrients as following:

$$t_{lag} = \frac{\gamma r}{3}\left(\frac{l}{P_h \Delta P} + \frac{1}{D\varphi}\right) \quad (2)$$

Where, D is dissolving coefficient; φ is mass ratio of a certain nutrient in total granule mass. This model gave a better modeling of lag period for nutrients release from polymer coated compound CRFs, but need further verification.

For the linear period ($t^* - t_{lag}$) the following rate equation was proposed for a single polymer coated granule:

$$R_{\text{lin}}(r,l,t) = \frac{3P_s C_{\text{sat}}}{rl\rho_s}(t - t_{\text{lag}}), t_{\text{lag}} \leqslant t \leqslant t^* \tag{3}$$

Where, P_s is solute permeability, cm²/d; C_{sat} is saturation concentration, g/cm³, standing for the difference between solute concentration within the granule, $C_{\text{in}} = C_{\text{sat}}$, and outside the coated granule, $C_{\text{out}} \sim 0$; ρ_s is fertilizer density, g/cm³; t^* is the time when the linear release ends. According to Shaviv et al. (2003b), such an expression can also be used for a population of granules provided that a narrow range of granule coating thickness and radii is used.

To account for the overall dependence of the release on temperature, an Arhenius type relation is offered, which should allow the estimation of the overall activation energy of the release from a coated granule during the linear period of release. Two parameters in Eq. 3 are likely to be temperature dependent: the permeability P_s, and the saturation concentration C_{sat}. Therefore the following temperature dependence of R_{lin} is proposed:

$$R_{\text{lin}} \propto C_{\text{sat}} \times P_s = C_{\text{sat}}^0 \exp\left(-\frac{EA_c}{RT}\right) \times P_s^0 \exp\left(-\frac{EA_{P_s}}{RT}\right) \tag{4}$$

Where, C_{sat}^0 and P_s^0 are reference (standard) values of the solubility and permeability respectively; EA_c is the energy of activation associated with the solubilization of the fertilizer, kJ/mol; EA_{P_s} is the energy of activation of its permeation through the membrane. By rearranging, one gets:

$$R_{\text{lin}} \propto C_{\text{sat}}^0 \times P_s^0 \exp\left(-\frac{EA_{P_s} + EA_c}{RT}\right) \tag{5}$$

Plotting ln R_{lin} against $1/(RT)$ for experimental data in which the temperature is changed, should thus allow the estimation of the overall energy of activation of the release $EA_{\text{rel}} = EA_c + EA_{P_s}$.

1.1.3 Experiment

1. Materials

Two polymer coated fertilizers ("polyurethane-like" coating): F_1, 19︰6︰13 (N︰P︰K), with a coating thickness of 0.0065 cm and F_2, 18︰6︰12 (N︰P︰K), with a coating thickness of 0.0096 cm were used.

2. Methods

Release was tested in three different systems: i, free water (common procedure, e.g., Shaviv, 2000); ii, water saturated sand packed in columns; and iii, sand at field capacity moisture.

Release in system i. 5 g polymer coated fertilizers were immersed in 33.3 mL water, which were then incubated at three different temperatures (20℃, 30℃ and 40℃). Each treatment was in four replicates. The supernatant solutions were sampled at predetermined time intervals and replaced by fresh de-ionized water. Nutrients in the samples were determined by a Lachat auto-analyzer (nitrate, ammonium, phosphate) and an Optima 1000 ICP (potassium).

Release in system ii. A glass column (2.8 cm inner diameter, and 16 cm height) with a funnel-shaped bottom was packed with 125 g silica sand (passing 160 mesh sieve). The sand was uniformly mixed with 5 g polymer-coated fertilizer and carefully placed in the column to which 33.3 mL of de-ionized water were added to saturate the sand. A plastic tube connected to the funnel-shaped bottom was used to keep the water level at the surface of the sand. The samples were then incubated at the three temperatures like those of system i. The columns were rinsed with 100 mL de-ionized water every few days and nutrient concentrations determined like in system i. Each treatment was run in four replicates.

Release in system iii. 5 g polymer-coated fertilizer were uniformly mixed with 125 g silica sand (passing 160 mesh sieve) and placed in a 200 mL plastic jar with a wide cover to which 20 mL distilled water were added to bring it to a moisture content slightly above field capacity (FC). The incubation was performed at 30℃. About 5 g of the moist sand were gently sampled every few days, extracted in de-ionized water and used for determination of nutrient concentration like in system i. On day 50 of the incubation the remaining sand was washed by adding 200 mL solution to remove the accumulated nutrients in the sand, after which the sand was separated from the solution, re-dried to FC and the incubation continued. To standardize the estimation of the duration of the lag periods of release for each nutrient, the lag was defined as the time at which the fractional release was less than 1%. The energy of activation of the release, EA_{rel}, was calculated on the basis of estimates of the rate of the release (% released per day) during the linear period of release R_{lin}. These were obtained from the release curves for the best fits to a straight line with $R^2 = 0.99$ (considering that the lines start at the end of the lag period).

1.1.4 Results and Discussion

1. *Release of individual nutrients*

Table 1 shows the individual rates of release of the different ions for the linear period and the 3 different temperatures. Fig. 1 and Table 1 clearly show the differences in the release rates between the different nutrients. Nitrate is released the fastest, followed by ammonium and potassium and phosphate is significantly slower. The rate of P release into water is in most cases 45% to 70% as compared to nitrate and it reduces to 35% to 50% in the saturated sand. The differences are even more striking when comparing the lag periods (Table 2). Little differences were obtained for nitrate, ammonium and potassium all ranging between 2 to 10 days under the varying experimental conditions, whereas for P they were almost one order of magnitude larger, ranging between 10 to 40 days. It is noteworthy that the differences in the lag increase with temperature.

Considering the fact that the volume of available water in the granule is limited (Shaviv et al., 2003a), it is thus expected that the first ions to dissolve are those with the higher solubility. Since the phosphates in the mixture have the lowest solubility it is expected that they will start dissolving only after a significant proportion of the other ions (particularly nitrate and ammonium) have been released leaving in the granule more available water for P dissolution.

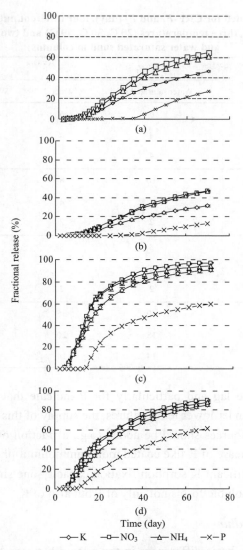

Fig. 1 Release of potassium, nitrate, ammonium and phosphate from: CRF-F_1 ((a)and (c)) and CRF-F_2 ((b) and (d)) into free water, at 20℃ ((a) and (b)) and 40℃ ((c) and (d))

Table 1 Release rates (% per day) during the linear period, obtained for CRFs-F_1 and F_2, four different nutrients (nitrate, ammonium, potassium and phosphate), three temperatures (20℃, 30℃, 40℃) and two release media (free water and water saturated sand in columns)

Nutrients	Release medium	CRF-F_1			CRF-F_2		
		20℃	30℃	40℃	20℃	30℃	40℃
NO_3-N	Sand column	1.05	1.62	2.08	0.81	1.16	1.77
	Water	1.04	2.04	3.32	0.76	1.35	2.24
NH_4-N	Sand column	0.85	1.55	2.83	0.67	1.16	2.01
	Water	0.98	1.85	3.26	0.75	1.49	2.24
K	Sand column	0.67	1.12	1.92	0.44	0.77	1.22
	Water	0.75	1.43	2.59	0.51	0.93	1.56
P	Sand column	0.39	0.63	0.91	0.28	0.49	0.84
	Water	0.77	0.89	2.1	0.32	0.65	1.03

Table 2 Lag periods estimated for CRF-F_1 and F_2 (day), four different nutrients (nitrate, ammonium, potassium and phosphate), three temperatures (20°C, 30°C, 40°C) and two release media (free water and water saturated sand in columns)

Nutrients	Temperature/°C	CRF-F_1		CRF-F_2	
		Water saturated sand	Water	Water saturated sand	Water
NO$_3$	20	6	6	8	6
	30	4	4	4	4
	40	4	2	4	4
K	20	10	8	10	8
	30	4	4	4	4
	40	2	2	2	2
NH$_4$	20	8	8	8	8
	30	4	4	4	4
	40	2	2	2	2
P	20	40	35	35	35
	30	18	20	14	16
	40	14	14	10	10

Such differences in the lag and particularly for P indicate that in early stages of plant growth, and particularly under lower temperatures, the supply of this nutrient may be too slow and thus more available sources should be added (e.g., a fraction of noncoated P or a thinly coated source with fast release of P, like coated mono-ammonium-phosphate). Furthermore, in soils and media with significant K sorption/fixation capacity one should consider applying a significant proportion of soluble forms not only of P but also of K.

2. Effect of release medium

Fig. 2 demonstrates the basic differences in the release between the three different media: free water, saturated sand in columns and sand at FC. Expectedly, the release into the free water is the fastest and that into the sand at FC the slowest. The rates of the linear release of nitrate, potassium and ammonium are generally about 5% to 20% slower in the saturated soil as compared to those in free water (Table 1). For phosphate the reduction is much more significant ranging between 25% to 55 % (Table 1).

Interestingly, in case of the sand at FC the relative reduction in the rates of nitrate, ammonium and potassium (e.g., Fig. 2) in the first 2-3 weeks is small. As the release goes on the rate significantly slows, as expected from the accumulation of the nutrients in the medium causing a significant reduction in the driving force of the release (i.e. Eq. (3), when the assumption $C_{out} \sim 0$ is no more valid).

In case of P, the reduction in the release rate is large (25% to 55%) when shifting from free water to water saturated sand and very drastic for FC conditions.

The effect of the medium, or eventually the degree of saturation or water availability, on the lag is relatively small (Fig. 2). This is acceptable considering that the total water potential difference between free water to sand at FC is not expected to change significantly. Therefore, the driving force for water entry (ΔP in Eq. (1)), which is a major factor to affect the lag, is not expected to change significantly between the media.

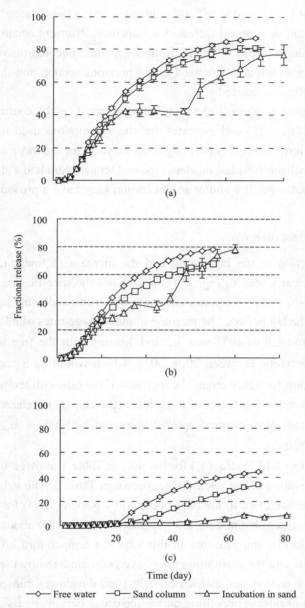

Fig. 2   Release of nitrate (a), potassium (b), and phosphate (c) from CRF-F_1 at 30℃ into: free water, water saturated sand (in column) and sand at field capacity

3. *Effect of coating thickness on release characteristics*

The average release rates into free water for all ions of CRF-F_1 where 0.89, 1.55 and 2.82 (% nutrient per day) for 20℃, 30 ℃ and 40℃, respectively and only 0.59, 1.11 and 1.77 for

CRF- F_2 at 20℃, 30℃ and 40℃, respectively. The ratios between the rates obtained for CRF-F_2 : CRF-F_1 range between 0.63 to 0.73 and agree well with the expected from Eq .(2) for the inverse-ratio of the coating thickness l (lF_1 : lF_2) corresponding with 0.68. The ratio slightly increases to 0.73-0.75 in case of the release into the water saturated sand and farther increases to 0.75-0.84 in the sand at FC. This indicates that in those cases the release was also affected by the medium or external factors that impede it. Nutrient accumulation outside the granules, when leaching or diffusion away from the granule become restricted is expected to impede the release as shown and discussed in the previous section and thus the effect of the coating thickness is slightly reduced.

The lag period (Table 2) seems to be almost non-affected by the coating thickness, which does not conform with Eq. (1) and indicates that the assumptions used for modeling the lag from a single coated fertilizer (e.g., urea) might not be valid for the more complex system of a compound N-P-K fertilizer. Besides, nutrient type and temperature had a significant impact on the lag, reiterating that solubility and/or solubilization may have a profound role in this case (Eq. (2)).

4. Temperature effect on release

Overall, the reduction in the lag period and the increase in linear rate of release with temperature can be clearly seen in Fig. 3. Table 1 shows the specific linear rates as obtained for each ion for CRF-F_1 and F_2 when releasing into free water and water saturated sand. Table 2 does the same for the lag period. The increase in the average rates of all four ions where the temperature raised from 20 to 40℃ was 3.2 and 3.0 times for the free water and the water saturated sand, respectively. Between 20 to 30℃, which would be a common range of the rhizosphere temperature for many crops, the increase of the rate with temperature is 1.8 to 2.0 times. This conforms to the Q_{10} factor of the release (the change in release rate at an increase of 10℃, which matches plant nutrient uptake) presented by Shoji et al., (1991) and further discussed by Shaviv (2000).

Plotting lnR_{lin} against $1/(RT)$ (Eq. (5)) for the data in Table 1 allowed the estimation of the overall energy of activation of the release EA_{rel} shown in Table 3. The values obtained for the release into free water are close to the ones obtained by Raban (1994) for urea with a similar type of coating. The values obtained for the water saturated sand are about 15% to 20% lower. This is assumed to further indicate that in this case the temperature effects on the release through the membrane and the dissolution may have been hindered by the medium itself (i.e. tortuosity, lower water content and thermal capacity) and the slower transport of the ions in it. Since the mechanism of release prevailing during the decay period are assumed to be the same as for the linear release the energy of activation estimated above should thus apply for the decay period as well.

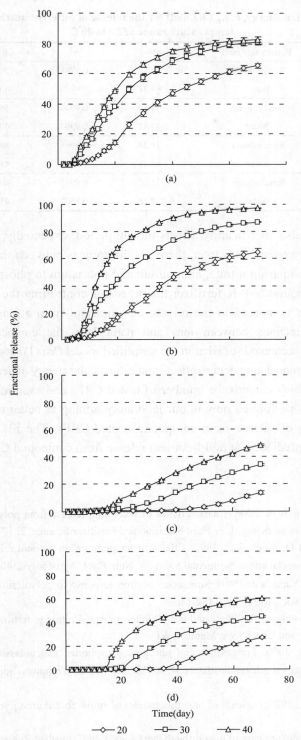

Fig. 3　Release of nitrate ((a) and (b)) and phosphate ((c) and (d)) from CRF- F_1 into free water ((b) and (d)) and water saturated sand ((a)and (c)) as affected by temperature (20℃, 30℃, 40℃)

Table 3 The activation energy, EA_{rel} (kJ/mol) for the release of different nutrients obtained for the temperature range of 20 to 40°C

Nutrients	Release medium	CRF-F_1	R^2	CRF-F_2	R^2
NO_3-N	Sand column	25.52	0.9788	29.13	0.997
	Water	43.31	0.9928	40.31	0.9992
NH_4-N	Sand column	44.80	0.9999	36.24	0.9955
	Water	44.81	0.9994	40.84	0.981
K	Sand column	39.24	0.9995	38.04	0.9977
	Water	46.21	0.9997	41.69	0.9988
P	Sand column	31.61	0.9953	40.96	1
	Water	37.28	0.8502	43.62	0.987

The temperature also significantly affects the lag period: expectedly, the lag gets shorter as the temperature increases. However, there are differences between the different ions and particularly when comparing nitrate, ammonium and potassium to phosphate. In case of the release from a compound N-P-K fertilizer the processes controlling the lag are expected to be more complex as compared to the release of a nutrient from a single coated fertilizer (e.g., Eq. (2)). Interactions between ions and particularly the competition on water for solubilization are not accounted for what in the simplified model (Eq.(1)) and therefore it was felt pre-mature to try and model temperature effects and estimate the energy of activation at this stage.

Ongoing work, which extends the number of tested CRFs and examines the release under various conditions is performed now in our laboratory aiming at better understanding of the complex mechanism of release from compound coated-CRFs. The results are expected to allow modeling and prediction of multi-nutrient release from compound CRFs.

References

Du C W, Zhou J M, Shaviv A. 2006. Characteristics of potassium release from polymer-coated controlled-release fertilizer and its modeling. J. of Plant Nutrition and Fertilizer Science, 2: 179-182

Engelsjord M E, Fostad O, Singh B R. 1997. Effects of temperature on nutrient release from slow-release fertilizers. I. Commercial and experimental products. Nutr. Cycl. Agroecosys., 46: 179-187

Gandeza A T, Shoji S, Yamada I. 1991. Simulation of crop response to polyolefin coated urea: I. Field dissolution. Soil Sci. Soc. Am. J., 55: 1462-1467

Hauck R D. 1985. Slow-release fertilizers and bio inhibitor-amended nitrogen fertilizers. In: Engelstad O P. Fertilizer Technology and Use. SSSA, Madison WI

Huett D O, Gogel B J. 2000. Longevities and nitrogen, phosphorus, and potassium release patterns of polymer-coated controlled-release fertilizers at 300°C and 400°C. Commun. Soil Sci. Plant Anal., 31: 959-973

Jarrell W M, Boersma L. 1980. (Release of urea by granules of sulfur coated urea.) Soil Sci. Soc. Am. J., 44: 418-422

Lu S M, Lee S F. 1992. Slow release of urea through latex film. J. of Controlled Release, 18: 171-180

Peoples M B, Freney J R, Moiser A R. 1995. Minimizing gaseous losses of nitrogen. In: Bacon P E. Nitrogen Fertilization in the Environment. New York: Marcel Dekeker Inc.: 565-601.

Raban S. 1994. Release mechanisms of membrane coated fertilizers. Master's Thesis, Agric. Eng. Technion-IIT. (Hebrew; contents, extended abstract, figures and tables in English)

Shaviv A. 1999. Preparation methods and release mechanism of controlled release fertilizers: agronomic efficiency and environment significances. International Fertiliser Society, York, UK, (431): 1-35

Shaviv A. 2000. Advances in controlled-release fertilizer. Advances in Agronomy, 71: 1-49

Shaviv A. 2001. Fertilizers and resource management for food security, quality and the environment. International Fertiliser Society, York, UK, (469): 1-23

Shaviv A, Smadar R, Zaidel E. 2003a. Model of diffusion release from polymer coated granular fertilizers. Envi Sci. & Tech., 37: 2251-2256

Shaviv A, Smadar R, Zaidel E. 2003b. Statistically based model for diffusion release from a population of polymer coated controlled release fertilizers. Envi Sci. & Tech., 37: 2257-2261

Shoji S, Gandeza A T. 1992. Controled release fertilizers with polyolefin resin coating. Controled release fertilizers with polyolefin resin coating. Konno Printing Co. Sendai, Japan

Shoji S, Gandeza A T, Kimura K. 1991. Simulation of crop response to polyolefin coated urea: II. Nitrogen uptake by corn. Soil Sci. Soc. Am. J., 55: 1468-1473

Smith G L, Robbins J A, Wallick W L. 1998. Kinetics of controlled release fertilizers. Book of Abstracts, 216th ACS National Meeting, Boston, August 23-27, American Chemical Society, Washington, DC

Tzika M, Alexandridou S, Kiparissides C. 2003. Evaluation of the morphological and release characteristics of coated fertilizer granules produced in a Wurster fluidized bed. Powder Technology, 132:16-24

Wilson F N, Chem C. 1988. Slow release true or false? A case study for control. Fertiliser Society of London

1.2 Enhancement of phosphorus solubility by humic substances in ferrosols

1.2.1 Introduction

Phosphorus (P) is an essential nutrient for plant growth and is often the limiting nutrient in agricultural ecosystems owing to its low availability in soils. Phosphate fixed by Fe, Al, and Ca in soils is a major cause of low phytoavailability (Ozanne, 1980; Ding and Pan, 2005; McBeath et al., 2005). P efficiency can be enhanced by increasing P solubility in soil solution or by reducing P fixation in soils. Humic substance interacting with P in soils may reduce P fixation (Saunders, 1965; Ding et al., 2005; Guppy et al., 2005), and make P more available to plants.

Numerous studies have been conducted on P phytoavailability. Perrott (1978) found that organic matter extracted from humified clover reduced P retention on hydrous alumina and amorphous synthetic aluminosilicate. The pot experiments of different independent studies demonstrated that humic substance was able to increase the corn dry matter yield, P uptake, and the labile P levels in soils (Easterwood and Sartain, 1990; Hue et al., 1994; Sharif et al., 2002). Using commercial humic-fulvic acid mixtures as soil amendments, Delgado et al. (2002) reported that these materials improved the utilization efficiency of P fertilizer as $NH_4H_2PO_4$.

The objective of this study was to investigate the effects of humic substance on the bioavailability of applied P in an acidic soil.

1.2.2 Materials and Methods

The soil used in this study was an Argi-Ustic Ferrosol according to Chinese Soil Taxonomy (CRGCST, 2001) or a Haplic Acrisol according to the soil classification developed by the

Food and Agriculture Organization of the United Nations. The soil sample was collected from the Red Soil Ecological Experimental Station, Chinese Academy of Sciences, Yingtan (28°15′30″ N, 116°55′30″ E), Jiangxi Province in southern China. Kaolinite and hydrous mica are the dominant clay minerals in this soil, and the soil also contains a small amount of vermiculite. Only the topsoil (0-15 cm) was sampled. After removing visible plant residual, stone, and soil fauna, the soil was ground and sieved through 0.4 mm. Some properties of the soil are detailed in Table 1.

Table 1 Physical and chemical properties of the surface layer (0-15 cm) of the Argi-Ustic Ferrosol

PH[①]	Organic matter (g/kg)	Total P[②] (g/kg)	Olsen P (mg/kg)	Free Fe_2O_3[③] (g/kg)	Free Al_2O_3 (g/kg)	Particle size composition[④](%)		
						Clay[⑤] (<2 μm)	Silt (2-50 μm)	Fine sand (>50 μm)
4.57	11.7	0.59	23.08	7.94	1.58	30.2	52.4	8.8

Notes:① Water : soil = 2.5 : 1. ② Total elemental composition by inductively coupled plasma-atomic emission spectrometer (ICP-AES) following $LiBO_2$ fusion (Lu, 1999). ③ Sample exacted with sodium citrate-sodium bicarbonate-sodium hydrosulfite followed by ICP-AES determination. ④ American Soil Science Society classification. ⑤ Sample examined using an LS230 laser diffractometer (Beckman Coulter, USA).

The humic substance (HS) used in this study is a biochemical product that contains a negligible amount of total P (Jufeng Chemical Technology Ltd., Shanghai). HS contains several functional groups, such as —OH, —COOH, and —NH_2, and these groups are responsible for its special chelating properties. P was used as monocalcium phosphate (MCP, analytical reagent grade), which is the main component of superphosphate fertilizer.

The batch equilibration experiment was conducted to investigate the mobilization of MCP in soil treated with various levels of HS. The soil (2.5 g), HS (0, 0.5, and 2.5 g), and P solution (0.31 and 1.25 g P/kg soil) were mixed with 50 mL distilled water in 80 mL polypropylene flasks and replicated three times. Suspensions were shaken on an end-over-end shaker at 25 ± 1℃ and then centrifuged at 2790 × g for 10 min. The supernatants were analyzed for P, Fe, and Al. After the residual soil samples were washed twice with alcohol, 50 mL of 0.01 mol/L $CaCl_2$ was added to the soil samples. The mixtures were then shaken for 1 h and equilibrated for 24 h at 30 ± 1℃. The supernatants were again extracted for P analysis. Finally, the same samples were washed as described above, and extracted with 0.5 mol/L $NaHCO_3$ (pH 8.5) for Olsen P.

Two different sequences of adding HS and P were used. In the first treatment, HS was first mixed with soil for 24 h, and then P was added to the suspension. The mixture was shaken for another 48 h. The samples of this sequence were designated HS-P. In the second treatment, P was first adsorbed onto the soil for 48 h, and then HS was added to the suspension. The mixture was shaken for another 24 h and designated P-HS. Phosphorus in the solution was determined by the ascorbic acid-molybdophosphate blue method (Olsen and Sommers, 1982) and Fe and Al by ICP-AES (Lu, 1999).

The analysis of variance (ANOVA) and multiple comparisons with Tukey's test at $P < 0.05$ were performed using SPSS 11.5 (Statistical Package for the Social Science) and Microsoft Office Excel.

1.2.3 Results

1. *Phosphorus in solution*

The addition of HS increased the P concentrations in solution at both low and high P levels with either sequence of HS addition (Fig. 1(a)). The P concentration in solution increased as the HS increased. For the low P level treatment, adding 0.5 g HS increased the P in solution by 3.6 times as compared to no addition of HS; further increasing HS to 2.5 g further raised the P in solution by 5 times. At the high P level, however, addition of 0.5 g HS increased the P in solution by 70%, and a high rate HS (2.5 g) further enhanced the P by 90% as compared to no addition of HS.

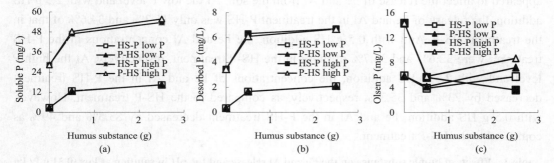

Fig. 1 The influence of humic substance (0, 0.5, and 2.5 g) on the P adsorption (a), P concentration extracted with $CaCl_2$ (b), and Olsen P concentration (c) in acidic soil at low (0.31 g P/ kg soil) and high P levels (1.25 g P/kg soil)

HS-P: humic substance was added to 2.5 g soil with deionized water prior to adding P; P-HS: P was added before humic substance was added to 2.5 g soil. Error bars represent standard deviations ($n = 3$)

2. *Residual P release with $CaCl_2$ extracted*

Phosphorus is extracted with 0.01 mol/ L $CaCl_2$ solution via an ion-exchange reaction between chloride and phosphate anions in the soil and HS mixtures. As seen in Fig. 1(b), the addition of HS significantly increased the concentration of desorbed P (DPC), similar to the P concentration in water solution. The DPC increased with the amount of HS. At the low P level, DPC in the treatment with 0.5 g HS was 5 times that in the treatment without HS, while 2.5 g HS enhanced DPC to 6.4 times. At the high P level, the DPC in the treatment with 0.5 g HS increased by approximately 80% as compared to that in the treatment with no HS, and the DPC in the treatment with 2.5 g HS increased by about 90%. As seen in Fig. 1(b), the mixing order of HS and P did not generate a significant difference in DPC.

3. *Olsen P release*

The addition of HS resulted in a significant decrease of Olsen P from the residual sample (Fig. 1(c)). The Olsen P concentration was least in the treatment with 0.5 g HS, except for the treatment with 2.5 g HS added at the high P level. The mixing sequence had no effect on the Olsen P concentration in the treatment with 0.5 g HS, but there was a significant difference in the treatment with 2.5 g HS. When P was added before HS (P-HS), Olsen P was considerably higher than that in the treatment with reverse mixing sequence (HS-P).

4. *Fe and Al release and pH in solution*

The addition of HS increased the concentrations of Fe and Al in solution (Table 2). As the ratio of HS to soil increased, the Fe concentration increased, while Al decreased at the low P level. However, at the high P level, Fe and Al increased with HS increase. When P was adsorbed prior to HS addition, the amounts of Fe and Al released in the treatment with 2.5 g HS were 9.5 and 5.8 times higher than those in the treatment with 0.5 g HS, respectively. In the reverse mixing sequence, Fe and Al in the treatment with 2.5 g HS increased to 4.8 and 3.6 times higher than those in the treatment with 0.5 g HS. The mixing order of HS and P appeared to affect the release of Fe and Al from the soil. At the low P level and with 2.5 g HS addition, the release of Fe and Al in the treatment P-HS was only 15.3% and 34.6% of that in the treatment HS-P. And with 0.5 g HS addition, the Fe and Al concentrations in the P-HS treatment were 32.6% and 35.7% of those in the HS-P treatment, respectively. At the high P level and with 2.5 g HS addition, the concentrations of Fe and Al in the P-HS treatment decreased by 73% and 35.3%, respectively, as compared to the HS-P treatment. However, with 0.5 g HS addition, Fe and Al in the P-HS treatment decreased by 83.5% and 49% as compared to the HS-P treatment.

Table 2   Effects of humic substance on the Fe and Al release and the pH in solution at low (0.31 g P/ kg soil) and high (1.25 g P/ kg soil) P levels

Treatment[1]	Humic substance (g)	Low P level			High P level		
		Fe (mg/L)	Al (mg/L)	pH	Fe (mg/L)	Al (mg/L)	pH
P-HS	2.5	6.26±0.072[2]c[3]	4.06±0.136c	2.69±0.042c	5.68±0.071b	4.49±0.040b	2.66±0.014c
	0.5	5.19±0.276c	4.86±0.266c	3.72±0.184b	0.60±0.047c	0.77±0.005d	3.50±0.007b
HS-P	2.5	41.0±1.425a	11.74±0.585b	2.66±0.007c	21.03±2.215a	6.94±0.168a	2.67±0.007c
	0.5	15.9±0.230b	13.61±0.185a	3.63±0.064b	3.63±0.364b	1.51±0.046 c	3.55±0.113b
Control	0.0	1.22±0.025d	2.63±0.094d	4.77±0.032a	0.015±0.002d	0.226±0.016e	4.41±0.050a

Notes: [1] P-HS: P was added before humic substance addition to 2.5 g soil; HS-P: humic substance was added to 2.5 g soil with deionized water prior to adding P; control: treatment for P adsorbed with no humic substance addition. [2] Means±standard deviations (n = 3). [3] Means in each column followed by the same letter are not significantly different by Tukey's test at 0.05 probability level.

The solution pH value decreased with increasing amount of HS with respect to the treatment without HS, and the mixing sequence of HS and P had no effect on the pH values.

1.2.4 Discussion

Humic substance can affect the soluble P and the easily extractable P (Guppy et al., 2005). This effect has been commonly ascribed to the competition between HS and P for soil sorption sites, especially at low pH (Hingston et al., 1972; Lopez-Hernandez et al., 1979; Sibanda and Young, 1986; Traina et al., 1986). As seen in Fig. 1(a), the more HS added in the suspension, the more P remaining in the solution. A small amount of P was sorbed onto the soil in the treatments of 2.5 g HS addition. Leaver and Russell (1957) pretreated a Tanzanian

Oxisol with HS as fulvic acid, and found 27%-63% reduction in the P sorption. Sibanda and Young (1986) also observed a significant reduction in the P sorption on gibbsite, goethite, and two highly weathered soils after the addition of humic acids or fulvic acids. Heng (1989) demonstrated that the addition of a small amount of HS reduced the P sorption (< 10% lower) in highly weathered soils of Malaysia. Wang et al. (1995) found that the addition of humic acids to alkaline soil with ammonium dihydrogen phosphate significantly increased the amount of water soluble P, and strongly retarded the formation of occluded phosphate.

Our results demonstrated that P could be released into solution from soil when HS was added. HS contains functional groups, such as carboxyl, phenolic hydroxyls, and alcoholic hydroxyls. The dissociation of protons from these groups results in lower pH values. In contrast, the complexation of P and HS contributes to the increase of soluble P or labile P (Bedrock et al., 1997). In this study, the reduction of pH and the increase of Fe and Al concentrations in solution (Table 2) elucidated the phenomenon. As a result, the concentrations of P in water and in 0.01 mol/L $CaCl_2$ solution increased with increasing amounts of HS, as shown in Fig. 1(a) and 1(b). In a batch study, the amount of P released increased with the shaking time and the application rates of humic and fulvic acids, probably because the solubility of $AlPO_4$ and $FePO_4$ increased owing to the humic and fulvic acid amendment (Lobartini et al., 1998).

References

Bedrock C N, Cheshire M V, Shand C A. 1997. The involvement of iron and aluminum in the bonding of phosphorus to soil humic acid. Commun. Soil Sci. Plant Anal., 28(11&12): 961-971

CRGCST (Cooperative Research Group on Chinese Soil Taxonomy). 2001. Chinese Soil Taxonomy. Beijing: Science Press: 203

Delgado A, Madrid A, Kassem S, et al. 2002. Phosphorus fertilizer recovery from calcareous soils amended with humic and fulvic acids. Plant Soil, 245: 277-286

Ding C P, Pan Y H. 2005. Exchange-adsorption characteristics of aluminium and manganous ions by red soil. Ⅳ. Chemical phenomenon for exchange of calcium/potassium ion by aluminium/manganous ions. Acta Pedologica Sinica, 42(1): 64-69 (in Chinese)

Ding Y Z, Li Z A, Zou B. 2005. Low-molecular-weight organic acids and their ecological roles in soil. Soils, 37(3): 243-250 (in Chinese)

Easterwood G W, Sartain J B. 1990. Clover residue effectiveness in reducing orthophosphate sorption on ferric hydroxide coated soil. Soil Sci. Soc. Am. J., 54: 1345-1350

Guppy C N, Menzies N W, Moody P W, et al. 2005. Competitive sorption reactions between phosphorus and organic matter in soil: a review. Aust. J. Soil Res., 43: 189-202

Heng L C. 1989. Influence of some humic substances on P-sorption in some Malaysian soils under rubber. J. Nat. Rubber Res., 4: 186-194

Hingston F J, Posner A M, Quirk J P. 1972. Anion adsorption by goethite and gibbsite. Ⅰ. The role of the proton in determining adsorption envelopes. J. Soil Sci., 23: 177-192

Hue N V, Ikawa H, Silva J A. 1994. Increasing plant-available phosphorus in an Ultisol with a yard-waste compost. Commun. Soil Sci. Plant Anal., 25(19&20): 3219-3303

Leaver J P, Russell E W. 1957. The reaction between phosphate and phosphate-fixing soils. J. Soil Sci., 8: 113-126

Lobartini J C, Tan K H, Pape C. 1998. Dissolution of aluminum and iron phosphate by humic acids. Commun. Soil Sci. Plant Anal., 29(5&6): 535-544

Lopez-Hernandez D, Flores D, Siegert G, et al. 1979. The effect of some organic anions on phosphate removal from acid and calcareous soils. Soil Sci., 128(6): 321-326

Lu R K. 1999. Analytical Methods of Soil Agrochemistry. Beijing: China Agricultural Science and Technology Press: 638 (in Chinese)

McBeath T M, Armstrong R D, Lombi E, et al. 2005. Responsiveness of wheat (*Triticum aestivum*) to liquid and granular phosphorus fertilizers in southern Australian soils. Aust. J. Soil Res., 43: 203-212

Olsen S R, Sommers L E. 1982. Phosphorus. In: Page A L, Miller R H, Keeney D R. Methods of Soil Analysis. Part 2. Chemical and Microbiological Properties. Madison: ASA and SSSA Publishing: 403-430

Ozanne P G. 1980. Phosphate nutrition of plants-A general treatise. In: Khasawneh F E, Sample E C, Kamprath E J. The Role of Phosphorus in Agriculture. Madison: ASA, CSSA and SSSA Publishing: 559-585

Perrott K W. 1978. The influence of organic matter extracted from humified clover on the properties of amorphous aluminosilicate. II. Phosphate retention. Aust. J. Soil Res., 16: 341-346

Saunders W M H. 1965. Phosphate retention by New Zealand soils and its relationship to free sesquioxide, organic matter and other soil properties. New Zeal. J. Agr. Res., 8: 30-57

Sharif M, Khattak R A, Sarir M S. 2002. Effect of different levels of lignitic coal derived humic acid on growth of maize plants. Commun. Soil Sci. Plant Anal., 33(19&20): 3567-3580

Sibanda H M, Young S D. 1986. Competitive adsorption of humus acids and phosphate on goethite, gibbsite and two tropical soils. J. Soil Sci., 37: 197-204

Traina S J, Sposito G, Hesterberg D, et al. 1986. Effects of pH and organic acids on orthophosphate solubility in an acidic, montmorillonitic soil. Soil Sci. Soc. Am. J., 50: 45-52

Wang X J, Wang Z Q, Li S G. 1995. The effect of humic acids on the availability of phosphorus fertilizers in alkaline soils. Soil Use Manage, 11: 99-102

1.3 Evaluation of water-borne coating for controlled release fertilizer using Wurster fluidized bed

1.3.1 Introduction

The effectiveness of fertilizers in increasing nutrients use efficiency and reducing environment risks mainly depends on their function of supplying an appropriate concentration of nutrients within root zone for a desired period(Shaviv, 2000). However, the active components in conventional fertilizer usually diminish rapidly before sufficient plant uptake as a result of degradation (e.g., chemical, photochemical and biological), leaching, runoff, volatilization, absorption or soil immobilization(Pérez-García et al., 2007). To maintain or even increase crop yields excessive fertilizer dosages are often applied, thus causing not only resources wasted but also potential environment threats.

Compared with conventional fertilizers, the use of controlled release fertilizer (CRF) demonstrates many advantages, such as decreasing the rate of nutrients removal from soil by rain or irrigation water, increasing the nutrients use efficiency, lowering the labor cost in fertilizer application, sustaining the supply of nutrients for a prolonged time and minimizing the negative toxic effects associated with overdose(Byung-Su et al., 1996; Al-Zahrani, 2000). Therefore CRF, as a promising direction, offers an excellent option to improve nutrients management and reduce environment hazards. Currently a variety of materials have been discovered to be suitable for CRF coatings, such as paraffin wax(Tomaszewska and Jarosiewicz, 2002), polyolefins (Kosuge and Tobataku, 1989), polyethylene(Salman, 1989; Salman et al., 1989; Posey and Hester, 1994), polystyrene(Garcia et al., 1996), kraft pine lignin(Cabrera, 1997), polyacrylamide(Rajsekharan and Pillai, 1996), polysulfone(Posey and Hester, 1994), and ethylcellulose(Pérez-García et al., 2007). However, organic solvents should be used to dissolve the above materials during coating process. Organic solvents are relatively expensive, and most of them are toxins or pollutants. Hence, more and more attentions are paid to water-borne coating, which has the advantage of non-toxicity, non-flammability, low price and good quality in comparison with the traditional organic coating(Ahmad et al., 2008), and a trend that shifts from solvent-based formulations to more environmentally friendly options has taken place in the area of general coatings industry(Tang et al., 2004).

For the water-borne coating materials, reacted layer technology has been widely used to improve water resistance of coating, and crosslinker is commonly used to modify this property since crosslinking is a very important process in the formation of network structure(Dušek and Dušková-Smrčková, 2000). Basically, any crosslinker employed must first be compatible with the polymer to produce homogeneous membrane with desirable properties.

In the present work, the influences of crosslinker on the structures and properties of the CRF coating as well as nutrients release rate from the CRF were systematically studied, and the main purpose was to assess the feasibility of using water-borne polymer as coating material for the CRF.

1.3.2 Experimental Section

1. Preparation of acrylic model membranes

Organic silicon modified acrylate latex, provided by Doctor Hydrophilic Chemicals Co. Ltd. (Yizheng, China), contains more than 65% of acrylic polymer. Distilled water (100 mL) and certain ratios of crosslinker aziridine (0, 1 and 2 wt%) was slowly added to acrylate latex (100 mL) at room temperature under continuous stirring for 15 min, then distributed the above latex into a 9 cm^2 leveled glass-plate mold, dried in oven at different temperatures (60, 70 and 80℃) for 24 h, and formed the model membranes on the glass-plate surface. The model membranes were removed from the mold, and stored in 4℃ refrigerator for use.

2. Preparation of polymer coated fertilizers

The conventional compound fertilizers (NPK, 15-12-15) were provided by Fulilong Chemical Co. Ltd., China. 80 kg of the fertilizer granules with 3-4 mm in diameter were loaded into a fluidized-bed coater (LDP-5, Jiafa Mechanic Co. Ltd., China) assembled with a Wurster bed on a pilot scale. The bed temperature was set at 30±5℃. After preheating at this temperature for 10 min, 16 kg of acrylate latex diluted with 16 kg of water and added different ratios (1 and 2 wt%) of crosslinker was sprayed as coating material through a nozzle at an atomizing pressure of 0.4 MPa, and the spray rate was controlled by a peristaltic pump at the speed of 0.20 L/ min. Then the coated fertilizers were dried at different temperatures (60, 70 and 80℃) in oven for 24 h to complete crosslinking reaction.

3. Characterization of model and coated membranes

There was 1 g of accurately weighed model membrane immersed in glass bottle containing 100 mL of distilled water and kept at 25℃ for 48 h until completely swollen. Then take it out with forceps, remove excess surface water with filter paper and weigh the swollen membrane to calculate its swelling capacity(Karadag et al., 2000).

The glass transition temperature (T_g) of model membrane was assayed using a differential scanning calorimetery (DSC, Pyris-DSC Perkin-Elmer, USA). Each sample was placed in a standard aluminum pan and heated at a rate of 20℃/min with indium and zinc as the calibration. As a rule, two successive scans were made for every sample. The midpoint of the total change in heat capacity was designated the glass transition. All calculations were performed on the second heating cycle. The settings applied were identical for every case studied: the initial temperature was –50℃ and the final one was set at 100℃.

The membranes removed from coated fertilizers were observed by an optical microscope (Olympus BH2) to test their uniformity and porosity. Fourier transform mid-infrared spectroscopy (FTIR-PAS, Nicolet 380) with a photoacoustic accessory (MTEC model 300, USA) was used to detect the structure information of coated membranes: the scans were conducted in the wavenumber range of 500-4000 cm^{-1} with a resolution of 4 cm^{-1} and a mirror velocity of 0.31 cm/s, and 32 successive scans were recorded.

4. Release profile of nutrients from polymer coated fertilizers

The release of nutrients as a function of time (day) for each formulation was performed in water medium. 2 g of coated fertilizers were placed in a glass bottle containing 100 mL deionized water and kept in oven at (40±1)℃ in three replicates. The nutrients relative content was evaluated by the solution conductivity(Yang et al., 2007), and the conductivity was measured every 24 h using electrical conductivity apparatus (DDS-320, China). On the 9th day of study, the coated fertilizers were grounded to determine the content of residual nutrients. The release profile was estimated as the cumulative release percentage versus time.

1.3.3 Results and Discussion

1. Characteristics of model membranes

Comparing with the crosslinker-free coating, the addition of crosslinker led to a significant decrease in swelling capacity, and the membrane contained 2 wt% of crosslinker exhibited lower swelling capacity than that containing 1wt% (Table 1). The addition of crosslinker could enhance the hydrophobicity of membrane. Furthermore, membrane-forming temperature can also influence swelling capacity. Membrane formed at 80℃ had the significantly ($P \leqslant 0.05$) lower swelling capacity than the other two temperature treatments for higher temperature promoted crosslinking reaction and thereof strengthened the membrane hydrophobicity. For the membrane used as fertilizer coating, its low swelling capacity can result in very slow nutrients release rate even no nutrients release while high swelling capacity will lead to very fast nutrients release even no controlled effect. Thus, the controlled release time can be adjusted by swelling capacity, and usually the swelling capacity should be less than 10%(Shen et al., 2009). According to the above results, it can be inferred that formulations with 2 wt% of crosslinker, especially those formed at 80℃ in oven thereafter, can decrease the rate of nutrients release in comparison with other formulations.

Table 1 Swelling capacity of model membranes containing different ratios of crosslinker and forming under different temperatures for 24 h

Forming-temperature (℃)	Crosslinker Ratio (wt%)		
	0	1	2
60	9.20 ± 0.65 A (a)	8.41 ± 1.23 A (ab)	6.92 ± 0.71 A (b)
70	8.96 ± 0.46 A (a)	7.04 ± 1.06 AB (b)	6.65 ± 0.09 A (b)
80	7.08 ± 0.49 B (a)	5.84 ± 0.28 B (b)	5.54 ± 0.32 B (b)

Notes: Means with the same letter (capital letters for temperature treatments and small letters in bracket for crosslinker treatments) are not significantly different at $P \leqslant 0.05$ level by SPSS 13.0.

The T_g of membrane with 1 wt% of crosslinker increased by 1℃ but the T_g of membrane with 2 wt% of crosslinker decreased slightly (Table 2). However, the fluctuation was mild, and no regular impact of crosslinker on the T_g of these materials could be deduced. Below the T_g the polymer is rigid, hard, brittle and glass like. Above the T_g the polymer is soft, rubbery and flexible(Barbour et al., 1996). For too high T_g the fertilizer granules would stick to each other, which would hamper particle fluidization and encapsulation; for too low T_g the membrane would be too hard, which would hinder the crosslinking reaction. The T_g ranging from 5℃ to 15℃ is suitable for the water-borne coating in fluidized bed(Shen et al., 2009). Generally, crosslinking reaction can raise T_g, but in this system only membrane with 1 wt% of crosslinker elevated the T_g slightly, indicating that such crosslinker posed little influence on the T_g of acrylate latex, and similar results have been reported(Xie et al., 2002). Therefore, this kind of crosslinker can be used to improve membrane property without great effect on the T_g, and the inner molecular mechanism needs further research.

Table 2 The glass transition temperature (T_g) and thermal capacity of model membranes formed at 80℃ for 24 h

Coating Latex	Sample Weight (mg)	T_g (℃)	Delta C_P [J/(g · ℃)]
Acrylate + 0%	6.49	6.82	019
Acrylate + 1%	6.26	7.88	0.23
Acrylate + 2%	6.99	6.13	0.22

Notes: 0%, 1% and 2% denotes 0 wt%, 1 wt% and 2 wt% crosslinker added in the coating formulations, respectively.

2. Characteristics of fertilizer coatings

The microphotographs of fertilizer coatings formed at 80℃ differed substantially with the quantity of crosslinker increasing from 1 wt% to 2 wt%: many microscopic pores disappeared, and the coating became more uniform (Fig. 1). It indicated that crosslinking reaction played an important role in the coating formation, which acted as a barrier for mass transfer and thus closely connected with nutrients release(Tzika et al., 2003). Consequently, it could be deduced that the release rate of nutrients from fertilizer coated by coating formulation with 2 wt% of crosslinker was slower than that with 1 wt% of crosslinker.

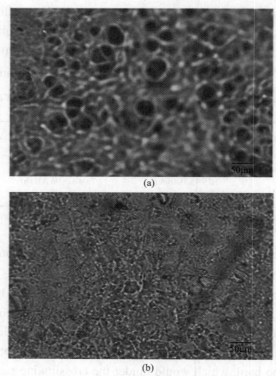

Fig. 1   Electron micrographs of coating surface on encapsulated fertilizer (formed at 80℃ for 24 h) incorporated: (a) 1 wt% and (b) 2 wt% of crosslinker

The FTIR-PAS spectra of fertilizer coatings showed a broad band due to O—H stretching vibration (3250-3550 cm^{-1}), and another band due to aliphatic C—H stretching vibration (2800-2950 cm^{-1}), including asymmetric and symmetric vibration (Fig. 2). The absorption band at 1710-1730 cm^{-1} proved the presence of C=O while the bands near 1100 cm^{-1} represented C—O and Si—O vibration(Han et al., 2009). The main difference between the two FTIR-PAS spectra was the weakness in the O—H vibration range (3250-3550 cm^{-1}) while strengthening in the remaining range (600-3200 cm^{-1}) for the coating added 2 wt% of crosslinker, which resulted from stronger interaction between the functional group (—COOH) and crosslinker. Such change modified the molecular structure, especially O—H. For the coating with 2 wt% of crosslinker, the ratio of absorption band denoting hydrophilic group O—H (3250-3550 cm^{-1}, band A) to that representing hydrophobic group aliphatic C—H (2800-2950 cm^{-1}, band B) remarkably decreased (Fig. 2), meaning the hydrophilicity of membrane was lowered(Ruth et al., 2009; Zhao et al., 2010). The results of FTIR-PAS spectra can verify that the increase of crosslinker amount performed obvious function on the improvement of coating hydrophobicity meanwhile complied with the result of swelling capacity of model membranes.

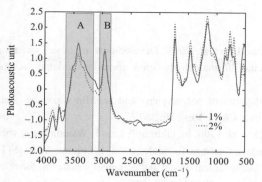

Fig. 2 Fourier transform mid-infrared photoacoustic spectra of fertilizer coatings (formed at 80℃ for 24 h) incorporated 1 wt % and 2 wt% crosslinker

3. *Nutrients release from polymer coated fertilizers*

The effects of crosslinker ratio on the release behavior of nutrients are shown in Fig. 3. The CRF whose coating formulation combined 2 wt% of crosslinker exhibited a much slower release of nutrients than that mixed with 1 wt% of crosslinker. In the 9 days of test, the former just delivered around 40% of the total nutrients while the latter nearly released completely, proving the great function of crosslinker percentage on release rate (Fig. 3). This fact could be explained by the membrane formation of more compact three dimension structure in the case of the coating formulation containing 2 wt% of crosslinker, as it has been observed in the EM pictures (Fig. 1). It was reported that such structure could control the nutrients diffusion better (Dušek and Dušková-Smrčková, 2000). Moreover, considering the results of swelling capacity and FTIR-PAS analysis, the enhancement of hydrophobicity of fertilizer coating was expected to contribute to the slow release of nutrients due to double amount of crosslinker employment. Hence it can be concluded that the reason why the coating formulation containing 2 wt% of crosslinker produced a longer duration of nutrients release is associated with both structure and property of coated membrane. The quantitative descriptions of nutrients release have been reported for polymer coated fertilizers(Du et al., 2004, 2008; Basu and Kumar, 2008; Fujinuma et al., 2009), and for this water-borne coating technology the nutrients release models may have to be improved or modified involving the factor of crosslinker; thus further investigations will be needed to optimize the controlled release parameters.

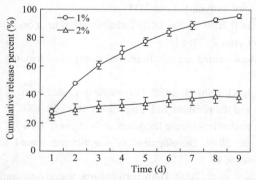

Fig. 3 Effect of crosslinker percentage (1%: 1 wt%; 2%: 2 wt%) on the cumulative release of nutrients from coated fertilizers heated at 80℃ in 40℃ static water

References

Ahmad S, Ashraf S M, Riaz U, et al. 2008. Development of novel waterborne poly(1-naphthylamine)/poly(vinylalcohol)-resorcinol formaldehyde-cured corrosion resistant composite coatings. Prog. Org. Coat., 62: 32

Al-Zahrani S M. 2000. Utilization of polyethylene and paraffin waxes as controlled delivery systems for different fertilizers. Ind. Eng. Chem. Res., 39: 367

Barbour M, Clarke J, Fone D, et al. 1996. In: Oldring P, Lam P. Waterborne & solvent based acrylics and their end user applications. Surface Coatings Technology Vol. 1. London: Wiley/SITA Technology

Basu S K, Kumar N. 2008. Mathematical model and computer simulation for release of nutrients from coated fertilizer granules. Math. Comput. Simulat., 79: 634

Byung-Su K, Young-Sang C, Hyun-Ku H. 1996. Controlled release of urea from rosin-coated fertilizer particles. Ind. Eng. Chem. Res., 35: 250

Cabrera R I. 1997. Let the nutrients flow slowly. Am. Nurseryman., 3: 32

Du C W, Tang D Y, Zhou J M, et al. 2008. Prediction of nitrate release from polymer-coated fertilizers using an artificial neural network model. Biosyst. Eng., 99: 478

Du C W, Zhou J M, Shaviv A, et al. 2004. Mathematical model for potassium release from polymer-coated fertilizer. Biosyst. Eng., 88: 395

Duśek K, Duśková-Smrčková M. 2000. Network structure formation during crosslinking of organic coating systems. Prog. Polym. Sci., 25: 1215

Fujinuma R, Balster N J, Norman J M. 2009. An improved model of nitrogen release for surface-applied controlled-release fertilizer. Soil Sci. Soc. Am. J., 73: 2043

Garcia M C, Diez J A, Vallejo A, et al. 1996. Use of kraft pine lignin in controlled-release fertilizer formulation. Ind. Eng. Chem. Res., 35: 245

Han X Z, Chen S S, Hu X G.2009. Controlled-release fertilizer encapsulated by starch/polyvinyl/alcohol coating. Desalination., 240: 21

Karadag E, Saraydin D, Caldiran Y, et al. 2000. Swelling studies of copolymeric acrylamide/crotonic acid hydrogels as carriers for agricultural uses. Polym. Adv. Technol., 11: 59

Kosuge N, Tobataku K. 1989. Coated granular Fertilizers., EP 030331

Pérez-García S, Fernández-Pérez M, Villafranca-Sánchez M, et al. 2007. Controlled release of ammonium nitrate from ethylcellulose coated formulation. Ind. Eng. Chem. Res., 46: 3304

Posey T, Hester R D. 1994. Developing a biodegradable film for controlled release of fertilizer. Plast. Eng., 1: 19

Rajsekharan A J, Pillai V N. 1996. Membrane-encapsulated controlled-release urea fertilizers based on acrylamide copolymer. J. Appl. Polym. Sci., 60: 2347

Ruth H E, Horst H G, Christian B.2009. In situ DRIFT characterization of organic matter composition on soil structural surfaces. Soil Sci. Am. J., 73: 531

Salman O A. 1989. Polyethylene-coated urea. 1. Improved storage and handing properties. Ind. Eng. Chem. Res., 28: 630

Salman O A, Hovakeemian J, Khraishi N. 1989. Polyethylene-coated urea. 2. Urea release as affected by coating material, soil type and temperature. Ind. Eng. Chem. Res., 28: 633.

Shaviv A. 2000. Advances in controlled-release fertilizers. Adv. Agron., 71: 1

Shen Y, Du C, Zhou J, et al. 2009. Development of water-borne coated fertilizer using reacted layer technology. Soil Fert. China., 6: 47

Tang L S, Zhang M, Zhang S F, et al. 2000. High performance waterborne aminoacrylic coatings from the blends of hydrosols and latexes. Prog. Org. Coat., 49: 54

Tomaszewska M, Jarosiewicz A. 2002. Use of polysulfone in controlled-release NPK fertilizer formulations. J. Agr. Food Chem., 50: 4634

Tzika M, Alexandridou S, Kiparissides C. 2003. Evaluation of the morphological and release characteristics of coated fertilizer granules produced in a Wurster fluidized bed. Powder Technol., 132: 16

Xie F, Liu Z H, Wei D Q. 2002. Properties and curing kinetic of acrylic resin cured with aziridine crosslinker. Chin. J. Synt. Chem., 10: 120 (in Chinese)

Yang Y, Zhang M, Ma L, et al. 2007. Fast measurement of nutrients release rate of coated controlled release fertilizer. Plant Nutr. Fert. Sci., 13: 730

Zhao C, Shen Y Z, Zhou J M, et al. 2010. Hydrophobicity characterization of polymer coating for controlled-release fertilizer using Fourier transform infrared photoacoustic spectroscopy. Chinese J. Anal. Chem., 38: 1186

1.4 Aqueous polyacrylate/poly (silicone-co-acrylate) emulsions-coated fertilizers for slow nutrient release application

1.4.1 Introduction

Controlled release fertilizers (CRFs) have been aiming at improving plant nutrient use efficiency and minimizing nutrient losses, thus reducing environmental threats and health problems often associated with poor fertilization management (Shaviv and Mikkelsen, 1993; Trenkel, 1997; Hanafi et al., 2000; Shaviv, 2000; Shaviv et al., 2003). Fertilizers coated with hydrophobic polymers are the major categories of CRFs as they have excellent nutrient release profiles (El-Refaie and Sakran, 1996; Al-Zahrani, 2000; Shaviv et al., 2003; Qiu et al., 2012).

In recent years, the waterborne coating technique has been applied widely since it can remove the possible secondary pollution caused by volatile organic solvent-based coating (Abraham and Pillai, 1996; Guo et al., 2012). Aqueous polyacrylate emulsion has the advantage of excellent film-forming characteristic, appropriate viscosity and low price (Qin et al., 2008; Zhang and Wang, 2008). However, hydroxyl and carboxylic groups in aqueous polyacrylate make aqueous polyacrylate polymer hydrophilic and swelling dramatically in wet environment. The low tolerance of aqueous polyacrylate to water would reduce the nutrient release duration of CRFs. Our previous research in which, cross-linker was used to modify the polyacrylate, showed that the preliminary solubility rate of nutrient was 25% and the cumulative nutrient release was 40% in 9 days at 40℃(Zhao et al., 2010). The nutrient release was further slowed through additional measures in the study. Firstly, the n-butyl acrylate /methyl methacrylate (BA/MMA) ratios had an important effect on the physicochemical properties of the emulsion (Cao et al., 2007; Dashtizadeh et al., 2011). Moreover, the hydrophilicity of aqueous polyacrylate had been reduced by organic silicone due to the stronger water repellency(Lim et al., 1999; Lee et al., 2003; Lin et al., 2005; Zhang et al., 2007; Naghash and Mohammadrahimpanah, 2011; Rüttermann et al., 2011). The poly (silicone-co-acrylate) emulsions were prepared by three methods: seeded polymerization, copolymerization and miniemulsion polymerization. In seeded polymerization and copolymerization, the principal locus of particle nucleation is in the monomer-swollen micelles. It is necessary for monomer to diffuse from monomer droplets to growing polymer particles, which causes highly premature cross-linking due to easily hydrolysis and condensation of organic silicone(Cao et al., 2007). In order to suppress hydrolysis and self-condensation of VTES, EG was used when preparing the poly (silicone-co-acrylate) by seeded polymerization and copolymerization. The miniemulsion is a relatively stable oil-in-water dispersion, which is typically obtained by shearing a system containing monomer, water, surfactant and a costabilizer(Qi et al., 2006). Monomer droplets in a miniemulsion become the dominant site for particle nucleation, which allows for organic silicone avoiding contact with water and suppress the organic silicone from hydrolysis and condensation(Zhang et al., 2009).

In this paper, waterborne polyacrylate and poly(silicone-co-acrylate) emulsions were synthesized, and the influences of BA/MMA ratio and polymerization technique on the properties of the coating and nutrient-release profiles of CRFs were investigated. The suitable emulsion would be well used to coat fertilizer and further slow the nutrient release of CRFs.

1.4.2 Experimental

1. Materials

The commercial granular compound fertilizers (NPK, 15-12-15) were provided by Fulilong Guangdong Fertilizer Co., Ltd. Methyl methacrylate (MMA, AR), *n*-butyl acrylate (BA, CP) and methacrylic acid (MAA, CP), and ethylene glycol (EG, AR) were from Nanjing Chemical Regent Co., Ltd. Sodium dodecylbenzenesulfonate (SDBS, AR) was from Chengdu Kelong Chemical Reagent Co., Ltd. Nonyl phenyl polyoxyethylene ether-10 (OP-10, LP) was received from Hebei Xingtai Kewang Auxiliary Agent Co., Ltd. Potassium persulfate (KPS, CP) and hexadecane (HD, CP) were supplied by Sinopharm Chemical Reagent Co., Ltd. Vinyltriethoxysilane (VTES, LP) was received from XFLP Silicones Co., Ltd. All the reagents were used as received. Deionized water was used to prepare all the solutions and emulsions.

2. Synthesis of polyacrylate and poly(silicone-co-acrylate) emulsions

The polyacrylate emulsions with different BA/MMA ratios were synthesized by semi-continuous polymerization. The emulsions polymerization was carried out in a 1000 mL three-neck flask equipped with a mechanical stirrer, reflux condenser and dropping funnel. The aqueous phase was prepared by dissolving 8.24 g OP-10 and 4.12 g SDBS in 248 g water. The organic phase was prepared by mixing the desired amount of BA and MMA as well as 3.5 g MAA. The total amount of BA and MMA was 200 g. Both phases were vigorously stirred for 30 min and the temperature was raised to 80℃ until the end of polymerization. 25 wt% of the oil-water mixture in the flask was used as the initial charge. The rest of mixture and the initiator solution (52 mL, 0.01 g/ mL, KSP) were fed alternately in 4 doses over 3 h, and then the polymerization was conducted under air atmosphere for additional 3 h.

The poly(silicone-co-acrylate) emulsions were prepared using three methods: seeded polymerization, copolymerization and miniemulsion polymerization (Fig. 1). VTES was prone

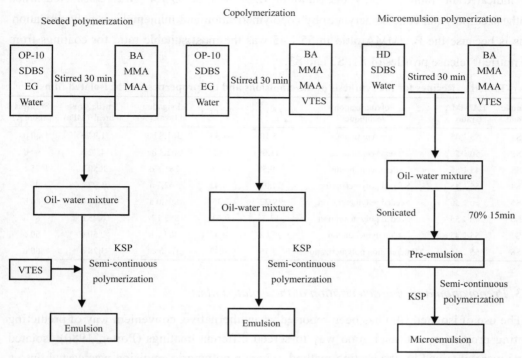

Fig. 1 Procedures for the synthesis of poly(silicone-co-acrylate) emulsions by seeded polymerization, copolymerization, and miniemulsion polymerization

to hydrolysis and polycondensation with itself. In order to suppress hydrolysis and self-condensation of VTES, EG was used when preparing the poly (silicone-co-acrylate) by seeded polymerization and copolymerization. However, EG was absent when preparing the poly(silicone-co-acrylate) by miniemulsion, for VTES was protected from the aqueous phase by a waterproof oil droplet and the hydrolysis reactions can be dramatically reduced in miniemulsion. The same polymerization parts as polyacrylate emulsion were no longer discussed and the different parts were only underlined as follows. For seeded polymerization, 31 g EG was added to water phase at the beginning and 5 g VTES (2.5wt% of the total acrylate monomer mass) was injected to the flask after the completion of the rest oil-water mixture addition(Naghash et al., 2006; Naghash and Mohammadrahimapanah, 2011); for copolymerization, 31 g EG was added to water phase, and the oil phase was prepared by mixing 5 g VTES with the acrylate monomer from the very beginning; for miniemulsion polymerization(Asua, 2002; Koukiotis and Sideridou, 2008; Ramos-Fernández et al., 2012), 8.24 g HD displaced OP-10 and was used as co-stabilizer, meanwhile, the oil phase was prepared by mixed 5 g VTES with the acrylate monomers. Both phases were mixed and stirred vigorously for 30 min to get the pre-emulsions. The pre-emulsion was then sonicated for 15 min at 70% output with Ultrasonic (KQ-500, Kun Shan Ultrasonic Instruments Co., Ltd.). The resultant miniemulsion was transferred into a 1000 mL three-neck flask and the polymerization was performed just as polyacrylate emulsion mentioned above. Eight emulsions were synthesized and the solid content of all emulsions was about 40%. The details of emulsions are indicated in Table 1. Only one ratio of BA/MMA in 55∶45 was considered when synthesized poly (silicone-co-acrylate) by copolymerization and miniemulsion polymerization. This is because the BA/MMA ratio in 55∶45 was the most suitable ratio for coatings from the nutrient release profiles of S1, S2 and S3.

Table 1 Recipes for the synthesis of the emulsions and the properties for the isolated films

Sample code	BA/MMA ratio	Polymerization technique	T_g^a (℃)	T_g^a (℃)	Elongation at break (%)	Tensile stress at break (MPa)	Stiffness (Shore A)
S1	55∶45	Semi-continuous	5.10	−1.33	361.8 bc	1.97 e	60 d
S2	50∶50	Semi-continuous	11.09	5.82	362.3 bc	4.72 c	60 d
S3	45∶55	Semi-continuous	19.58	13.36	281.5 d	8.50 a	75 c
S4	55∶45	Seeded polymerization	4.02	−4.42	862.7 a	1.97 e	45 f
S5	50∶50	Seeded polymerization	9.22	2.36	804.0 a	3.49 d	80 b
S6	45∶55	Seeded polymerization	13.73	9.53	383.1 b	6.25 b	85 a
S7	55∶45	Copolymerization	5.62	−4.42	407.7 b	2.50 e	50 e
S8	55∶45	Miniemulsion polymerization	5.75	−4.42	311.5 cd	3.78 d	60 d

3. Preparation and characterization of the isolated films

The use of isolated film has been reported as an alternative, convenient way of predicting coating properties, and as a good way to screen different coatings (Porter, 1980). Isolated films were obtained by the casting method, where a polymeric emulsion was casted onto a non-stick substrate and water was evaporated completely in an oven at 80℃ for 8 h. The swelling degree was determined by the following procedure: the weighed film (W_{f1}) was immersed into deionized water and remained sinking at 25℃, followed by removing the swollen films periodically from water and wiping off the surface water with a piece of filter paper to determine the weight (W_{f2}). The swelling degree was defined as ($W_{f2} - W_{f1}$)/ W_{f1} × 100 (Chen et al., 2010). The lower of swelling degree indicated the better of water-resistance property of the isolated films.

Circa 10 mg of isolated film was weighed and differential scanning calorimetry (DSC, Perkin-Elemr Pyris1, USA) was taken at a heating rate of 20℃/min. The thermal behavior for the samples was examined under the nitrogen atmosphere between −100℃ and 150℃. The glass transition temperature (T_g) was taken at the onset of the corresponding heat capacity jump. As a rule, two successive scans were made for every sample. All calculations were performed on the second heating cycle.

Tensile properties of isolated films were measured using a universal testing machine (CMT 5254, Shenzhen SANS Testing Machine Co., Ltd., China) according to the procedures outlined in ASTM D 638-03. A dumb-bell shaped die wide (type A2) was cut from the isolated film. The initial gauge length was 10 mm, and the measuring speed was 200 mm/min. Elongation at break and tensile stress at break were measured based on three independent drawing experiments performed at the same conditions. Hardness tests (Shore A) were conducted on a hardness tester (XL-A, Jiangdu Mingzhu Testing Machine Co., Ltd., China), and referred to ASTMD 2240-03 at (23±2)℃ and relative humidity of 50% (Xiang et al.,

2011). All hardness data were the average values of three runs.

4. *Preparation of coated fertilizers*

The fertilizers granules were coated in a Wurster fluidized bed equipped with a bottom-spray pneumatic nozzle (LDP-3, Changzhou Jiafa Granulation Drying Equipment Co., Ltd.). The process parameters were: product temperature 45-50℃, spray rate 2.5 g/min, and atomization pressure 0.1 MPa. The amount of coating emulsions was 75 g per 300 g original fertilizer granules with 2-3 mm in diameter. The average coating thickness was about 100 μm. In all the cases, the coated granules were tray-dried in an oven at 80℃ for 8 h before further evaluation.

5. *FTIR-PAS characterization and detection release profiles of CRFs*

A FTIR spectrometer (Nicolet 380, USA) equipped with photoacoustic accessory (Model 300, MTEC, USA) was used for the determination of the spectra of CRFs coated with S1, S3 and S4. The spectra were recorded in wavenumber range of 500-4000 cm^{-1}, and the mirror velocity was set to 0.63 cm/s, 32 successive scans were conducted with a resolution of 4 cm^{-1}.

Five grams of coated fertilizer were immersed in 100 mL of deionized water at 25℃. 100 mL solution was removed periodically (to determine the amount of nutrient release) and replaced by deionized water (100 mL) with three replicates; the nutrient relative content was evaluated by solution conductivity (Dai et al., 2007), measured with an electrical conductivity apparatus (DDS-320, China). On the 30th day of release, the coated fertilizers were ground to determine the content of residual nutrient. The release profiles were estimated as the cumulative release percentages versus time.

1.4.3 Results and Discussion

1. *Effect of BA/MMA ratios on swelling degree of the isolated films*

The swelling degree of isolated films is an important evidence of water resistance performance, and the lower of swelling degree indicates stronger of water resistance performance. As shown in Fig. 2, the increasing of MMA caused stronger water resistance performance on both polyacrylate (Fig. 2(a)) and poly(silicone-co-acrylate) films (Fig. 2(b)). The polymer chains were arranged closely and the free space between polymer chains was decreased with increasing of MMA. It improved water resistance of the film.

2. *Effect of synthesis methods on swelling degree of the isolated films*

The swelling degrees of poly(silicone-co-acrylate) emulsions prepared by different ways are showed in Fig. 3. In the case of the BA/MMA ratio of 55∶45, poly(silicone-co-acrylate) emulsions prepared by miniemulsion polymerization suggested the lowest swelling degree while the other methods revealed similar swelling degree. That is because hydrophilic EG was absent when preparing poly(silicone-co-acrylate) by minioemulsion.

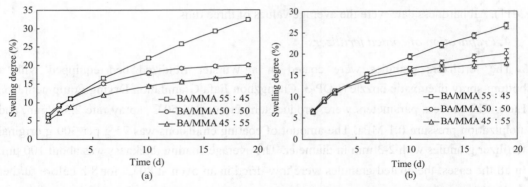

Fig. 2 Effect of the BA/MMA ratios on the swelling degrees of isolated films obtained by (a) polyacrylate and (b) poly(silicone-co-acrylate) emulsions prepared by seeded polymerization

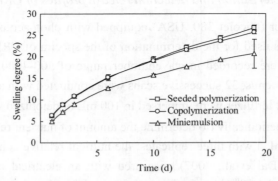

Fig. 3 Effect of the synthesis methods on the swelling degrees of the isolated films

3. *Glass transition temperatures of isolated films*

Glass transition temperature, which is closely related to the film formation states of CRFs' coating, can be used as an indirect indicator for selecting coatings of CRFs. As shown in Table 1, the measured T_g's increased with increasing of MMA, whereas decreased with increasing of BA irrespective of polyacrylate and poly(silicone-co-acrylate) emulsions. The effects of VTES on glass transition temperature were associated with the synthesis techniques. When poly(silicone-co-acrylate) emulsion was prepared by seeded polymerization, the T_g's were lower than those of the corresponding polyacrylate isolated films with the same BA/MMA ratios. When poly(silicone-co-acrylate) emulsions were prepared by copolymerization and miniemulsion, the T_g's were higher. If the ethoxy group of VTES did not hydrolyze, the glass transition temperature would be lower because the T_g of poly (VTES) is –90℃(Maitra et al., 2003). In addition, the glass transition temperature could be increased with crosslinks in copolymer films. These two opposite aspects could therefore affect the final glass transition temperature(Guo et al., 2005). VTES was added to the flask after the completion of the oil-water mixture in seeded polymerization, while it was injected to the flask at the beginning in copolymerization and miniemulsion. The addition of VTES in the later period of polymerization reaction would avoid the prolonged contacting between the alkoxysilane and water, thus alleviating hydrolysis processes. Normally T_g of coatings suitable for CRFs was 5-15℃(Shen et al., 2009), thus neither S3 nor S4 was not suitable for CRFs coating. The T_g's were predicted according to Fox equation (Table 1). The measured T_g's were all higher than predicted one, because Fox equation assumed all monomer were random copolymerization and ignored the influence of segment distribution on T_g's of the copolymer. In copolymerization system of poly (MMA-co-BA), the obvious difference of the reactivity ratio between MMA (1.88) and BA (0.43) caused more block copolymerization(Cao et al., 2007).

4. Mechanical properties of isolated films

It is necessary to develop CRFs coating combining the flexibility, strength and stiffness. Elongation at break is a measure of film flexibility, whereas tensile stress is a measure of film's strength and the stiffness is characterized by Shore A hardness. Table 1 demonstrates the mechanical properties of isolated films. Increasing MMA content led to a strain hardening coating with low flexibility irrespective of polyacrylate and poly(silicone-co-acrylate) emulsions. Compared with S1, the poly(silicone-co-acrylate) emulsion prepared by seeded polymerization (S4) showed higher flexibility and lower stiffness. The strain softening coating could be explained by lower T_g of poly(silicone-co-acrylate) film prepared by seeded polymerization. The tensile strength for the film of miniemulsion polymerization presented higher value compared with the film of seeded polymerization and copolymerization. EG, besides suppressing hydrolysis and self-condensation of VTES, also has the plasticizing function. It decreased the intermolecular interactions, therefore the strength was lowered(He, 1990). EG was absent when preparing the poly(silicone-co-acrylate) by miniemulsion, while it was present when preparing the poly(silicone-co-acrylate) by seeded polymerization and copolymerization.

5. FTIR-PAS characterization of the CRFs

The FTIR-PAS spectra of CRFs coated with S1, S3 and S4 are displayed in Fig. 4. The wide absorption at 3250-3550 cm^{-1} (O—H stretching vibration) suggested lower intensity in S3 than S1, which resulted from higher monomer reactivity ratios of MMA than BA. We

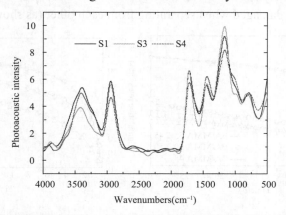

Fig. 4 FTIR–PAS spectra of CRFs coated with S1, S3, and S4

S1 and S3 denote polyacrylate emulsions with BA/MMA ratios of 55∶45 and 45∶55, espectively. S4 represents the poly(silicone-co-acrylate) emulsion prepared by seeded polymerization with a BA/MMA ratio of 55∶45（color figure can be viewed in the online issue, which is available at wileyonlinelibrary.com）

could note the absence of C=C at 1645 cm^{-1} in S4, which manifested that the vinyl groups in VTES did join the reaction and were consumed during the emulsion polymerization. Besides, the spectra of S4 in the range of 990-1136 cm^{-1} were broader compared to S1 and S3 because of Si—O—Si and Si—O—C asymmetric stretching. Coupled with high content of EG in poly(silicone-co-acrylate) emulsions, all above provided evidences for the copolymerization of VTES onto polyacrylate chain(Zou et al., 2005).

6. *Nutrient cumulative release profiles*

Nutrient release profile is essential to evaluate whether the coatings are suitable for CRFs. The final nutrient release profiles of CRFs are directly related to the physicochemical properties of the coating, i.e., water resistance performance, glass transition temperature and mechanical properties. The effect of BA/MMA ratio on the nutrient release behavior of CRFs is displayed in Fig. 5. For polyacrylate emulsions (Fig. 5(a)), an increase in BA content resulted in a decrease in nutrient release rate. Nutrient release profile of S1 was similar to that of S2, while the nutrient release of S3 was sharply accelerated. The water resistance performance of S3 was improved, which was of benefit to the nutrient slow release. However the T_g (19.58℃) was much higher than that of coatings suitable for CRFs (5-15℃). Additionally, the stiffness of S3 was greatly increased and the flexibility was significantly decreased, which led to brittle coating. So CRFs coated with S3 were more vulnerable to release nutrient through "fail mechanism"(Shaviv, 2000). For poly(silicone-co-acrylate) emulsions prepared by seeded polymerization (Fig. 5(b)), the nutrient release was the slowest with the BA /MMA ratio in 50∶50 (S5), while the nutrient release of ratio in 55∶45 (S4) was intermediate and that of ratio in 45∶55 (S6) was the fastest. The effects of BA/MMA ratio on the nutrient release profile were not accordant with polyacrylate emulsion. In comparison with S5, T_g of S4 dropped from 5.1℃ to 4.02℃ and hardness decreased from 60 to 45. All these caused sticky coatings, thus twins coated granules were formed and the coatings integrity was damaged when separating from each other (as shown in Fig. 6).

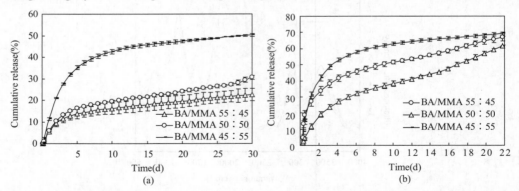

Fig. 5　Effect of the BA/MMA ratios on the nutrient-release profiles of CRFs coated with (a) polyacrylate and (b) poly(silicone-co-acrylate) emulsions prepared by seeded polymerization

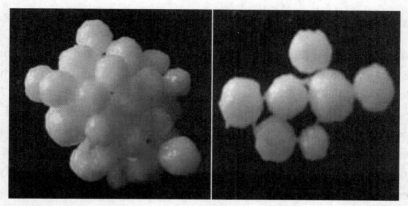

Fig. 6 Optical pictures of the sticky coating of CRFs. The CRFs were coated with poly(silicone-co-acrylate) emulsions prepared by seeded polymerizationwith a BA/MMA ratio of 55 : 45 (S4)

The effect of the synthesis methods on the nutrient release profiles of CRFs is demonstrated in Fig. 7. In the case of BA/MMA ratio in 55 : 45, CRFs coated with poly(silicone-co-acrylate) emulsion prepared by miniemulsion polymerization exhibited the slowest nutrient release. CRFs just delivered about 15% of the total nutrient in the 30 days of test, whereas CRFs coated with poly (silicone-co-acrylate) emulsions prepared by copolymerization and seeded polymerization released 36% and 78% of the total nutrient, respectively. Seeded polymerization (S4) resulted in lower glass transition temperature and a strain softening effect; lower glass transition temperature and strain softening effect led to a sticky coating (as shown in Fig. 6), accelerating the nutrient release. In comparison with copolymerization (S7), miniemulsion polymerization (S8) resulted in stronger water resistance performance, higher glass transition temperature and a strain hardening effect. It suggested slower nutrient release of CRFs. Moreover, high water resistance performance was beneficial to slow nutrient release of CRFs. Nevertheless, it was observed that the nutrient release rate was not directly related to water resistance performance, and the mechanical properties of the coating had a greater influence on the nutrient release in this study.

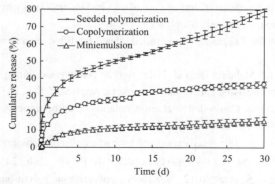

Fig. 7 Effect of the synthesis methods on the nutrient-release profiles of CRFs

1.4.4 Conclusions

In order to slow nutrient release of CRFs coated with aqueous polyacrylate emulsion, the effects of BA/MMA ratio, VTES and synthesis technique on the coating properties and the nutrient release behavior had been investigated. Water resistance performance, T_g, strength and stiffness increased with increasing the MMA content. VTES improved the water resistance performance of coating, but the effects on glass transition temperature, mechanical properties and nutrient release behavior varied with different synthesis techniques. When the BA/MMA ratio was 55∶45, the poly(silicone-co-acrylate) prepared by seeded polymerization (S4) resulted in low glass transition temperature and a strain softening coating, accelerating the nutrient release; the poly(silicone-co-acrylate) prepared by copolymerization (S7) led to higher glass transition temperature and lower stiffness; while miniemulsion polymerization (S8) gave higher glass transition temperature and a strain hardening effect, slowing nutrient release. The preliminary solubility rate of nutrient was about 3% and the cumulative nutrient release did not exceed 15% within 30 days at 25℃. The nutrient release rate was greatly decreased compared to that of CRFs prepared in our previous study (the preliminary solubility rate was 25% and the cumulative nutrient release was 40% in the 9 days at 40℃)(Zhao et al., 2010). In summary, waterborne poly(silicone-co-acrylate) emulsion synthesized by miniemulsion at the BA/MMA in 55∶45 could be well used in CRFs coatings.

References

Abraham J, Pillai, R V N. 1996. Membrane-encapsulated controlled- release urea fertilizers based on acrylamide copolymers. J. Appl. Polym. Sci., 60: 2347-2351

Al-Zahrani S M. 2000. Utilization of polyethylene and paraffin waxes as controlled delivery systems for different fertilizers. Ind. Eng. Chem. Res., 39, 367-371

Asua J M. 2002. Miniemulsion polymerization. Prog. Polym. Sci., 27: 1283-1346

Cao T, Liu Q, Hu J. 2007. The synthesis principle, properties and applications of polymer emulsion. Beijing: Chemical Industry Press

Chen F, Ye F, Chu G. 2010. Synthesis of acrylate modified vinyl chloride and vinyl isobutyl ether copolymers and their properties. Prog. Org. Coat., 67: 60-65

Dai J, Fan X, Yu J, et al. 2007. Longevity of controlled release modified resin fertilizer. Sci. Agr. Sin. CHED., 40: 966-971

Dashtizadeh A, Abdouss M, Mahdavi H, et al. 2011. Acrylic coatings exhibiting improved hardness, solvent resistance and glossiness by using silica nano-composites. Appl. Surf. Sci., 257: 2118-2125

El-Refaie K, Sakran A A.1996. Controlled release formulation of agrochemicals from calcium alginate. Ind. Eng. Chem.Res., 35: 3726-3729

Guo T Y, Xi C, Hao, G J, et al. 2005. Preparation and properties of room temperature self-crosslinking poly (MMA-co-BA-co-St-co-VTES) latex film. Adv. Polym. Tech., 24: 288-295

Guo Y H, Li S C, Wang G S, et al. 2012. Waterborne polyurethane/poly(n-butyl acrylate-styrene) hybrid emulsions: particle formation, film properties, and application. Prog. Org. Coat., 74: 248-256

Hanafi M M, Eltaib S M, Ahmad M B. 2000. Physical and chemical characteristics of controlled release compound fertilizer. Eur. Polym. J., 36: 2081-2088.

He M J. 1990. Polymer Physics. Shanghai: Fudan University Press

Koukiotis C, Sideridou I D. 2008. Preparation of high solids stable translucent nanolatexes of MMA/BA copolymers and MMA/BA/Veova-10 terpolymers with low MFFT using green industrial surfactants. Prog. Org. Coat., 63: 116-122

Lee Y J, Akiba I, Akiyama S J. 2003. The study of surface segregation and the formation of gradient domain structure at the blend of poly (methyl methacrylate)/poly(dimethyl siloxane) graft copolymers and acrylate adhesive copolymers. J. Appl. Polym. Sci., 87: 375-380

Lim K T, Webber S E, Johnston K P. 1999. Synthesis and characterization of poly(dimethyl siloxane)-poly(alkyl (meth) acrylic acid) block copolymers. Macromolecules, 32: 2811-2815

Lin M, Chu F, Guyot A. 2005. Silicone-polyacrylate composite latex particles. Particles formation and film properties. Polymer, 46: 1331-1337

Maitra P, Ding J, Huang H, et al. 2003. Poly (ethylene oxide) silananted nanosize fumed silica: DSC and TGA characterization of the surface. Langmuir, 19: 8994-9004

Naghash H J, Mallakpour S, Mokhtarian N. 2006. Synthesis and characterization of silicone-modified vinyl acetate -acrylic emulsion copolymers. Prog. Org. Coat., 55: 375-381

Naghash H J, Mohammadrahimpanah R. 2011. Synthesis and characterization of new polysiloxane bearing vinylic function and its application for the preparation of poly (silicone-co-acrylate)/montmorillonite nanocomposite emulsion. Prog. Org. Coat., 70: 32-38

Porter S C. 1980. The effect of additives on the properties of an aqueous film coating. Pharm. Technol., 4: 67-75

Qi D, Bao Y, Weng Z, et al. 2006. Preparation of acrylate polymer/silica nanocomposite particles with high silica encapsulation efficiency via miniemulsion polymerization. Polymer, 47: 4622-4629

Qin Y, Tang S, Huang H. 2008. Preparation of new types of controlled release fertilizers and their controlled release capability. Chinese Journal of Tropical Agriculture, 28: 29-33

Qiu X Y, Tao S M, Ren X Q, et al. 2012. Modified cellulose films with controlled permeatability and biodegradability by crosslinking with toluene diisocyanate under homogeneous conditions. Carbohydr. Polym., 88: 1272-1280

Ramos-Fernández J M, Beleña I, Romero-Sánchez M D, et al. 2012. Study of the film formation and mechanical properties of the latexes obtained by miniemulsion co-polymerization of butyl acrylate, methyl acrylate and 3-methacryloxypropyltrimethoxysilane. López-Buendía, Á.M. Prog. Org. Coat., 75: 86-91

Rüttermann S, Trellenkamp T, Bergmann N. 2011. A new approach to influence contact angle and surface free energy of resin-based dental restorative materials. Acta Biomater, 7: 1160-1165

Shaviv A, Mikkelsen R L. 1993. Controlled-release fertilizers to increase efficiency of nutrient use and minimize environmental degradation. Fert. Res., 35: 1-12.

Shaviv A, Raban S, Zaidel E. 2003. Model of diffusion release from polymer coated granular fertilizer. Environ. Sci. Technol., 37: 2251-2256

Shaviv A. 2000. Advances in controlled-release fertilizers. Adv. Agron., 71: 1-49

Shen Y Z, Du C W, Zhou J M, et al. 2009. Development of water-borne polymercoated fertilizer using reacted layertechnology. Soil and Fertilizer Sciences in China, 6, 47-51

Trenkel M E. 1997. Slow-and controlled-release and stabilized fertilisers: an option for enhancing nutrient use efficiency in Agriculture. IFA: Paris

Xiang X Q, Chen S J, Zhang J, et al. 2011. Photodegradation of plasticized poly (vinyl chloride) stabilized by different types of thermal stabilizers. Polym. Eng. Sci., 51: 624-631

Zhang B, Liu B, Deng X. 2007. A novel approach for the preparation of organic-siloxane oligomers and the creation of hydrophobic surface. Appl. Surf. Sci., 254: 452-458.

Zhang F, Wang Y. 2008. Current situation and development trend of slow/controlled- release fertilizer in China. Soil Fertilizer in China, 4: 1-4

Zhang S W, Liu R, Jiang J Q, et al. 2009. Film formation and mechanical properties of the alkoxysilane-functionalized poly (styrene-co-butyl acrylate) latex prepared by miniemulsion copolymerization. Prog. Org. Coat., 65: 56-61

Zhao C, Shen Y Z, Du C W. 2010. I Evaluation of waterborne coating for controlled-release fertilizer using

Wurster fluidized bed. Ind. Eng. Chem. Res., 49: 9644-9647

Zou M, Wang S, Zhang Z. 2005. Preparation and characterization of polysiloxane-poly (butyl acrylate-styrene) composite latices and their film properties. Eur. Polym. J., 41: 2602-2613

1.5 Biodegradation of a biochar-modified water-borne polyacrylate membrane coating for controlled release fertilizer and its effects on soil bacterial community profiles

1.5.1 Introduction

Fertilizers are extremely important for crop yield. Controlled release fertilizers (CRFs) coated with a polymer control the release of nutrients so their availability coincides with the crop's requirement, and show great potential in agriculture (Shaviv, 2001). In contrast to conventional fertilizers, CRFs are advantageous because they reduce nutrient leaching by rain or irrigation and save labor-associated costs of fertilizer application (Zhao et al., 2010; Han et al., 2009), especially for rice production (Choudhury and Kennedy, 2005; Li et al., 2009; Yang et al., 2012).

Control over the rate of nutrient release from coated fertilizer is mainly decided by the polymer coating. Currently there are two main kinds of coating polymers, and they are distinguished by the type of solvent used in CRF production. There are the organic solvent dissolvable polymers, such as polyacrylamide (Rajsekharan and Pillai, 1996) or polystyrene (Garcia et al., 1996), which are expensive and toxic to the environment in production and in application. The other type of coating polymer is the water dissolvable polymer, such as the water-borne polymers, which are relatively cheap and environmentally non-toxic during production (Zhao et al., 2010). Consequently, increasing interest is focused on water-borne polymer coatings.

Degradation of the CRF polymer coating in soil is a factor that is highly considered for control over the rate of nutrient release and also for environmental safety. Degradation of the polymer is influenced by soil temperature, soil moisture, and especially by soil biological activity (Kennedy and Smith, 1995; Shah et al., 2008). Degradation of the polymer refers not only to bond scission and chemical transformation, but also to the formation of new functional groups (Celina, 2013). Water-borne polyacrylate contains ester groups and carboxyl groups, which are sensitive to moisture (Zhang et al., 2012). These groups offer opportunities to tune the biomechanical properties within a broad range of desired properties and to accelerate polymer degradation (Zhang et al., 2012; Decker and Zahouily, 1999). Variation in polymer degradation is increased by the complex soil environment, which may result in polymer coating failure and unguaranteed control over the rate of nutrient release. On the other hand, the degradation products of synthetic polymers may impact on microorganisms that play a central role in nutrient cycling and provide an important ecosystem service (Hadad et al., 2005; Albertsson and Karlsson, 1990; Costanza et al., 1987). Furthermore, changes in soil bacterial community profiles are important to the soil environment (Kirk et al., 2004). For example, Liu et al. (2011) detected that one kind of CRF resin coating increased the quantity of bacteria and actinomyces, determined using plate count and enzyme activity in soil, and Ikeda et al. (2014) found that urea-formaldehyde, the oldest slow release N fertilizer, markedly increased bacterial diversity. Despite water-borne polymer coated CRFs having the most potential, there is little published information about changes to soil bacterial profiles in response to their degradation in soil.

In our previous study, we found that water-borne polyacrylate emulsion could be applied to CRF development; furthermore, biochar-modified water-borne polyacrylate material enhanced the mechanical strength and increased the release period, hence demonstrating great potential for water-borne polymer coated CRFs (Zhou et al., 2013). The biochar involved was derived from locally available wheat residues (Lehmann, 2007; Huggins et al., 2014), and the effects of the biochar on soil microbial profiles were studied (Zimmerman et al., 2011). However, the degradation of biochar-modified water-borne polyacrylate membranes and its impact on soil microbes are unknown.

The objectives of this study were: ① to detect the degradation of the biochar-modified water-borne polyacrylate membrane in paddy soil using Fourier transform infrared photoacoustic spectroscopy (FTIR-PAS); and ② to explore the effects of the biochar-modified water-borne membrane on soil bacterial community and activity by polymerase chain reaction-denaturing gradient gel electrophoresis (PCR-DGGE) and Biolog EcoPlates.

1.5.2 Materials and methods

1. *Materials*

Water-borne polyacrylate emulsion (Doctor Hydrophilic Chemicals Co. Ltd., Yizheng, China) was used as a coating material to control the rate of nutrient release. The emulsion contained butyl acrylate, methyl methacrylate, methyl acrylic acid, and the cross-linker aziridine. The biochar was from wheat straw that was pyrolyzed at 400 ℃; detailed information has been presented elsewhere (Xu et al., 2013). Biochar (1% wt/wt) was added to the water-borne polyacrylate emulsion for both chemical and physical modification. The detailed information about making membranes was as follows: 55 g water-borne polyacrylate latex or 55.275 g biochar-modified water-borne polyacrylate latex (0.275 g crushed biochar was added into 55 g latex and then stirred for 15 min at room temperature) was distributed into a polytef culture dish (internal diameter, 10 cm) dried in an oven at 80℃ for 24 h, and formed into the membranes of 1 mm thickness on dish surface, then the model membranes removed from the mold tailored into circles with 1 cm diameter and stored in a 4℃ refrigerator for use.

Paddy soil was collected from the Ecological Station of Red Soil, Chinese Academy of Sciences, in Yingtan City, Jiangxi Province of China (28°15′N, 116°55′E). Soil agro-chemical properties were as follows: pH (H_2O), 5.2; organic carbon, 22.13 g/kg; total nitrogen, 1.92

g/kg; total phosphorous 0.61 g/kg; and total potassium 6.01 g/kg. The soil texture using in the experiment was loam soil (Ultisols and Oxisols in US Soil Taxonomy) contained 38% sand, 42% silt, and 20% clay. The soil was air-dried at room temperature and passed through a 2 mm sieve.

2. *Soil incubation experiment*

Twenty seven of polyethylene cups (height, 13 cm; internal diameter, 8.5 cm) were filled with 500 g of soil, and four pieces of membrane (about 0.7 g) were buried horizontally in the soil at 1 cm depth from the soil surface. The soil was waterlogged and a 2 cm layer of water was maintained on the surface. Deionized water was used for all water treatments. The cups were covered with plastic film perforated with holes for gas exchange and incubated at 28 ± 1℃. Water was added to the cups every two days to maintain the water layer.

Bifactorial design with the factors treatment and time each replicated three times was set. The first factor was membrane treatment which included control without membranes in soil (CK), unmodified water-borne polyacrylate (UP) membranes and biochar-modified water-borne polyacrylate (BP) membranes. The second factor is sampling position treatment which included the contacted soil layer (top layer, 0-2 cm distance from the buried membrane surface) and the noncontacted soil layers (bottom layer, 8-10 cm distance from the buried membrane surface). Soil samples and buried membranes were both sampled after 2, 6 and 12 months.

The sampled membranes were washed with water and dried at 80℃, weighed, and then measured using Fourier transform infrared photoacoustic spectroscopy (FTIR-PAS). The fresh soil samples were subdivided into two subsamples. One was used to determine the functional diversity of microbial communities by Biolog EcoPlates, and the other was stored at −20℃ prior to DNA extraction and then the genetic structures of bacteria were evaluated using PCR-DGGE.

3. *Recording of FTIR-PAS spectra*

FTIR-PAS was based on the photoacoustic effect while the photoacoustic signal generation was affected by the physical properties of soils (Rosencwaig and Gersho, 1976), and the technique has been previously demonstrated to be very suitable to analyze polymeric materials, especially the heterogeneous compound polymers without sample pretreatments when compared with conventional transmission and reflection techniques (Almeida et al., 2002; Du et al., 2010; Zhang et al., 2012). FTIR-PAS spectra of the membranes were recorded using a FTIR spectrometer (Nicolet 6700) with a photoacoustic accessory (MTEC model 300, USA). The scans were conducted in the wavenumber range of 500-4000 cm^{-1} with a resolution of 4 cm^{-1}, using 32 scans and mirror velocities of 0.16, 0.32, 0.64 and 1.28 cm s^{-1}. For the spectra recording, a piece of the membrane was put into the photoacoustic accessory cell, and the cell was purged with dry helium for 10 s prior to scanning. Black carbon was used as reference (Du et al., 2010).

The profiling depth of the membrane was obtained using FTIR-PAS according to Eq. (1):

$$\mu = \sqrt{\frac{D}{\pi v \gamma}} \tag{1}$$

where μ is the profiling depth (μm), D is the thermal diffusivity of sample, v is the moving mirror velocity (cm/s), and γ is the wavenumber (cm^{-1}). The thermal diffusivity of most polymeric materials was about 0.01×10^{-5} m^2/s (Zhang et al., 2012), and the profiling depths were calculated under different moving mirror velocities and wavenumbers (Table 1).

Table 1 The profiling depth of polymer membrane using FTIR-PAS calculated using typical thermal diffusivity, wavenumber and moving mirror velocity

Assignment	Band position (cm^{-1})	Moving-mirror velocity (cm/s)	Thermal diffusion distance (μm)
C—H stretching vibration	~ 2900	0.16	8.21
		0.32	5.81
		0.64	4.11
		1.28	2.90
C=O stretching vibration	~ 1730	0.16	10.72
		0.32	7.58
		0.64	5.36
		1.28	3.79
C—H bending vibration	~ 1450	0.16	11.72
		0.32	8.28
		0.64	5.86
		1.28	4.14
C—O stretching vibration	~ 1100	0.16	13.10
		0.32	9.26
		0.64	6.55
		1.28	4.63

Notes: The depth profiling function of photoacoustic spectroscopy is $\mu = \sqrt{\frac{D}{\pi v \gamma}}$, in which μ is the profiling depth (μm), D is the thermal diffusivity of sample (0.01×10^{-5} m^2/s), v is the moving-mirror velocity (cm/s), and γ is the wavenumber (cm^{-1}).

Principal component analysis was conducted on the FTIR-PAS spectra, and the Euclidean distances using the first 11 components were used to determine dissimilarities between membranes before incubation and membranes in each incubation period. Different spectral ranges, i.e., 500-4000 cm^{-1}, 2800-3200 cm^{-1}, 1500-1900 cm^{-1}, 1300-1500 cm^{-1}, 1000-1300 cm^{-1}, and 500-1000 cm^{-1} were selected according to functional group to calculate the Euclidean distance. A higher Euclidean distance meant greater compositional difference between membranes.

4. Biolog EcoPlate analysis

Biolog EcoPlates (Biolog Inc., USA) have been widely used to determine the total activity and functional diversity of soil culturable microbial communities (Harch et al., 1997). The 96-well EcoPlates contained 31 different carbon sources, replicated three times, plus three blank wells without any carbon source. A 5 g sample of fresh soil was suspended in 50 mL of 0.85% (w/v) sterile NaCl solution, shaken for 30 min, and then allowed to settle for 5 min. The supernatant was diluted 10-fold. A 150 μL aliquot of the diluted sample was inoculated onto the EcoPlate and incubated at 28℃ in the dark. Substrate utilization on the plate was monitored by measuring the absorbance at 590 nm wavelength every 24 h for 168 h using a Bio-Rad Microplate Reader (Bio-Rad, USA). The total culturable microbial activity and functional diversity were measured by the average well color development (AWCD) and three diversity indices (Shannon-Weaver index H', Simpson index $1/D$, and McIntosh index U) according to Fang et al. (2012).

5. PCR-DGGE analysis

To assess changes in the community structure of the entire soil bacteria, a PCR-DGGE method was used (Muyzer et al., 1993). Soil DNA was extracted from each homogenized sample (approximately 0.5 g) following the manufacturer's protocols using the FastDNA® SPIN Kit for soil (MP, USA). The extracted soil genomic DNA was dissolved in 50 μL TE buffer, and stored at −20℃ prior to use. The DNA samples were purified using PowerClean® DNA Clean-Up Kit (MO BIO, Inc., USA) and PCRs were run using 50 μL reaction volumes. The PCR procedures were as follows: an initial 94℃ denaturation for 5 min, followed by 35 cycles at 94℃ for 30 s, 60℃ for 30 s, 72℃ for 30 s, and a final extension step at 72℃ for 7 min, and then held at 4℃. Using these conditions, the PRBA338F (5'-ACT CCT ACGGGA GGC AGC AG-3') and PRUN518R (5'-ATT ACC GCG GCT GCT GG-3') primers were used to amplify the 338 to 518 rDNA region, as described by Nakatsu et al. (2000) and to obtain products of about 200 bp, while forward primers contained a 40-bp GC-clamp (5'-CGC CCG CCG CGC GCG GCG GGG GCG GGG GCA CGG GGG G-3') attached to the 5' end (Muyzer et al., 1993). Amplicons were checked by electrophoresis on 1% agarose gel. We then performed 16S rDNA-DGGE using the DCode System (BIO-RAD, USA). A 10 μL aliquot of amplicons was loaded (top filling method) onto 8% acrylamide-bisacrylamide gel containing a denaturant gradient of 45% to 65% at 70 V for 16 h in 1× TAE running buffer at 60℃, followed by 0.5 h coloration using Gel-Red nucleic acid gel stain (Biotium, USA). The gels were visualized and digitalized by using a Gel Doc™ EQ imager (Bio-Rad, USA) combined with Quantity One 4.4.0 (Bio-Rad, USA). Species richness was calculated from the band numbers per sample. H' and evenness were calculated using the number of bands and peak intensities (Yu and Morrison, 2004).

6. Statistical analysis

All spectral data was processed using MATLAB 2009b. Digital information for DGGE was determined using Quantity One 4.4.0. All bacterial community diversity data were analyzed by a three-way analysis of variance (ANVOA). One-way ANOVA with Tukey test at the 0.05 probability was used to determine Euclidean distance over the difference of incubation and the microbial community diversity among three treatments. Independent-sample T tests was used to detect the microbial community diversity difference between soil top layer and bottom layer using SPSS 16.0, and the graphics were plotted using SigmaPlot 12.5.

1.5.3 Results and discussion

1. *The change of water-borne membranes over the soil incubation period*

Fig. 1(a) shows the weight loss of water-borne membranes buried in the waterlogged paddy soil during the 12 months incubation. Compared with the control, the weight losses for the UP and BP membranes after two months were about 1.01% and 0.99%, respectively, and increased to 1.24% and 1.15% after 12 months, respectively. This demonstrates that the membranes were degraded at a relatively low rate, thus they remained integrated and could function effectively as controlled release membranes. Fig. 1(b) further demonstrates that the UP and BP membranes remained totally intact. The color of the UP membranes became darker over the incubation period, suggesting that some small molecules in the soil might diffuse into the membranes and that the membranes were compatible with soil. The color of the BP membranes remained black due to the biochar component, and only a small amount of biochar was lost from the membrane border according to the membrane color, suggesting that BP was a more stable coating than UP for controlled nutrient release.

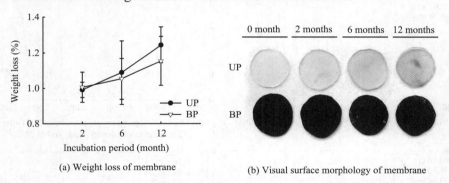

(a) Weight loss of membrane   (b) Visual surface morphology of membrane

Fig. 1 Changes of water borne membranes over 12 months incubation in paddy soil at 28 ℃

UP, unmodified ployacrylate membrane; BP, biochar modified polyacrylate membrane

Fig. 2 shows the spectra of all samples over the incubation period, and several similar function groups are apparent, such as 2900, 1730, 1450, and 1200 cm^{-1} which represent the C—H stretching vibration, the C=O stretching vibration, the C—H bend vibration, and the C—O stretching vibration, respectively. However, there were numerous variances in the spectra between the treatments.

According to Eq. (1), the profiling depths (μm) were calculated at different moving mirror velocities (Table 1). Fig. 2 shows the FTIR-PAS spectra of UP and BP membranes at different moving mirror velocities, and indicates that both UP and BP membranes were heterogeneous. For the UP membranes (Fig. 2(a)), the surface layer (2.92-4.65 μm, with a moving velocity of 1.28 cm/s) showed the greatest difference to the next three deeper layers (4.14-13.10 μm, with a moving velocity of 0.16, 0.32 and 0.64 cm/s, respectively). The most significant difference occurred for absorption around 1030 cm^{-1}, which is assigned to CO—O—C or C—O—C vibration (Movasaghi et al., 2008), indicating that more bonds of CO—O—C or C—O—C were observed in the surface layer (0-4.65 μm). After 12 months of incubation, the UP membranes were still heterogeneous, and the difference mainly occurred in the surface layer, and the vibration intensity of CO—O—C or C—O—C became significantly weaker (Fig. 2(b)), suggesting that some of these bonds were broken. Although similar results were observed for the BP membranes (Fig. 2(c)), the vibration intensity of CO—O—C or C—O—C showed an opposite trend compared with that for the UP membranes. The vibration of CO—O—C or C—O—C in the surface layer of the BP membranes was prevented by the involvement of biochar, and after 12 months of incubation, the prevention might be removed due to the loss of biochar from the surface layer, as indicated by the ~50 cm^{-1} shift of the absorption band towards the direction of lower wavenumbers: 980 cm^{-1} for BP membranes versus 1030 cm^{-1} for UP membranes (Fig. 2(d)). Therefore, less degradation of CO—O—C or C—O—C bonds occurred for the BP membranes over 12 months of incubation, which demonstrated that they were more stable than the UP membranes.

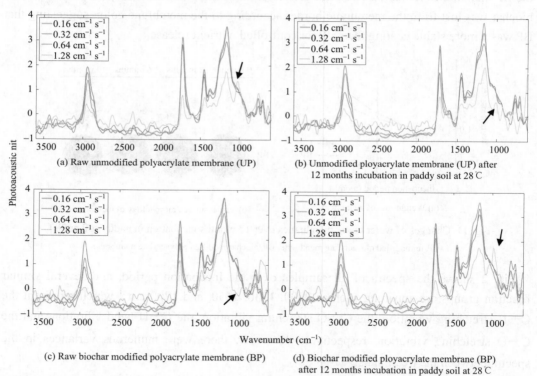

Fig. 2 FTIR-PAS depth profiling spectra of water borne membranes with the moving mirror velocity of 0.16, 0.32, 0.64, and 1.28 cm/s

The arrow showed the significant change around 1050 cm^{-1} of membrane after 12 months incubation in paddy soil

FTIR-PAS spectra based Euclidean distance was used to judge the changes in the membranes during different incubation times. Six regions (500-4000, 500-1000, 1000-1300, 1300-1500, 1500-1900, and 2800-3200 cm^{-1}) were selected to calculate the Euclidean distances. Because the main changes occurred in the surface layer (the spectra with moving mirror velocity of 1.28 cm/s), the spectra at this depth were used to calculate the Euclidean distances (Table 2). Significant differences were observed for the UP membranes in the total 500-4000 cm^{-1} region over 12 months of incubation. The main contribution resulted from the fingerprint region of 500-1000 cm^{-1}, followed by two other regions of 1300-1500 cm^{-1} and 1500-1900 cm^{-1}, and the remaining regions showed less contribution. However, for the BP membranes, although there was some change in the 500-1000 cm^{-1} region, no significant difference was found in the total 500-4000 cm^{-1} region, which further verified that the BP membranes were more stable than the UP membranes.

Table 2 Euclidean distances calculated from FTIR-PAS spectra data over the incubation time at the surface layer with the moving mirror velocity of 1.28 cm/s

Middle-infrared regions (cm^{-1})	Treatments	2 months	6 months	12 months
500-4000 (whole spectral region)	UP	15.08 ± 2.10 c	21.51 ± 0.24 b	37.98 ± 0.99 a
	BP	22.20 ± 4.18 a	17.80 ± 3.57 a	18.45 ± 3.84 a
2800-3200 (C—H stretching vibration)	UP	5.39 ± 0.76 a	6.53 ± 1.77 a	5.95 ± 0.82 a
	BP	5.33 ± 0.21 a	5.17 ± 1.25 a	5.55 ± 0.64 a
1500-1900 (C=O stretching vibration)	UP	3.70 ± 0.21 b	5.23 ± 0.52 a	3.13 ± 0.14 b
	BP	4.85 ± 0.88 b	5.73 ± 0.81 ab	7.41 ± 0.83 a
1300-1500 (C—H bending vibration)	UP	3.38 ± 1.15 b	5.65 ± 0.18 a	3.42 ± 0.43 b
	BP	3.19 ± 0.74 a	4.24 ± 0.92 a	3.59 ± 1.09 a
1000-1300 (C—O stretching vibration)	UP	3.13 ± 0.56 a	4.14 ± 1.75 a	2.79 ± 0.54 a
	BP	3.04 ± 0.98 a	4.45 ± 0.60 a	4.63 ± 1.83 a
500-1000 (fingerprint region)	UP	6.05 ± 3.17 b	9.68 ± 2.38 b	17.17 ± 4.75 a
	BP	10.53 ± 3.31 a	8.82 ± 2.71 a	10.32 ± 1.59 a

Notes: ① UP, unmodified polyacrylate membrane; BP, biochar modified polyacrylate membrane. ② Different capital letters in the same row indicate significant differences ($P < 0.05$).

Combining the results of the membrane weight loss, morphology, and FTIR analysis, biodegradation of the water-borne membranes mainly occurred in the surface layer, and the biodegradation rate of the UP membranes was significantly greater than that of the BP membranes. The surface layer was directly subjected to the environment, and was thus more easily degraded, especially the C—O—C groups that were easily broken. The involvement of biochar might form a thin coating outside the C—O—C groups, which could protect the bonds from degradation, although the groups were released after 12 months of incubation due to removal of biochar from the surface layer. In addition, it is possible that the biochar products contained some toxic substance that suppressed microbial activity, which reduced the polyacrylate biodegradability by soil microorganism (Zimmerman et al., 2011).

2. Effects of water-borne polyacrylate membranes on soil bacterial community profiles

The absorbance values from the Biolog EcoPlates after 96 h incubation were used to evaluate the soil culturable bacterial functional communities based on AWCD value and three diversity indices (Shannon-Weaver index H', Simpson index $1/D$, and McIntosh index U). The

AWCD, which reflects the oxidative capacity on 31 kinds of carbon sources in Biolog EcoPlates, is used as an indicator of overall culturable microbial activity (Bossio and Scow, 1995; Garland, 1996). The AWCD values of all samples on 96 h incubation were sharply reduced at early 6 months of incubation, and then gradually decreased (Fig. 3). It is possible that aerobic microorganisms predominated in the early incubation phase, and then were suppressed while anaerobic microorganisms predominated during the later incubation period as oxygen was rapidly depleted (Liesack et al., 2000). In addition, products of anaerobic metabolism in soil, such as H2S, NH3, or volatile fatty acids may have inhibited microbial activity (Sahrawat, 2004) and the soil microorganisms gradually adapted to the conditions during the later phase.

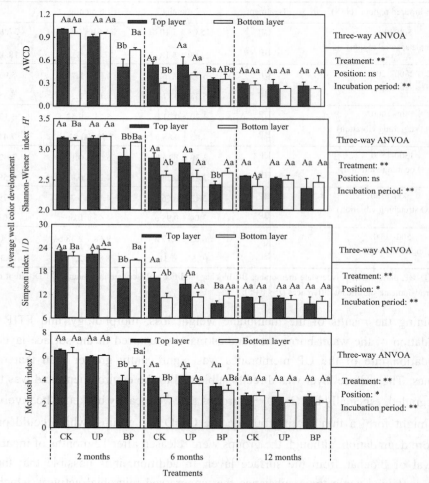

Fig. 3 Diversity indices of microbial communities in paddy soil using Biolog EcoPlate over the soil incubation

All data were subjected to a three-way analysis of variance (three-way ANOVA) after testing the normality and homogeneity. Treatment: CK, soil without polyacrylate membrane; UP, soil with unmodified waterborne polyacrylate membrane; BP, soil with biochar modified polyacrylate membrane. Position: top layer, the contacted soil layer (soil with 0-2 cm distance from the buried membrane surface); bottom layer, the noncontact soil layer (soil with 0-2 cm distance from the buried membrane surface). Incubation period: 2 months, 6 months, and 12 months. ns, not significant difference ($P > 0.05$); *, significant difference ($P < 0.05$); **, significant difference ($P < 0.01$). The same letter in each box (capital letter for membrane treatments and small letter for position) meant not significantly different ($P > 0.05$)

In the 2nd month, the AWCD values for both the top and bottom layer in the CK and UP treatments were not significantly different, but they were significantly higher than the BP treatment, and the AWCD value for the top layer was higher than the bottom layer for the BP treatment, which means that the BP membrane suppressed the microbial activity both in the top layer and the bottom layer in the 2nd month, and the suppression in bottom layer was alleviated. In the 6th month, the soil microbial activity was still suppressed by the BP treatment compared with in top layer of the CK treatment based on the AWCD values, but AWCD value of BP treatment was relatively less suppressed than that in the 2nd month, and the AWCD values in the bottom layer for all treatments showed no significant difference, indicating that the soil microbial activity in the bottom layer of the BP treatment could be recovered. In the 12th month, the AWCD values in the top and bottom layers for all treatments were not significantly different, indicating that the suppression of soil microbial activity by biochar disappeared.

Three diversity indices (Shannon-Weaver index H', Simpson index $1/D$, and McIntosh index U) were used to assess the richness, dominant population, and evenness of soil microorganisms, respectively (Fang et al., 2012). Fig. 3 shows the changes of three diversity indices between treatments over the incubation period. For the top layer, each of the three indices of soil microbial communities in the BP treatment were significantly lower in the 2nd month, but the Simpson index $1/D$ recovered in the 6th month, and the the McIntosh index U and Shannon-Weaver index H' recovered in the 12th month, indicating that richness, dominant population, and evenness of soil microorganisms in the 2nd month decreased in the BP treatment, but the suppression of dominant population disappeared in the 6th month, and the suppression of evenness and richness of soil microorganisms disappeared in the 12th month. For the bottom layer, McIntosh index U was significantly lower in BP treatment while other two indices, i. e. Shannon-Weaver index H' and Simpson index $1/D$, showed no significant difference from CK treatment in all bottom layers in the 2nd month, suggesting that the evenness of soil microorganisms was affected but no differences were observed for the index of the richness and dominant population. Furthermore, there were no significant differences within these three indices between the BP and CK treatments since the 6th month.

There are two reasons why the BP lowered the microbial carbon utilization during the early phase. First, a small amount of the membrane soluble fraction dissolved into the soil, such as ammonium persulfate catalyst, which lowered the carbon source utilization (Li et al., 2013). Second, the biochar products contain some substances, such as dioxins, furans, phenols, and polyaromatic hydrocarbons, which could reduce the microbial activity (Zimmerman et al., 2011). The suppression of microbial activity recovered over the 12-month incubation period, suggesting that a short duration of suppression due to the water-borne membrane involvement might result in the proliferation of soil microorganisms.

The effects of water-borne polyacrylate membranes on the bacterial community composition was verified by molecular analysis, which was performed on DGGE gel using PCR amplification of 16S rDNA genes from soil DNA of each sample and separated by electrophoresis. Fig. 4 shows bacterial results for the DGGE gel in the 2nd, 6th, and 12th

months. All samples showed that numerous bacterial groups appeared to be ubiquitous (strong and weak), thus indicating a polymicrobial community. The intensities and numbers of bands for all samples reduced in DGGE gel, and the bands of DGGE lower gel became clear over the incubation period. These results may be attributed to aerobic microorganisms predominating during the early incubation time, which were then suppressed while anaerobic microorganisms predominated over the incubation period as oxygen was rapidly depleted (Liesack et al., 2000). The DGGE bands were digitalized using Quantity One software for extracting more information. Band numbers as well as intensities and patterns provided information about the richness (S), Shannon-Wiener diversity index (H'), and evenness (E), which were calculated as follows: OTU richness (S) was determined from the number of bands in each lane, and Shannon-Wiener diversity index (H') was calculated from $H' = -\sum P_i \ln P_i$, and evenness (E) was calculated as $E = H'/H'_{max}$, where $H'_{max} = \ln S$ (Yu and Morrison, 2004). The indices for each treatment are listed in Fig. 5. During the three sampling times, the OTU richness (S) values, Shannon-Wiener index (H'), and evenness (E) values generally reduced as a function of incubation time for all treatments. The anaerobic environment suppressed aerobic microorganisms (Liesack et al., 2000) and might contribute to the reduced bacterial community diversity in the three treatments over the incubation period. And in each sampling time (2nd month, 6th month, or 12th month), three community composition diversity values, i. e., S, H', and E values of BP and UP treatments showed no significantly different from the CK for both the top and bottom layers.

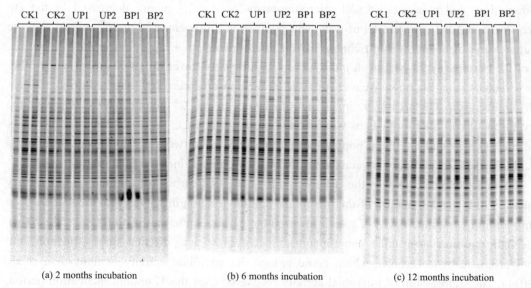

(a) 2 months incubation   (b) 6 months incubation   (c) 12 months incubation

Fig. 4 DGGE gel of bacterial community in paddy soil from different treatments over the soil incubation

CK, soil without waterborne polyacrylate membrane UP, soil with unmodified membrane; BP, soil with biochar modified membrane; CK1, UP1, BP1: the contacted soil layer (soil with 0-2 cm distance from the buried membrane surface) from treatment of CK, UP and BP, respectively; CK2, UP2, BP2: the noncontact soil layer (soil with 8-10 cm distance from the buried membrane surface) from treatment of CK, UP and BP, respectively

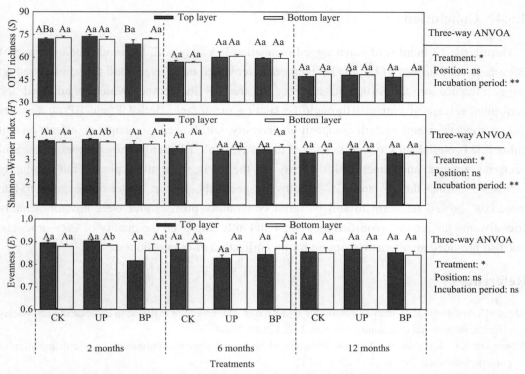

Fig. 5   Composition characterization of bacterial community diversity in paddy soil using indexes of OTU richness (S), Shannon-Wiener index (H') and evenness (E)

All data were subjected to a three-way analysis of variance (three-way ANOVA) after testing the normality and homogeneity. Treatment: CK, soil without polyacrylate membrane; UP, soil with unmodified waterborne polyacrylate membrane; BP, soil with biochar modified polyacrylate membrane. Position: top layer, the contacted soil layer (soil with 0-2 cm distance from the buried membrane surface); bottom layer, the noncontact soil layer (soil with 0-2 cm distance from the buried membrane surface). Incubation period: 2 months, 6 months, and 12 months. ns, not significant difference ($P > 0.05$); *, significant difference ($P < 0.05$); **, significant difference ($P < 0.01$). The same letter in each box (capital letter for membrane treatments and small letter for position) meant not significantly different ($P > 0.05$)

The results obtained by Biolog EcoPlates and PCR-CGGE analyses are not necessarily in contrast, which is similar to the results of others (Vestergard et al., 2008; Abbate et al., 2013; Bushaw-Newton et al., 2012), because the Biolog EcoPlates was used to assess the functional diversity of the soil bacterial community while the PCR-DGGE was focused on the composition diversity of the bacterial community, and the activity of some bacterial community was susceptible prior to the community composition (Mijangos et al., 2009). Consequently, two microbial methods used together provided a comprehensive understanding of the effects of BP on soil bacterial community profiles. Considering the Biolog EcoPlates and PCR-DGGE results for UP treatment, the bacterial functional diversity and the composition diversity showed no significant different from CK in the whole incubation period. For the BP treatment, the soil bacterial composition diversity was no different from the control, whereas soil culturable bacterial activity and functional diversity were lower than the control in early incubation phase, and they recovered to the control level in the 6th month and the 12th month, respectively.

1.5.4 Conclusions

During the 12 months of waterlogged incubation in paddy soil, the biodegradation rate of BP was lower than that of UP, the BP membrane remained more intact, and the soil showed no large influence on the structural integrity of the BP membrane, which guaranteed the controlled release of nutrients through the coating membrane. Both UP and BP membranes did no impacts on soil bacteril composition diversity. UP showed no negative effects on soil culturable bacterial activity and functional diversity, while BP membranes suppressed the soil bacterial activity and functional diversity at the early incubation phase, but gradually recovered during the 6th and 12th months, respectively. Hence, water-borne polyacrylate materials showed no harm to soil bacterial community profiles and were environmentally friendly, and the biochar-modified membrane not only improved the quality of CRF products, but also provided an alternative option for the utilization of crop residues.

References

Abbate C, Ambrosoli R, Minati J L, et al. 2013. Metabolic and molecular methods to evaluate the organoclay effects on a bacterial community. Environ. Pollut., 179: 39-44

Albertsson A C, Karlsson S. 1990. The influence of biotic and abiotic environments on the degradation of polyethylene. Prog. Polym. Sci., 15: 177-192

Almeida E, Balmayore M, Santos T. 2002. Some relevant aspects of the use of FTIR associated techniques in the study of surfaces and coatings. Prog. Org. Coat., 44: 233-242

Bossio D A, Scow K M. 1995. Impact of carbon and flooding on the metabolic diversity of microbial communities in soils. Appl. Environ. Microb., 61: 4043-4050

Bushaw-Newton K L, Ewers E C, Velinsky D J, et al. 2012. Bacterial community profiles from sediments of the Anacostia River using metabolic and molecular analyses. Environ. Sci. Pollut. Res., 19: 1271-1279

Celina M C. 2013. Review of polymer oxidation and its relationship with materials performance and lifetime prediction. Polym. Degrad. Stabil., 98: 2419-2429

Choudhury A T M A, Kennedy I R. 2005. Nitrogen fertilizer losses from rice soils and control of environmental pollution problems. Commun. Soil Sci. Plant Anal., 36: 1625-1639

Costanza R, d'Arge R, de Groot R, et al. 1987. The value of the world's ecosystem services and natural capital. Nature, 387: 253-260

Decker C, Zahouily K. 1999. Photodegradation and photooxidation of thermoset and UV-cured acrylate polymers. Polym. Degrad. Stab., 64: 293-304

Du C W, Zhou G Q, Wang H Y, et al. 2010. Depth profiling of clay-xanthan complexes using step-scan mid-infrared photoacoustic spectroscopy. J. Soils Sediments, 10: 855-862

Fang H, Tang F F, Zhou W, et al. 2012. Persistence of repeated triadimefon application and its impact on soil microbial functional diversity. J. Environ. Sci. Health Part B-Pestic. Contam. Agric. Wastes, 47: 104-110

Garland J L. 1996. Analytical approaches to the characterization of samples of microbial communities using patterns of potential C source utilization. Soil Biol. Biochem. ,28: 213-221

Garcia M C, Diez J A, Vallejo A, et al. 1996. Use of kraft pine lignin in controlled-release fertilizer formulations. Ind. Eng. Chem. Res., 35: 245-249

Hadad D, Geresh S, Sivan A. 2005. Biodegradation of polyethylene by the thermophilic bacterium *Brevibacillus borstelensis*. J. Appl. Microbiol., 98: 1093-1100

Han X Z, Chen S S, Hu X G. 2009. Controlled-release fertilizer encapsulated by starch/polyvinyl alcohol coating. Desalination, 240: 21-26

Harch B D, Correll R L, Meech W, et al. 1997. Using the Gini coefficient with BIOLOG substrate utilisation data to provide an alternative quantitative measure for comparing bacterial soil communities. J. Microbiol. Methods, 30: 91-101

Huggins T, Wang H M, Kearns J, et al. 2014. Biochar as a sustainable electrode material for electricity production in microbial fuel cells. Bioresour. Technol., 157: 114-119

Ikeda S, Suzuki K, Kawahara M, et al. 2014. An assessment of urea-formaldehyde fertilizer on the diversity of bacterial communities in onion and sugar beet. Microbes Environ., 29: 231-234

Kennedy A C, Smith K L. 1995. Soil microbial diversity and the sustainability of agricultural soils. Plant Soil, 170: 75-86

Kirk J L, Beaudette L A, Hart M, et al. 2004. Methods of studying soil microbial diversity. J. Microbiol. Methods, 58: 169-188

Lehmann J. 2007. A handful of carbon. Nature, 447: 142-144

Li F L, Liu M, Li Z P, et al. 2013. Changes in soil microbial biomass and functional diversity with a nitrogen gradient in soil columns. Appl. Soil Ecol., 64: 1-6

Li H, Liang X Q, Lian Y F, et al. 2009. Reduction of ammonia volatilization from urea by a floating duckweed in flooded rice fields. Soil Sci. Soc. Am. J., 73: 1890-1895

Liesack W, Schnell S, Revsbech N P. 2000. Microbiology of flooded rice paddies. Fems Microbiol. Rev., 24: 625-645

Liu M, Zhang M, Yang Y C, et al. 2011. Effects of controlled-release fertilizer coating residual on soil microbial quantity and enzyme activity. J. Plant Nutr. Fert. Sci., 17: 1012-1017

Mijangos I, Becerril J M, Albizu I, et al. 2009. Effects of glyphosate on rhizosphere soil microbial communities under two different plant compositions by cultivation-dependent and independent methodologies. Soil Biol. Biochem., 41: 505-513

Movasaghi Z, Rehman S, Rehman I U. 2008. Fourier transform infrared (FTIR) spectroscopy of biological tissues. Appl. Spectrosco. Rev., 43: 134-179

Muyzer G, Wall E C D, Witterlinden A G. 1993. Profiling of complex microbial populations by denaturing gradient gel electrophoresis analysis of polymerase chain reaction-amplified genes coding for 16S rRNA. Appl. Environ. Microbiol., 59: 695-700

Nakatsu C H, Torsvik V, Ovreas L. 2000. Soil community analysis using DGGE of 16S rDNA polymerase chain reaction products. Soil Sci. Soc. Am. J., 64: 1382-1388

Rajsekharan A J, Pillai V N. 1996. Membrane-encapsulated controlled-release urea fertilizers based on acrylamide copolymer. J. Appl. Polym. Sci., 60: 2347-2351

Rosencwaig A, Gersho A. 1976. Theory of the photoacoustic effect with solids. J. Appl. Phys., 47: 64-69

Sahrawat K L. 2004. Organic matter accumulation in submerged soils. Adv. Agron., 81: 169-201

Shah A A, Hasan F, Hameed A, et al. 2008. Biological degradation of plastics: A comprehensive review. Biotechnol. Adv., 26: 246-265

Shaviv A. 2001. Advances in controlled release fertilizers. Adv. Agron., 71: 1-49

Vestergard M, Henry F, Rangel-Castro J I, et al. 2008. Rhizosphere bacterial community composition responds to arbuscular mycorrhiza, but not to reductions in microbial activity induced by foliar cutting. Fems Microbiol. Ecol., 64: 78-89

Xu Y P, Xie Z B, Zhu J G, et al. 2013. Effects of pyrolysis temperature on physical and chemical properties of corn biochar and wheat biochar. Soils, 45: 73-78

Yang Y C, Zhang M, Li Y C, et al. 2012. Controlled release urea improved nitrogen use efficiency, activities of leaf enzymes, and rice yield. Soil Sci. Soc. Am. J., 76: 2307-2317

Yu Z T, Morrison M. 2004. Comparisons of different hypervariable regions of rrs genes for use in fingerprinting of microbial communities by PCR-denaturing gradient gel electrophoresis. Appl. Environ. Microbiol., 70: 4800-4806

Zhang W R, Zhu T T, Smith R, et al. 2012. Non-destructive study on the degradation of polymer coating I :

Step-scan photoacoustic FTIR and confocal Raman microscopy depth profiling. Polym. Test, 31: 855-863

Zhao C, Shen Y Z, Du C W, et al. 2010. Evaluation of water-borne coating for controlled-release fertilizer using Wurster fluidized bed. Ind. Eng. Chem. Res., 49: 9644-9647

Zhou Z J, Du C W, Shen Y Z, et al. 2013. Development of biochar modified polyacrylate emulsions coated released fertilizers. J. Funct. Mater., 44: 1305-1308

Zimmerman A R, Gao B, Ahn M Y. 2011. Positive and negative carbon mineralization priming effects among a variety of biochar-amended soils. Soil Biol. Biochem., 43: 1169-1179

2 Evolution pathway and evaluation methods of soil fertility

2.1 Plants use alternative strategies to utilize nonexchangeable potassium in minerals

2.1.1 Introduction

Soil potassium reserves are generally large since K is the seventh most abundant element in the Earth's crust (Schroeder, 1978). Only a small portion of the soil K is in solution or exchangeable forms that can be easily used by plants, with the majority of soil K retained in minerals in nonexchangeable or structural forms that are slowly or less available to plants (Sparks, 2000; Huang, 2005). In countries lacking K fertilizer resources, there is great potential to improve the use of soil K by plants. Nonexchangeable K (NEK) is the K held between adjacent tetrahedral layers of dioctahedral and trioctahedral micas, vermiculites and intergrade clay minerals and not bonded covalently within the crystal structures of soil minerals particles (Martin and Sparks, 1985). The NEK content of soils varies from 2% to 8 % of total K as evaluated with the traditional boiling HNO_3 method (Xie et al., 2000). The total amount of soil NEK released in tetraphenylboron sodium solution can reach up to 20%–55% of total K (Zhou and Wang, 2008). This indicates that NEK reserves in soils are quite large, and the majority of the soil K is not recovered by the boiling HNO_3 method but exploited by K efficient plants in many studies might initially be in the nonexchangeable form. Thus NEK plays a much more important role in soil K supply to plants than structural K.

Plant species or genotypes within species may differ in their capacity to use NEK in soils or minerals (Coroneos et al., 1996; Rengel and Damon, 2008). Ryegrass could mobilize more K from gneiss than pak-choi, while alfalfa could not utilize K from gneiss (Wang et al., 2000a). Sugar beet was more effective in mobilization of low available K in the rhizosphere than wheat and barley (Dessougi et al., 2002). Different cultivars of many plant species such as amaranth (Tu et al., 2000), potato (Trehan et al., 2005), rice (Jia et al., 2008) and maize (Lv and Zhang, 2010) differ in their K uptake efficiencies. However, breeding new K-uptake-efficient genotypes has so far been sporadic, because the exact or determinative mechanisms by which plants sequester K efficiently from soils were unknown. Differential exudation of organic compounds to facilitate release of NEK was considered as one of the mechanisms of differential K uptake efficiency (Rengel and Damon, 2008). Organic acids such as oxalic acid for maize or grain amaranth (Krafffczyk et al., 1984; Tu et al., 1999), tartaric acid for pak-choi or radish (Chen et al., 1999; Zhang et al., 1997), malic acid for oilseed rape (Zhang et al., 1997) and citric acid for maize (Krafffczyk et al., 1984) could be released in root exudates under K sufficient or insufficient conditions. These acids facilitate release of NEK or mineral K to various extents according to data from batch or incubation experiments (Song and Huang, 1988; Zhu and Luo, 1993; Wang et al., 2000b; Li et al., 2006; Wang et al., 2007). In addition to mineral K activation by root exudates, both root morphology, including root length, biomass and root hairs (Hogh-Jensen and Pedersen, 2003; Li and Ma, 2004), and root K uptake activity measured in terms of rates of K uptake or affinity of K uptake systems (Hinsinger and Jaillard, 1993; Trehan and Sharma, 2002) have been investigated with regard to plant K uptake efficiency (Rengel and Damon, 2008). Most studies hypothesize possible mechanisms through morphological and physiological differences in K uptake efficiency, but very few make detailed assessments of the observed differences at a mechanistic level. Uncertainty of the key mechanism prevents us from taking a strategic approach towards breeding efficient K uptake genotypes through conventional breeding or genetic engineering.

Efficient K uptake plant species or genotypes may have specific physiological mechanisms to promote NEK release in the rhizosphere. Release of NEK from soils or minerals is reliant on or influenced by factors including concentrations of ions and K in solution (Martin and Sparks, 1985; Huang, 2005). In plant-soil systems, root secretion of organic acids or depletion of solution K to a low level in the rhizosphere would be the main mechanisms that facilitate NEK release. However, the relative importance of NEK activation by organic acids in root exudates and the K-sequestration capacity of roots in facilitating NEK release and utilization by plants is still obscure.

The current study aimed to ① compare the K-sequestration capacity from various K-bearing minerals of the two plant species, ② explain the differences between the K-sequestration capacity from various K-bearing minerals by the two plant species.

2.1.2 Materials and Methods

1. Minerals

Five minerals (biotite, vermiculite, muscovite, phlogopite and feldspar) with varying K availabilities were used in this study. The mineral samples were collected from Linshou County, Hebei Province, and their mineral and chemical compositions showed some impurities (Table 1). The mineral samples were ground, passed through a 60 mesh screen, and washed three times with 0.5 mol / L $CaCl_2$ to remove labile K and with deionized water to remove free $CaCl_2$. The samples were then air-dried.

Table 1 Location and basic properties for the soils tested

Soils	Location	Soil type	Parent material	Cropping system	CEC (cmol /kg)	Organic matter (g/kg)	pH	$CaCO_3$ (%)	Clay content <0.002mm (%)	Available K (mg/kg)	Slow available K (mg/kg)	Total K (%)	Main clay minerals
LY	Laiyang, Shandong	Aquic inceptisol	Alluvial deposit	Wheat-maize	9.20	10.0	6.80	-	12.2	93.0	1068	1.63	HM*, VC, KK, SM
WC	Wangcheng, Hunan	Ultisol	Quaternary red clay	Double rice	9.97	40.2	5.14	-	30.5	73.1	334	1.41	HM, SM
FQ	Fengqiu, Henan	Calcic aquic inceptisol	Alluvial deposit	Wheat-maize	8.31	6.5	8.65	7.36	21.8	126.0	1092	2.18	HM, CH, VC, KK, SM
CS	Changshu, Jiangsu	Entisol	Lacustrine deposit	Wheat-rice	27.33	46.6	6.65	1.08	34.3	149.5	582	1.61	HM, SM, CH, VC, KK

Note: *HM - hydromica; SM - smectite; CH - chlorite; VC - vermiculite; KK - kaolinite; GT - goethite; MI - mica.

2. *Use of mineral K by plant species*

Ryegrass (*Lolium perenne* L.) has a comparatively high capacity for K uptake from minerals (Hinsinger and Jaillard, 1993). Grain amaranth (*Amaranthus* sp.) K12 is a K enrichment genotype selected by Li and Ma (2003) that is able to activate soil K (Li et al., 2006). Their capacities to sequester K were compared in a pot experiment. Each pot contained a mixture of 500 g quartz sand and 50 g mineral. Seven treatments were set up, comprising the five K-bearing minerals, quartz with enough K in nutrient solution, and quartz without K. Seeds of ryegrass or grain amaranth were sown on minerals, with three replicates of each. One hundred ml of half strength nutrient solution (Hoagland and Arnon, 1950) omitting K was added to each pot every day, while a nutrient solution containing K was only added to the quartz with K treatment. The surplus nutrient solution was discharged from the bottom of the pot to prevent accumulation of salts, including K, released from the minerals, thus the K used by the plants was predominantly the newly released K from minerals. The experiment was carried out in a greenhouse with ambient light and a temperature ranging from 15 to 30℃. After a growth period of 7 weeks, the above-ground part of the plants was harvested. Samples were oven-dried, weighed and digested using the $H_2SO_4 + H_2O_2$ method (Lu, 1999) for K concentration determination.

3. *Kinetics of root K uptake and root exudate collection*

Seedlings of ryegrass and grain amaranth were raised in quartz sand watered with deionized water daily and with nutrient solution once a week. After 20 days the seedlings were transferred to a hydroponic culture system. For each species, a 40 L PVC tank with a 50-hole cover was used. Two seedlings were supported with sponge in each hole. The nutrient solution was aired continuously, renewed once a week for the first two weeks and then every three days. The pH was adjusted daily to 6.0 with 0.1 mol/L HCl or NaOH.

After 6 weeks in hydroponic culture, 42 plants of each species were moved to K-free nutrient solution for 48 hours to induce K deficiency. After three rinses with 0.2 mmol/L $CaSO_4$, the plants were put into dark bottles (two plants per bottle) containing 100 mL of 0.2 mmol/L $CaSO_4$ and various levels of KCl (0.002, 0.005, 0.01, 0.1, 1.0, 5 or 10 mmol/L), with three replicates. The plants were harvested after six hours of K uptake, and their root length and fresh weight were measured. The length of the longest root was measured with a ruler. The K concentration and volume of solution in each bottle were measured to calculate the amount of K uptake by plants under the various K supply levels.

The remaining plants were cultured for another week in nutrient solution with the K level reduced to 0.125 mmol/L. After rinsing with deionized water, 18 plants of each species were put into 0.2 mmol/L $CaSO_4$ and 30 mg/L chloromycetin solutions for 120 and 30 min, respectively. Following this they were rinsed again with deionized water and then put into dark bottles containing 250 mL of deionized water (six plants per bottle) for 12 h to collect root exudates. For each bottle, the deionized water was renewed every 3 h, with four batches of root exudates being collected. These were combined and concentrated to 50 mL using a rotary-evaporator under vacuum at 40℃. The plants were harvested and the root fresh weights were recorded.

4. *Activation of mineral K by citric and oxalic acid*

Since the root exudates collected in the current study contained very high concentrations of K (data not shown) and were not suitable for a mineral K release experiment, 10 mL of deionized water or a solution containing 5 or 50 mmol/L of citric or oxalic acid was added to 5.00 g of minerals and incubated at 25℃ for a week to investigate the effect of organic acids in root exudates on the release of mineral K. After incubation, the available K in the minerals was determined using the 1 mol/L NH_4OAc method (Lu, 1999).

5. *Release of NEK from minerals in various solutions*

The release of NEK from the tested minerals was significantly influenced by various cations such as Ca^{2+}, Na^+, H^+ and NH_4^+ in solution (Shen et al., 2009). To investigate whether there is a threshold of K concentration for NEK release from minerals and how the threshold would be influenced by the presence of the various cations, the easily releasable NEK of the minerals was removed via a previous 60 d incubation. During the incubation, 50 mL of a solution of 0.01 mol/L $CaCl_2$, 0.02 mol/L NaCl or HCl, or water was added to 5 g of the minerals and was renewed daily (Shen et al., 2009). From the 61st day on, the solutions were not renewed and the minerals were incubated for another 50 d at 25℃. During the 50 d incubation period, 4.5 mL of solution was sampled periodically for K concentration, and 4.5 mL of the original solution was added to keep the solution volume constant in each tube. Three replicates were set up.

6. *Analysis*

The composition of K-bearing minerals was measured using an X'Pert PRO X-ray diffractometer with Cu K_a radiation (40 kV, 40 mA) and a graphite filter, from 3.0° to 60.0° with a scan speed of 4.0°/min. The chemical composition of the minerals was determined by ICP-AES (IRIS-Advantage, Thermo Elemental, MA, USA) after melting by the lithium metaborate method (Lu, 1999).

Determination of the potassium in solution was made using a flame photometer (Model HG-5, Beijing Detection Instrument Ltd.) using an internal standard procedure employing 3 mmol/L lithium chloride.

The organic acids in root exudates were detected by HPLC (LC-10AT VP, SHIMADZU, Tokyo, Japan) equipped with a Shim-pack SCR-102H column (8.0mm (i. d.)× 30 cm) according to Ma et al. (2002).

All data are means of three values. The differences among means was statistically evaluated with SPSS 10.0, using analyses of variance (ANOVA) taking a *P* value of <0.05 as significant. The parameters V_{max}, C_{min} and K_m of the Michaelis-Menten kinetics equation $V = V_{max} \times (C - C_{min}) / [K_m + (C - C_{min})]$ were evaluated with the nonlinear regression procedure of SPSS 10.0.

2.1.3 Results

1. *Growth of plant species with K supply from K-bearing minerals*

Grain amaranth and ryegrass differed greatly in their ability to grow using the K supplied by the different minerals used in the current study. As indicated by the biomass accumulation (Fig. 1), grain amaranth only grew well in the full K treatment with K in nutrient solution, followed by biotite as the second K-sufficient resource. With K supplied by vermiculite, phlogopite or muscovite, growth of grain amaranth was seriously stunted, ranking at the same level for the three minerals. Grain amaranth could not survive without external K when it grew on quartz or feldspar. Ryegrass showed better adaptability than grain amaranth to grow on K-bearing minerals, as indicated by biomass or relative biomass in Fig. 1. The most significant difference was found in the vermiculite treatment, where ryegrass had a relative biomass of 97%, while grain amaranth had only 20%, showing that ryegrass could grow well with K supplied by vermiculite but grain amaranth could not. The data in Fig. 1 also indicated that the biomass of ryegrass responded sensitively and proportionally to different K-bearing minerals; thus ryegrass was much more suitable than grain amaranth for use as an indicator species distinguishing the NEK availability of K-bearing minerals.

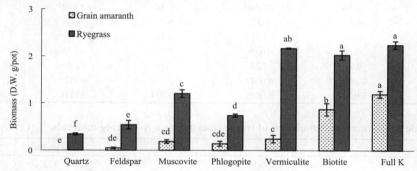

Fig. 1 Biomass accumulation of two species grown on K-bearing minerals

For each species, the same letter indicates no significant difference between treatments ($P<0.05$) by LSD

Grain amaranth was proposed as a K-enriched species because of its higher K concentration in tissue (Li and Ma, 2003). With K supplied by biotite or nutrient solution, K content in grain amaranth reached > 5%, reducing to about 1.5% in K insufficient treatments (Table 2). Grain amaranth had higher K contents than ryegrass under both K sufficient and insufficient conditions, except in the vermiculite treatment. Under K insufficient conditions, the higher tissue K content and lower relative biomass of grain amaranth (Fig. 1, Table 3) indicated that grain amaranth need higher tissue K content for normal growth. Thus grain amaranth could be termed as K-enriched, but compared with ryegrass was neither K-use efficient (related to tissue K concentration (Rengel and Damon, 2008)) nor K-uptake efficient (Table 2). With K supplied by vermiculite, the K content in ryegrass was 50% higher than that in grain amaranth, and the difference between the K uptakes of the two species was evident (Table 4). The data demonstrate that ryegrass was more K-uptake efficient than grain amaranth. An adequate supply of K is very important to distinguish the K-uptake efficiencies of species. A very high K supply level, such as that supplied by biotite containing K in very high availability, or a very low K supply level, such as that supplied by phlogopite, muscovite or feldspar containing K in very low availability, were not suitable for distinguishing K-uptake efficiencies of species. However, a medium or adequate level of K supply, such as the K supplied by vermiculite, was much better to distinguish K-uptake efficiency among plant species or genotypes.

Table 2 Composition of K-bearing minerals used

	Feldspar	Muscovite	Phlogopite	Vermiculite	Biotite
	Mineral composition				
	Orthoclase 30%-40% Plagioclase 35%-45% Quartz 15%-25% Mica little	Muscovite 75%-85% Quartz 10%-20% Ferruginous matter little	Phlogopite 60%-70% Orthoclase 15%-25% Involving biotite <10%	Vermiculite 15%-25% Quartz 35%-45% Orthoclase 15%-25% Achromate <10% Dolomite little	Biotite 55%-65% magnesium amphibole 25%-35% Quartz little
	Chemical composition (%)				
K_2O	5.76	10	10.71	3.87	3.72
Na_2O	5.02	0.61	0.61	1.46	1.82
Al_2O_3	14.86	28.94	31.06	14.02	12.52
CaO	0.44	0.12	0.29	4.63	9.8
MgO	0.25	1.15	1.18	5.05	6.9
SiO_2	76.36	50.68	49.94	58.73	42.56
Fe_2O_3	0.53	7.24	7.07	12.44	24.95
P_2O_5	0.0268	0.0747	0.0474	0.4523	1.9827
MnO	0.0045	0.0009	0.0034	0.1041	0.2141

Table 3 Biomass accumulation of two species grew on K-bearing minerals   (Unit: g/pot)

Treat	Grain amaranth			Ryegrass		
Quatz	0.00	(0%)	e	0.34	(15%)	f
Feldspar	0.05	(4.2%)	de	0.54	(24%)	e
Muscovite	0.19	(16%)	cd	1.20	(53%)	c
Phlogopite	0.14	(12%)	cde	0.73	(32%)	d
Vermiculite	0.24	(20%)	c	2.18	(97%)	ab
Biotite	0.86	(72%)	b	2.03	(90%)	a
Full K	1.19	(100%)	a	2.25	(100%)	a

Table 4 Potassium content and uptake by two species with K supply from K-bearing minerals

Treat	K content (%)				K uptake (mg/pot)					
	Grain amaranth		Ryegrass		Grain amaranth			Ryegrass		
Quatz	0.00	d	0.45	d	0.0	(0%)	e	1.5	(1.6%)	e
Feldspar	1.42	c	0.43	d	0.7	(1.1%)	e	2.3	(2.5%)	e
Muscovite	1.53	c	0.61	d	2.9	(4.4%)	d	7.4	(8.0%)	c
Phlogopite	1.77	bc	0.55	d	2.5	(3.8%)	d	4.0	(4.3%)	d
Vermiculite	2.05	b	3.00	c	5.1	(7.8%)	c	65.2	(71%)	b
Biotite	5.14	a	4.72	a	44.7	(68%)	b	95.8	(104%)	a
Full K	5.50	a	4.09	b	65.7	(100%)	a	92.0	(100%)	a

2. Root exudates and morphology

The weight and length of roots of the plants grown under hydroponic culture are compared in Fig. 2. The root weight of grain amaranth was 21.9% higher than that of ryegrass, while the root length of ryegrass was 13.5% higher than that of grain amaranth. By mass, the root morphology of the two species did not differ much. Data obtained under hydroponic culture may not represent what happens in pot culture, but could be compared with data that also collected under hydroponic culture.

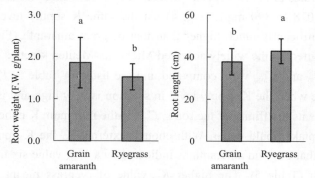

Fig. 2 Root weight and root length of two species after six weeks in hydroponic culture

Vertical bars represent standard deviation ($n=21$)

Significant amounts of citric (0.249 mmol/kg root/h) and oxalic acids (0.145 mmol/kg root/h) were detected in the root exudates of grain amaranth, but were not found in the root exudates of ryegrass. Research has previously shown that grain amaranth releases root exudates containing organic acids which activate mineral K and facilitate plant K uptake (Tu et al., 1999; Li et al., 2006). In the current study, grain amaranth was not as efficient as ryegrass at using the NEK in minerals even if organic acids were released by the roots of grain amaranth. The effect of organic acids on NEK release from minerals was largely dependent on the dose of organic acids. At low levels (5 mmol/L), the increase in available K caused by the addition of organic acids in biotite, vermiculite and feldspar was small (Fig. 3). Activation of

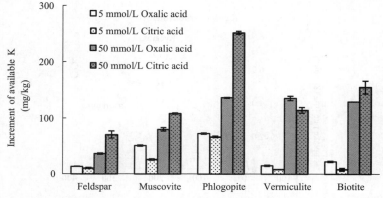

Fig. 3 Increment of NH_4OAc-extractable K in minerals after incubation with organic acids for one week

Calculated as the difference between the measured values obtained for each acid-mineral extraction and the corresponding water-mineral extraction. Vertical bars represent standard deviation ($n=3$)

NEK by organic acids could be significant when high levels of organic acids were added, and the effects were also related to the types of organic acids and minerals involved (Fig. 3). For ryegrass, organic acids in root exudates seemed to be unrelated to its higher K-uptake efficiency.

3. Root K uptake kinetics

The K uptake capacity of the two species was also compared under a hydroponic culture system. K uptake rates of the two species increased correspondingly as the K levels in solution increased from 0.078 to 390 mg/L (Fig. 4). At the same K supply level, K uptake rates of ryegrass were significantly much higher than that of grain amaranth (Fig. 4). The curves in Fig. 4 were well fitted by the widely accepted Michaelis-Menten kinetics. Kinetics parameters including V_{max}, K_m and C_{min} were computed and are listed in Table 3. V_{max} is the maximum root K-uptake rate when the K supply level in solution is very high and is not limiting root K uptake. K_m represents K-affinity of the roots. C_{min} is the minimum K concentration in solution at which net K-uptake could occur. With enough external K, the K uptake of ryegrass was much faster than that of grain amaranth as indicated by a V_{max} value six times higher than that of grain amaranth (Table 5). The higher K_m value of ryegrass meant it had a lower root K-affinity than grain amaranth. However, the value of K_m equals the external K concentration at which K-uptake rates reach up to half of V_{max}. The large difference between the V_{max} values

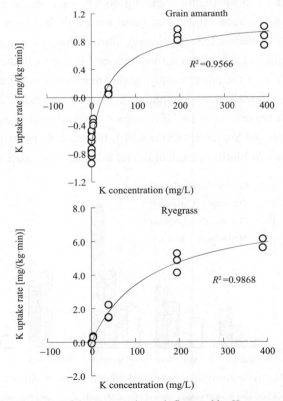

Fig. 4 Root K uptake rates of the two species as influenced by K concentration in solutions

of ryegrass and grain amaranth made K_m a less important parameter in the current study. The C_{min} of ryegrass was much lower than that of grain amaranth, suggesting a strong K-uptake capacity of ryegrass at low external K concentrations (Table 5). The K-uptake efficiency of plants was mostly considered when plants were grown in K insufficient conditions, thus C_{min} is the most important parameter to distinguish the K-uptake efficiency of the plants.

Table 5 The correlation coefficient and slope of simple linear regression equations describing the relationship of soil K removed by 1 to 8 crops of ryegrass and K extracted by different chemical methods

Crops	1 mol/L NH$_4$OAc		0.2 mol/L NaBPh$_4$					
			30 min		60 min		120 min	
	r^2	Slope	r^2	Slope	r^2	Slope	r^2	Slope
1	0.68*	0.43	0.84*	0.31	0.83*	0.27	0.84*	0.25
2	0.61*	0.79	0.84*	0.62	0.92*	0.55	0.94*	0.52
3	0.40*	0.83	0.68*	0.71	0.89*	0.70	0.91*	0.65
4	0.36*	0.82	0.64*	0.72	0.87*	0.72	0.89*	0.68
5	0.35*	0.86	0.64*	0.76	0.87*	0.76	0.89*	0.72
6	0.34*	0.88	0.63*	0.79	0.86*	0.79	0.89*	0.75
7	0.32*	0.89	0.60*	0.81	0.85*	0.82	0.88*	0.77
8	0.28*	0.88	0.57*	0.83	0.83*	0.85	0.86*	0.80

Note:* Correlation coefficient is significant at level of $P < 0.01$.

2.1.4 Discussion

The capacity of a plant to exploit soil NEK depends not only on the plant itself, but also relates to NEK release from soil or minerals. The mechanism for high plant K uptake-efficiency should facilitate soil NEK release, and thus closely link to the mechanism of NEK release. Ionic factors and K concentration are the main factors that determine NEK release from soil or minerals (Martin and Sparks, 1985; Huang, 2005).

Previous research has shown that the presence of other cations can increase K release not only from exchangeable sites but also from nonexchangeable sites (Rich and Black, 1963; Martin and Sparks, 1985; Shen et al., 2009). In the current study, the effects of various ions on NEK release were investigated after the exchangeable and easily releasable NEK of minerals was removed via a previous 60 d incubation in the same solution. The data in Fig. 5 shows that the promotive effect of ions on NEK release followed the order of $H^+ \gg Na^+$, $Ca^{2+} > H_2O$ (except that Na^+ and Ca^{2+} had no significant effect on K release from feldspar) (Fig. 5). The effect of Na^+ and Ca^{2+} on the release of NEK was related to mineral type. Although Na^+ promoted a greater release of NEK from phlogopite and muscovite than Ca^{2+}, Ca^{2+} promoted a greater release of NEK from biotite than Na^+. For vermiculite, Ca^{2+} increased the release of NEK more than Na^+ during the previous 60 d extraction (Shen et al., 2009), and Na^+ increased the release of NEK more than Ca^{2+} at a later stage than that reported in this paper. All the results demonstrated that Na^+ has a more positive effect on the release of initially recalcitrant NEK than Ca^{2+}, while Ca^{2+} is stronger than Na^+ at promoting the release of easily releasable NEK from minerals. The relative effectiveness of Na^+ on the release of initially recalcitrant NEK may be due to the small diameter of its hydrated ion (0.45 nm) which facilitates access to the interlayer K that cannot be reached by Ca^{2+} (0.6 nm) (Bolt et al., 1963). For easily releasable NEK in swelling biotite, no access hindrance existed, so the higher valence of the Ca^{2+} ion led to more NEK release as compared to Na^+ (Rich, 1964; Scott and Smith, 1966).

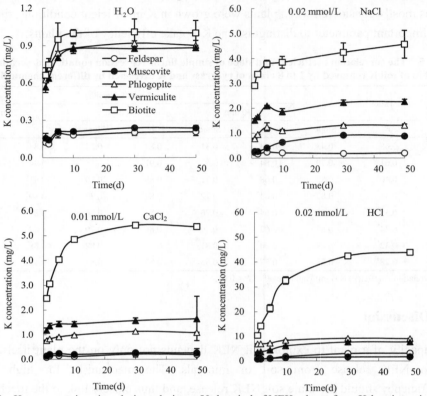

Fig. 5 K concentrations in solutions during a 50 d period of NEK release from K-bearing minerals
Vertical bars represent standard deviation (n=3)

As the contact time increased, the NEK release to all solutions was initially quick, but then slowed down and eventually almost stopped as the solution K concentration reached a certain level (Fig. 5). The results could be explained as NEK release from most minerals being a diffusion controlled process (Mortland and Ellis, 1959; Mortland, 1961; Feigenbaum et al., 1981), and the K concentration in solution being a key factor in determining NEK release from minerals (Scott and Smith, 1966; Martin and Sparks, 1985; Huang, 2005). The final K concentration in the solutions (Fig. 5) could be considered as the K thresholds for NEK release from the minerals, and are listed in Table 6. The data showed that K thresholds not only depended on mineral type, but were also significantly increased by the presence of Ca^{2+} or Na^+. The existence of K concentration thresholds for NEK release has been previously reported by many researchers (Datta and Sastry, 1988; Adhikari and Ghosh, 1993; Hinsinger and Jaillard, 1993; Schneider, 1997; Huang, 2005), and indicates that the promotive effect of other cations on NEK release through ion exchange can only happen when the K concentration in solution is below the threshold. In real soil-plant systems, some researchers reported that the presence of Ca^{2+}, Mg^{2+} or Na^+ in irrigation water or the rhizosphere could increase soil NEK availability to plants (Rahmatullah et al., 1994; Moritsuka et al., 2004). The results obtained in the current study suggest that Ca^{2+}, Na^+ and some other metal ions not tested in this study may increase soil K availability not only through their exchange effect on initial exchangeable K, but also through their positive effect on the release of initial NEK via elevating the thresholds of K concentration for NEK release.

Table 6 The correlation coefficient of simple linear regression equations (soil K vs. ryegrass leaf K), and the critical soil K estimated according to a critical ryegrass leaf K level of 18.88 g/(kg DM)

Soil	NH$_4$OAc method		NaBPh$_4$ 60-min method	
	r^2 of linear equation	Critical soil K (mg/kg)	r^2 of linear equation	Critical soil K (mg/kg)
LY	0.60*	48	0.89*	190
WC	0.96*	157	0.95*	211
FQ	0.85*	97	0.90*	202
CS	0.90*	132	0.93*	270

Note: * Correlation coefficient is significant at level of $P < 0.01$.

The effect of H^+ on NEK release was traditionally explained as being due to its weathering effect on minerals (Huang, 2005). However, Norrish (1973) proposed that in concentrations lower than 1 mmol, the H^+ ion behaves as other cations do in replacing interlayer-K and that only with higher concentrations of acid is the octahedral sheet attacked and its structure destroyed (Martin and Sparks, 1985). With 20 mmol/L HCl in the current study, the easy-dissoluble components of the minerals were removed during a prior 60 d incubation in the same solution, and the NEK released from minerals did not further increase after 30 d contact time (Fig. 5). Clear K thresholds were also obtained for NEK release from minerals in 20 mmol/L HCl, and the thresholds in HCl solution were much higher than in salts (Table 6). These results demonstrate that H^+ behaved like a metal cation in replacing NEK because under the weathering effect the release of NEK would not stop and no clear threshold of K concentration could be found, unlike in the present study.

The behaviour of H^+ as a metal cation in replacing NEK would also be an important mechanism in the rhizosphere, where organic acids are secreted by roots. Song and Huang (1988) reported that these organic acids can facilitate the weathering of minerals and rocks through the formation of metal-organic complexes, but showed that the weathering of the K-minerals by organic acids resulted in a higher proportion of K loss than of Si loss, which was attributed to partial K release by H^+ exchange. The ability of organic acids to facilitate release of NEK or mineral K was also proved by many researchers (Zhu and Luo, 1993; Wang et al., 2000b; Li et al., 2006; Wang et al., 2007), although the data could not clearly identify whether the NEK release was promoted by H^+ exchange or by weathering of the K-minerals. However, the dissolving effect of organic acids on K-minerals would be very slow, but the exchange of K^+ by H^+ would be much faster (Feigenbaum et al., 1981). Thus, the promotive effect of most organic acids on NEK release in the rhizosphere may also occur mainly through exchange of K^+ by H^+ and by elevating the thresholds of K concentration for NEK release.

H^+ acts as a metal cation and does better than other cations in replacing NEK from minerals. This conclusion has significant meaning in distinguishing what form of K could be activated or mobilized by acidification in real plant-soil system. Acidification of soils in cropland is a common phenomenon that may be resulted from high rate application of nitrogen fertilizer, acid rain, or H^+ secretion by plant roots etc. (Romheld et al., 1984; Barak et al., 1997; Liu et al., 2010) Presence of H^+ could activate K release in soils, but the soil K activated by H^+ is mainly from the nonexchangeable part but not from the structural part according to the results got in current study. It confirmed that NEK plays a much more important role in soil K supply to plants than structural K. Thus the soil NEK reserve, but not the total K reserve determines soil K fertility. Since the total amount of soil NEK can reach up to 20%-55% of total K (Zhou and Wang, 2008), how to evaluate the availability of soil NEK reserve and to exploit soil NEK reserve are substantial important for cropland K management in future.

Attributing the positive effect of H^+ (at a concentration of 20 mmol/L) on NEK release to ion exchange makes K concentration a much more important factor in determining NEK release in soils (Table 6). A very low concentration of K in solution could lead to a very quick NEK release from many kinds of K-minerals; this phenomenon could be easily proved by the high release rate of NEK during sodium tetraphenylboron extraction (Cox et al., 1999). In a plant-soil system, the ability of roots to reduce the K concentration in the rhizosphere would surely affect soil NEK release. The NEK in mineral can only be exploited by plant root when the C_{min}, the minimum external K concentration at which K is taken up by the plant root is lower than the K threshold for NEK release from the K-bearing mineral. Thus the C_{min}, a parameter reflecting the K-uptake capacity of roots at low K concentrations, would probably be a key parameter in determining plant NEK uptake efficiency. While the capacity to secrete organic acids into the rhizosphere is a secondary or minor mechanism for plant NEK uptake efficiency, since the release of organic acids to the rhizosphere is a common phenomenon for most plants, but does not inevitably result in a high NEK uptake efficiency in most plants.

Neither root exudates nor root morphology, but K-uptake capacity of roots at low K concentration determined plant NEK uptake capacity. This concept was physiologically consistent to recent discoveries of many transporters and channels involved in plant K uptake from the soil (Lebaudy et al., 2007; Chen et al., 2008; Lan et al., 2010). It was reported that K uptake and fluxes within the plant are mediated by several families of transporters and channels differing in their affinity for K (Markus and Pascal, 2007; Lebaudy et al., 2007). To improve the plant NEK uptake capacity, transporters controlling K uptake at low concentration are extremely important according to the results got in current study. Discovery and regulation of such transporters through conventional breeding or genetic engineering might help to breed genotypes with high NEK uptake capacity in future.

References

Adhikari M, Ghosh T K. 1993. Threshold levels of potassium for intermediate potassium release in relation to clay mineralogy and specifically held potassium of soils. Journal of the Indian Society of Soil Science, 41: 663-666

Barak P, Jobe B O, Krueger A R, et al. 1997. Effects of long-term soil acidification due to nitrogen fertilizer inputs in Wisconsin. Plant Soil, 197: 61-69

Bolt G H, Sumner M E, Kamphorst A. 1963. A study of the equilibria between three categories of potassium in an illitic soil. Soil Sci. Soc. Am. Proc., 27: 294-299

Chen K, Ma J, Cao Y. 1999. Exudation of organic acids by the roots of different plant species under phosphorus deficiency. Journal of China Agricultural University, 4(3): 58-62 (in Chinese)

Chen Y F, Wang Y, Wu W H. 2008. Membrane transporters for nitrogen, phosphate and potassium uptake in plants. Journal of Integrative Plant Biology, 50(7): 835-848

Coroneos C, Hinsinger P, Gilkes R J. 1996. Granite powder as a source of potassium for plants: a glasshouse bioassay comparing two pasture species. Fertilizer Research, 45: 143-152

Cox A E, Joern B C, Brounder S M, et al. 1999. Plant-available potassium assessment with a modified sodium tetraphenylboron method. Soil Sci. Soc. Am. J., 63: 902-911

Datta S C, Sastry T G. 1988. Determination of threshold levels for potassium release in three soils. J. India Soc. Soil Sci., 36: 676-681

Dessougi H E, Claassen N, Steingrobe B. 2002. Potassium efficiency mechanisms of wheat, barley, and sugar beet grown on a K fixing soil under controlled conditions. J. Plant Nutr. Soil Sci., 165: 732-737

Feigenbaum S, Edelstein R, Shainberg I. 1981. Release rate of potassium and structural cations from micas to ion exchangers in dilute solutions. Soil Sci. Soc. Am J., 45: 501-506

Hinsinger P, Jaillard B. 1993. Root-induced release of interlayer potassium and vermiculitization of phlogopite as related to potassium depletion in the rhizosphere of ryegrass. J. Soil Sci., 44: 525-534

Hoagland D R, Arnon D I. 1950. The water culture method for prowling plants without soil. Circular 347, California Agriculture Experimental Station, Berkeley, CA, USA: 142

Hogh-Jensen H, Pedersen M B. 2003. Morphological plasticity by crop plants and their potassium use efficiency. J. Plant Nutr., 26: 969-984

Huang P M. 2005. Chemistry of potassium in soils. In: Tabatabi M A, Sparks D L. Chemical Processes in Soils. SSSA Book Series, no 8. Soil Science Society of America, Madison, WI: 227-292

Jia Y, Yang X, Islam E, et al. 2008. Effects of potassium deficiency on chloroplast ultrastructure and chlorophyll fluorescence in inefficient and efficient genotypes of rice. Journal of Plant Nutrition, 31: 2105-2118

Krafffczyk I, Trolldenier G, Beringer H. 1984. Soluble root exudates of maize: influence of potassium supply and rhizosphere microorganisms. Soil Biology and Biochemistry, 16 (4): 315-322

Lan W Z, Wang W, Wang S M, et al. 2010. A rice high-affinity potassium transporter (HKT) conceals a calcium-permeable cation channel. Proc. Natl. Acad. Sci., 107(15): 7089-7094

Lebaudy A, Very A A, Sentenac H. 2007. K^+ channel activity in plants: genes, regulations and functions. FEBS Letters, 581: 2357-2366

Li T X, Ma G R. 2003. Screening grain amaranths for genotypes of the capability of enrichment in potassium. Plant Nutrition and Fertilizer Science, 9(4): 473-479 (in Chinese)

Li T X, Ma G R. 2004. Physiological and morphological characteristics of roots in grain amaranth genotypes enrichment in potassium. Acta Agronomica Sinica, 30(11): 1145-1151 (in Chinese)

Li T X, Ma G R, Zhang X Z. 2006. Root exudates of potassium enrichment genotype grain amaranth and their activation on soil mineral potassium. Chinese Journal of Applied Ecology, 17(3): 368-372 (in Chinese)

Liu K H, Fang Y T, Yu F M, et al. 2010. Soil Acidification in response to acid deposition in three subtropical forests of subtropical China. Pedosphere, 20: 399-408

Lu R K. 1999. Analytic Methods for Soil and Agro-chemistry. Beijing: Chinese Agricultural Technology Press (in Chinese)

Lv F T, Zhang X S. 2010. Study on the characteristic of potassium absorption and low potassium endurance mechanism for different genotypes maize. Journal of Maize Sciences, 18(1): 61-65 (in Chinese)

Ma J F, Shen R F, Zhao Z Q, et al. 2002. Response of rice to Al stress and identification of quantitative trait loci for Al tolerance. Plant and Cell Physiology, 43: 652-659

Markus G, Pascal M. 2007. Potassium transporters in plants—involvement in K^+ acquisition, redistribution and homeostasis. FEBS Letters, 581: 2348-2356

Martin H W, Sparks D L. 1985. On the behavior of nonexchangeable potassium in soils. Commun Soil Sci

Plant Anal., 16(2): 133-162

Moritsuka N, Yanai J, Kosaki T. 2004. Possible processes releasing nonexchangeable potassium from the rhizosphere of maize. Plant Soil, 258: 261-268

Mortland M M. 1961. The dynamic character of potassium release and fixation. Soil Sci., 91: 11-13

Mortland M M, Ellis B. 1959. Release of fixed potassium as a diffusion controlled process. Soil Sci. Soc. Am. Proc., 23:363-364

Norrish K. 1973. Factors in the weathering of mica to vermiculite. Proc. of the Internat Clay Conf, Madrid, Spain: 417-432

Rahmatullah, Shaikh B Z, Gill M A, et al. 1994. Bioavailable potassium in river-bed sediments and release of interlayer potassium in irrigated arid soils. Soil Use and Management, 10: 43-46

Rengel Z, Damon P M. 2008. Crops and genotypes differ in efficiency of potassium uptake and use. Physiologia Plantarum, 133: 624-636

Rich C I. 1964. Effect of cation size and pH on potassium exchange in Nason soil. Soil Sci., 98: 100-106

Rich C I, Black W R. 1963. Potassium exchange as affected by cation size, pH, and mineral structure. Soil Sci., 97: 384-390

Romheld V, Muller C, Marschner H. 1984. Localization and capacity of proton pumps in roots of intact sunflower plants. Plant Physiology, 76(3): 603-606

Schneider A. 1997. Short-term release and fixation of K in calcareous clay soils. Consequence for K buffer power prediction. Eur. J. Soil Sci., 48:499-512

Schroeder D. 1978. Structure and weathering of potassium containing minerals. Proc. Congr. Int. Potash Inst., 11: 43-63

Scott A D, Smith S J. 1966. Susceptibility of interlayer potassium in micas to exchange with sodium. Clays Clay Miner, 14: 69-81

Shen Q H, Wang H Y, Zhou J M, et al. 2009. Dynamic release of potassium from potassium bearing minerals as affected by ion species in solution. Soils, 41(6): 862-868 (in Chinese)

Song S K, Huang P M. 1988. Dynamics of potassium release from potassium-bearing minerals as influenced by oxalic and citric acids. Soil Sci. Soc. Am. J., 52: 383-390

Sparks D L. 2000. Bioavailability of soil potassium. In: Sumner M E. Handbook of Soil Science. Boca Raton, F L: CRC Press: 38-53

Trehan S P, El-Dessougi H, Claassen N. 2005. Potassium efficiency of 10 potato cultivars as related to their capability to use nonexchangeable soil potassium by chemical mobilization. Commun. Soil Sci. Plant Anal., 36: 1809-1822

Trehan S P, Sharma R C. 2002. Potassium uptake efficiency of young plants of three potato cultivars as related to root and shoot parameters. Commun. Soil Sci. Plant Anal., 33: 1813-1823

Tu S X, Huang M, Guo Z F, et al. 2000. Genotypic variations in kinetics of potassium absorption and utilization of soil and applied potassium by *Amaranthus* spp. Pedosphere, 10(4): 363-372

Tu S X, Sun J H, Guo Z F. 1999. The root exudation of grain amaranth and its role in release of mineral potassium. Acta Agriculturae Nucleatae Sinica, 13(5): 305-311 (in Chinese)

Wang D S, Liang C H, Du L Y. 2007. Effect of organic acids on K release from K-bearing minerals. Journal of Shenyang Agricultural University, 38(1): 65-69 (in Chinese)

Wang J G, Zhang F S, Cao Y P, et al. 2000a. Effect of plant types on release of mineral potassium from gneiss. Nutr. Cycl. Agroecosys., 56: 37-44

Wang J G, Zhang F S, Zhang X L, et al. 2000b. Release of potassium from K-bearing minerals: effect of plant roots under P deficiency. Nutr. Cycl. Agroecosys., 56: 45-52

Xie J C, Zhou J M, Hardter R. 2000. Potassium in Chinese Agriculture. Nanjing: Hehai University Press: 53 (in Chinese)

Zhang F S, Ma J, Cao Y P. 1997. Phosphorus deficiency enhances root exudation of low-molecular weight organic acids and utilization of sparingly soluble inorganic phosphates by radish (*Raghanus satiuvs* L.) and rape (*Brassica napus* L.) plants. Plant Soil, 196: 261-264

Zhou J M, Wang H Y. 2008. K forms and transformation in soils. In: Zhou J M, Magen H. Soil Potassium Dynamics and K Fertilizer Management. Nanjing: Hehai University Press: 3-9 (in Chinese)

Zhu Y G, Luo J X. 1993. Release of soil nonexchangeable K by organic acids. Pedosphere, 4(3): 269-276

2.2 Nutrient budget and soil nutrient status in greenhouse system

2.2.1 Introduction

Since the introduction of greenhouse cultivation technique in mid 1970s, greenhouse production has been widely popularized and applied in China. China rank first in terms of the area under greenhouse cultivation and total yield in the world. The production of vegetables contributes major share in total yield (Yu et al., 2007). The area under greenhouse cultivation increases rapidly, but the quality is low. There are many problems, which are different from those occurring in regular open field cultivation; especially improper fertilization leads to deterioration of soil quality and other related environmental problem. The greenhouse production begins relatively late in China and in-depth researches on production technique and cultivation management are insufficient. There is urgent need to develop environment-friendly and sustainable production technology which would enhance the yield with high quality of produce. The research on nutrient budget and utilization rate of fertilizer is important to find out the rationality of fertilization system, forecasting soil fertility trend and possible environmental impact, it also facilitates the industrialized and large scale greenhouse cultivation (Liu et al., 2002; Suo, 2008).

Currently, most of researches on soil quality under greenhouse cultivation are focused on salinization, change of nutrient content, heavy metal pollution and microorganism features (Wang et al., 2006; Huang et al., 2007; Yu et al., 2007; Liu et al., 2008). Studies on nutrient budgets mainly aimed at regional farmland ecosystem (Fang, 2008). Though greenhouse cultivation is a rising industry, which represents the development trend of modern agriculture, the nutrient budget and nutrient status of greenhouse cultivation are rarely reported.

According to previous researches, problems regarding greenhouse soil quality deterioration were closely related with rates and kinds of fertilizer application. This paper further discussed utilization rates of fertilizers, nutrients translocation and changes in soil nutrients. The 18 representative greenhouses in Shouguang, Shandong Province, China were selected as research objective to find out problems in nutrient input and the influence of these problems on soil nutrient by intensive survey and soil analysis.

2.2.2 Materials and Methods

1. Description of study area

Shouguang City of Shandong Province is the largest greenhouse vegetable production base in China. Now it boasts more than 300000 sunlight greenhouse and its vegetable planting area exceeds 60000 ha so it plays an important and exemplary role in popularizing its development mode. Further, its fertilizer application level and soil nutrient status under local greenhouse cultivation condition are representative in North China.

2. Nutrient budget of greenhouse

After collecting information on fertilizer input, cropping pattern, and crop yield (Table 1), we calculated the apparent balance of soil nutrient by subtracting the pure quantity of total input of both chemical and organic fertilizers from the gross nutrient uptake of crops.

Table 1　Fertilization level of greenhouse system in Shouguang, Shangdong Province

Sampling site		Area (m^2)	Yield of vegetable (kg/a)	Organic manure		Chemical fertilizer	
				Type and application rate (t/a)		Type and application rate (kg/a)	
Luocheng Town	1	650	Cucumber 5000、towel gourd 7500	Pig manure、duck manure、chicken manure	15	Compound fertilizer	950
	2	650	Cucumber 5000、towel gourd 7500	Pig manure、duck manure、chicken manure	15	Compound fertilizer	950
	3	630	Pimiento 3000、towel gourd 7500、Egg plant 7500	Pig manure、duck manure、chicken manure	14	Compound fertilizer	950
	4	650	Musk melon 2500、eggplant 7500、pimiento 2500	Pig manure	13	Compound fertilizer	950
	5	530	Pimiento 10000	Chicken manure	10	Compound fertilizer Diammonium phosphate	850 50
	6	480	Tomato 7000、marrow 7000	Chicken manure	9	Compound fertilizer	1000
Wenjia Town	1	325	Tomato 7500	Pig manure、wheat straw	3	Diammonium phosphate Potassium sulphate Superphosphate Aminophenol	35 75 150 700
	2	680	Tomato 12500、eggplant 7000	Pig manure、wheat straw	4	Diammonium phosphate Potassium sulphate Superphosphate Aminophenol	50 100 200 1000
	3	500	Tomato 12500、eggplant 13000	Pig manure、chicken manure	7	Compound fertilizer	850
	4	500	Tomato 12500、eggplant 15000	Pig manure、chicken manure	7	Compound fertilizer	850
	5	560	Tomato 5000、oil bean 2000	Pig manure、chicken manure	12	Compound fertilizer	1150
	6	600	Tomato 7500	Pig manure	2	Compound fertilizer	300
Sunji Town	1	500	Cucumber 5000、balsam pear 5000	Chicken manure	12	Diammonium phosphate Compound fertilizer	100 900
	2	320	Cucumber 2000、balsam pear 2500	Chicken manure	7	Diammonium phosphate Compound fertilizer	300 600
	3	640	Cucumber 7500、balsam pear 7500	Chicken manure	10	Compound fertilizer	1150
	4	500	Cucumber 6000、balsam pear 6000、tomato 2500	Chicken manure	12	Compound fertilizer	750
	5	390	Cucumber 5000、balsam pear 2500 marrow 2000	Chicken manure、duck manure	9	Compound fertilizer Diammonium phosphate	400 50
	6	390	Cowpea 1750、Pimiento 1750	Chicken manure	6	Ammonium carbonate Compound fertilizer	150 300

3. Nutrient input

The nutrient inputs (N, P_2O_5 and K_2O) mainly refer to the pure quantity of input in chemical and organic fertilizers, and were calculated by multiplying the input of each fertilizer with the percent of effective nutrient. The nutrient contents of chemical fertilizer were calculated according to specification on packing bag, and the nutrient contents of organic fertilizer were calculated according to the parameters specified in Anonymous (1999). The amount of nutrients absorbed through irrigation and crop seed accounts for minute percentage of total nutrient input, and therefore, they were neglected.

4. *Nutrient uptake*

It mainly refers to the gross absorption of the nutrients by the crop, considering its economic yield and nutrient uptake parameters (N, P_2O_5 and K_2O)(Table 2). The nutrient losses arising from leaching and runoff were not included in nutrient uptake because the greenhouse cultivation system was in long-term semi-closed condition and seldom leached by rains.

Table 2 Nutrient uptake of vegetable crops with 100 kg of yield
(Ma, 1997; Bai and Huang, 1999; Huang, 2001)

Vegetable species	N (kg)	P_2O_5 (kg)	K_2O (kg)
Cucumber	0.21	0.10	0.35
Towel gourd	0.41	0.13	0.45
Muskmelon	0.41	0.22	0.55
Balsam pear	0.53	0.18	0.69
Pimiento	0.58	0.11	0.74
Eggplant	0.31	0.07	0.49
Tomato	0.31	0.07	0.50
Oil bean	0.80	0.23	0.68
Marrow	0.39	0.21	0.73
Cowpea	0.90	0.22	0.70

5. *Sampling and analysis*

The samples from topsoil (0-20 cm) were collected from every 6 representative greenhouses located at Luocheng, Wenjia and Sunji towns of Shouguang, Shandong Province, China. The samples at the depth of 0-20, 20-40, 40-60, 60-80 to 80-100 cm in the greenhouse soil with longer cultivation period were also collected. The sampling was done in S-shaped manner in each greenhouse. As the soil and water table of greenhouse and the adjacent open-field were the same, the similar sampling method was employed to collect the topsoil and profile samples from adjacent open-field. The samples were then air-dried and sieved through 2 mm and 0.149 mm mesh for analysis. Organic matter was determined with dilution heat $K_2Cr_2O_7$ oxidation volumetric method, total nitrogen with Kjeldahl method, nitrate nitrogen and ammonium nitrogen with 2 mol/L KCl digestion and flow analyzer, available phosphorus with 0.5 mol/L pH 8.5 $NaHCO_3$ digestion and Mo-Sb colorimetry, available kalium with 1 mol/L pH 7.0 CH_3COONH_4 digestion and flame Photometer. The grading of the soil on the basis of its nutritional status for vegetable cultivation was done according to Shen and Zou (2004) (Table 3).

Table 3 Criteria of soil nutrient grade for soils under vegetable cultivation (Shen and Zou, 2004)

Fertility grade	Organic matter (g/kg)	Total nitrogen (g/kg)	Nitrate nitrogen (mg/kg)	Available phosphorus (mg/kg)	Available potassium (mg/kg)
High	>50	>2.0	>125	>120	>200
Relatively high	40-50	1.5-2.0	100-125	90-120	150-200
Medium	20-40	1.0-1.5	50-100	30-90	100-150
Low	15-20	0.8-1.0	25-50	15-30	75-100
Very low	<15	<0.8	<25	<15	<75

2.2.3 Results

1. Fertilization features

The investigation showed that the organic fertilizers used in greenhouse cultivation at Shouguang were mainly fresh pig manure and chicken manure while only a few cases duck manure and wheat straw were used. Organic fertilizers mainly used as basal fertilizer with annual average and the maximum consumption being 177 t/ha and 240 t/ha, respectively. Chemical fertilizers mainly included compound fertilizer, diammonium phosphate, potassium sulphate, ammonium carbonate, and super phosphate. The 56% of greenhouses received high dose of compound fertilizer (15-15-15). The annual average consumption of chemical fertilizer was 17 t/ha while the maximum consumption was 29 t/ha. It was found that the average annual input amount of N, P_2O_5 and K_2O were 4088 (CV=30%), 3655 (CV=35%) and 3437 kg/ha (CV=36%), respectively under greenhouse cultivation (Table 4). Further, the 63%, 61% and 66% of N, P_2O_5 and K_2O were supplied from chemical fertilizer with the input amount 2539, 2184 and 2202 kg/ha, respectively. It accounted for 64%, 48% and 44% higher than those provided by organic fertilizer.

Table 4 The fertilizer application of greenhouse system in Shouguang

Sampling site		Organic manure (kg/ha)			Chemical fertilizer (kg/ha)			Application rate (kg/ha)		
		N	P_2O_5	K_2O	N	P_2O_5	K_2O	N	P_2O_5	K_2O
Luocheng Town	1	1687	1675	1298	2131	2469	2515	3817	4144	3813
	2	1687	1675	1298	2131	2469	2515	3817	4144	3813
	3	1624	1596	1235	2198	2547	2595	3822	4143	3830
	4	1542	1478	1174	2131	2469	2515	3672	3947	3689
	5	1764	1640	1432	2594	2405	2405	4358	4045	3837
	6	1827	1676	1524	3125	3125	3125	4952	4800	4648
Wenjia Town	1	505	518	326	3640	1049	1154	4145	1567	1480
	2	322	330	208	2485	691	735	2807	1021	943
	3	1057	1017	798	2550	2550	2550	3606	3566	3348
	4	1057	1017	798	2550	2550	2550	3606	3566	3348
	5	182	187	118	750	750	750	932	937	868
	6	2093	2047	1524	3080	3080	3080	5173	5127	4604
Sunji Town	1	2137	2097	1828	1795	1538	1538	3931	3635	3367
	2	1587	1455	1324	1807	1154	1154	3395	2609	2477
	3	1612	1478	1344	2695	2695	2695	4307	4173	4039
	4	2477	2271	2065	2250	2250	2250	4726	4520	4314
	5	2477	2271	2065	3100	2700	2700	5576	4970	4764
	6	2258	2070	1882	4688	2813	2813	6945	4882	4695

2. Nutrient budget in greenhouse system

Heavy dose of fertilizer was found to be a common practice to obtain high yield under greenhouse cultivation. It was observed that the input ratio among N, P_2O_5 and K_2O (1∶0.9∶0.8) was quite different from the uptake ratio (1∶0.3∶1.4) (Table 5). It caused an excess N,

P_2O_5 and K_2O in the soil and the theoretical surplus were 3214, 3401 and 2322 kg/ha, respectively. The surplus of P_2O_5 was the highest with its maximum of being 4702 kg/ha. As various vegetables consumed the smallest amount of P_2O_5, its utilization ratio for current season was only 10% - 25% and most of P_2O_5 was remained in soil. The more attention should be paid to residual effect of fertilization and input less phosphate fertilizer to avoid waste of resources. Though, the input amount of N was the highest, its surplus was lower because of higher uptake of N by vegetables. The surplus of K_2O was the lowest, which was mainly due to its higher uptake and lower input.

Table 5 Nutrient budget of greenhouse system in Shouguang

Sampling site		Input of nutrient (kg/ha)			Uptake of nutrient (kg/ha)			N : P_2O_5 : K_2O		Surplus of nutrient (kg/ha)		
		N	P_2O_5	K_2O	N	P_2O_5	K_2O	Input ratio	Uptake ratio	N	P_2O_5	K_2O
Luocheng Town	1	3817	4144	3813	623	227	788	1 : 1.1 : 1.0	1 : 0.4 : 1.3	3194	3917	3025
	2	3817	4144	3813	623	227	788	1 : 1.1 : 1.0	1 : 0.4 : 1.3	3194	3917	3025
	3	3822	4143	3830	1121	290	1471	1 : 1.1 : 1.0	1 : 0.3 : 1.3	2701	3853	2358
	4	3672	3947	3689	738	208	1061	1 : 1.1 : 1.0	1 : 0.3 : 1.4	2934	3739	2628
	5	4358	4045	3837	1094	208	1396	1 : 0.9 : 0.9	1 : 0.2 : 1.3	3264	3838	2441
	6	4952	4800	4648	1021	408	1794	1 : 1.0 : 0.9	1 : 0.4 : 1.8	3931	4392	2854
Wenjia Town	1	4145	1567	1480	715	162	1154	1 : 0.4 : 0.4	1 : 0.2 : 1.6	3430	1406	326
	2	2807	1021	943	889	201	1423	1 : 0.4 : 0.3	1 : 0.2 : 1.6	1918	821	−481
	3	3606	3566	3348	1581	357	2524	1 : 1.0 : 0.9	1 : 0.2 : 1.6	2026	3209	824
	4	3606	3566	3348	1645	385	2720	1 : 1.0 : 0.9	1 : 0.2 : 1.7	1962	3181	628
	5	932	937	868	525	135	643	1 : 1.0 : 0.9	1 : 0.3 : 1.2	407	802	224
	6	5173	5127	4604	415	94	670	1 : 1.0 : 0.9	1 : 0.2 : 1.6	4758	5034	3935
Sunji Town	1	3931	3635	3367	949	359	1333	1 : 0.9 : 0.9	1 : 0.4 : 1.4	2983	3276	2033
	2	3395	2609	2477	447	167	622	1 : 0.8 : 0.7	1 : 0.4 : 1.4	2947	2442	1855
	3	4307	4173	4039	867	328	1219	1 : 1.0 : 0.9	1 : 0.4 : 1.4	3440	3845	2821
	4	4726	4520	4314	1043	371	1498	1 : 1.0 : 0.9	1 : 0.4 : 1.4	3683	4149	2817
	5	5576	4970	4764	631	274	987	1 : 0.9 : 0.9	1 : 0.4 : 1.6	4945	4696	3778
	6	6945	4882	4695	809	181	788	1 : 0.7 : 0.7	1 : 0.2 : 1.0	6136	4702	3907

3. *Accumulation and translocation of organic matter in topsoil*

The accumulation of organic matters in greenhouse soil was in little quantity. The results showed the content of organic matter in topsoil (0-20 cm) of open field at Shouguang was 8.0-14.7 g/kg and its average content amounted to 11.6 g/kg (CV=29%), which showed its serious shortage (Fig.1). Though, the annual input of organic fertilizer under greenhouse cultivation reached 153.9-240.0 t/ha, the major organic fertilizer was labile fresh chicken manure from local poultry farm. Moreover, the internal temperature of greenhouse was higher and the activity of microorganism was strong which resulted in rapid mineralization of organic matter.

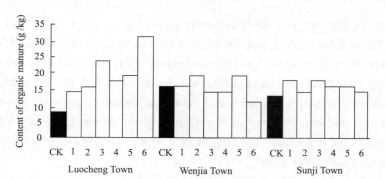

Fig. 1 The content of organic matter in the topsoil (0-20 cm) of greenhouse. CK: open field; 1-6: the number of greenhouse. The same as below

4. Accumulation and translocation of nitrogen in topsoil

The contents of total nitrogen, nitrate nitrogen and ammonium nitrogen were higher in green house topsoil than those in open field soil to different extent (Fig. 2). The respective averages were 2.6 g/kg (CV=23%), 254 mg/kg (CV=64%) and 8.9 mg/kg (CV=27%), which were 1.9, 21.2 and 1.6 times higher than those in open field soil. The rise in the nitrate nitrogen was the highest. Under greenhouse cultivation, the ratio of soil mineral nitrogen to total nitrogen raised from 0.9%-1.7% (open field) to 2.5%-23.4% indicating that fertilization was the key reason for available nutrient accumulation in soil. According to the criteria of soil nutrient grade for soils under vegetable cultivation, the content of total nitrogen (0.9-1.5 g/kg) in open field soils was at medium/lower level while the content of nitrate nitrogen (8.6-18.6 mg/kg) was at serious shortage. Under greenhouse cultivation, the contents of total nitrogen and nitrate nitrogen were 1.6-3.7 g/kg and 103-330 mg/kg, respectively, which were at relatively high level. In terms of accumulation of nitrogen in greenhouse soil profile, the contents of total nitrogen and nitrate nitrogen gradually decreased along with the depth of soil (Fig. 3). The contents of total nitrogen in topsoil (0-20 cm) of greenhouse differed greatly from that in open field soil but, have no significant difference with those in corresponding layers under 20 cm. It was also obvious that the content of NO_3^- accumulated mainly in surface soil and leached downwards as it was hard to be adsorbed by soil colloids for being carrying negative charge, and easily leached by soil water. It was observed that the content of mineral nitrogen in different soil layers of greenhouse was significantly higher than that in corresponding soil layers of open field (Fig. 3). Heavy amount of NO_3^- leaching downwards caused a higher potential risk to the quality of local groundwater. The analysis showed that the amount of nitrate nitrogen in groundwater was 9 mg/L, which was close to 10 mg/L as per the drinking water quality standard specified by both World Health Organization and China (Jin et al., 2004).

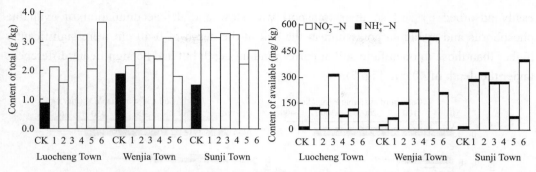

Fig. 2 Change in the total N and available N content in the topsoil of greenhouse

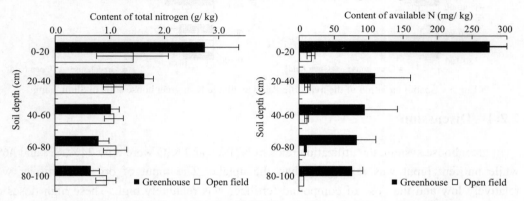

Fig. 3 Spatial variation of total N and available N content in greenhouse soils in Shouguang

5. *Accumulation and translocation of available nitrogen and available potassium in soil*

The contents of available nitrogen and available potassium in topsoil of open field were 33-72 and 102-165 mg/kg respectively, which were at medium/low level. Under greenhouse cultivation, the contents of available nitrogen and available potassium in topsoil (0-20 cm) increased greatly with the average contents being 248 (CV=31%) and 486 mg/kg (CV=37%), respectively, which were 5.4 and 3.7 times higher than those in topsoil of open field (Fig. 4). The overdose of fertilizer led the amount of available phosphorus in soil of greenhouses to far beyond the required. In greenhouse soil, the amount of available phosphorus was 135-377 mg/kg, while its required amount for vegetables was only 60-90 mg/kg. The amount of available potassium in greenhouse soil was 192-764 mg/kg and 94% of investigated greenhouses showed its higher level (>200 mg/kg). Phosphorus and potassium could be

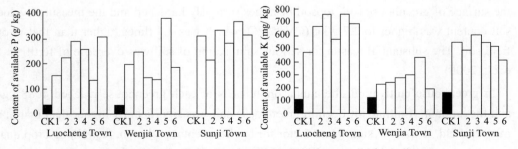

Fig. 4 Change in the available P and available K in the topsoil of greenhouse

easily adsorbed by soil so their transport was slow and the accumulation of available phosphorus and available potassium in the soil at the depth of 0-40 cm were significantly higher than those in open field soil at corresponding depth but had no significant differences under the depth of 40 cm (Fig. 5).

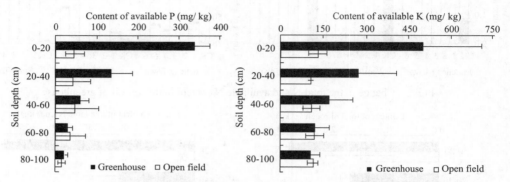

Fig. 5 Spatial variation of the available P and available K in greenhouse soils in Shouguang

2.2.4 Discussion

In greenhouse system, the utilization rates of N, P_2O_5 and K_2O were only 24%, 8% and 46% while nutrient input was far higher than the uptake. The input of organic fertilizer was relatively low and the dose of compound fertilizer was relatively high. These high doses of compound fertilizers could cause imbalance of nutrient, so they should not be used separately (Lu et al., 1996). However, under greenhouse cultivation, these compound fertilizers were used for basal dressing and also accounted for a large percent for top-dressing. For example, di-ammonium phosphate (consist of N (16%), and P_2O_5 (45%)) was applied in area under investigation. It was found that the amount of N was relatively low; after dressing, and the effect was not obvious in the beginning. Therefore, farmers believed that the input of fertilizer was low, and used more fertilizers which made excess dose of chemical fertilizer. It led to surplus of available nutrients and lack of organic matters in greenhouse soil. Considering uptake ratio of N, P_2O_5 and K_2O for most of vegetables, the input of P_2O_5 was high while the input of N and K_2O was low. The imbalance of nutrient input and selective absorption of nutrient arising from monocropping pattern in greenhouse intensified the imbalance between nutrient input and uptake which had resulted a series of problems such as secondary salinization of soil. On-site investigation indicated that salt efflorescence in large area was on the surface of greenhouse soil, topsoil becomes seriously hardened and the measured value of salt content was higher to 2.7 g/kg (CV=58%), which was 4.4 times higher than that in open field soil. The substantial reason for serious salinization of soil was overdose of fertilizer (Yu et al., 2006).

In greenhouse cultivation, the annual input of N at each greenhouse reached 4088 kg/ha, and the amount of nitrate nitrogen in greenhouse soil was significantly higher than that in soil of open field. It was the significant factor for environmental pollution, restraining crop quality improvement and hindering sustainable utilization soil. On the one hand, nitrate was easily uptaken by vegetables and 81% of nitrate absorbed by human body was from vegetables (Xiong et al., 2004). The research indicated that if the input of total nitrogen amounted to 450 kg/ha, the contribution ratio for accumulation of nitrate was higher than 85% (Qiu et al., 2004). Thus large accumulation of nitrate nitrogen in soil had no contribution to yield of crops, on the contrary, it would deteriorate crop quality and made a negative influence on human health. On the other hand, overdose of nitrogenous fertilizer, especially ammonium nitrogen fertilizer would lead to serious soil acidification and greatly increase the content of active aluminum in soil (Malhia et al., 1998; Xu and Coventry, 2002). It was found that the average pH value of greenhouse soil decreased from 7.5 (open field soil) to 6.9 and the lowest one was only 5.7, which proved that acidification of surface soil was significant (Yu et al., 2006, 2007). Furthermore, the content of NO_3^- in soil (at the depth of 80-100 cm) was significantly higher than that in open field soil at corresponding depth. It indicated that NO_3^- translocates downward significantly, and the pollution risk in groundwater would increase gradually.

Considering the aforesaid problems, the top priority for greenhouse cultivation should be given to increase the input of organic fertilizer. The rational dressing of chemical and organic fertilizers would not only facilitate a higher yield of vegetables, but also greatly decrease the accumulation of nitrate and the losses of nitrate nitrogen, ammonium nitrogen, and water-soluble total phosphorus via leaching. Thus, rational dressing is worthy to be popularized and applied to vegetable cultivation (Huang et al., 2009). According to nutrient uptake of crops and soil fertility, balanced fertilization is required. The soil nutrient diagnosis should therefore be made if necessary to increase utilization ratio of fertilizer and the yield of crops.

References

Anonymous. 1999. Station for popularizing agricultural techniques.In: Chinese Nutrient Records of the Organic Fertilizer. Beijing: China Agricultural Press: 6-59

Bai G Y, Huang D M. 1999. New Fertilizer Technical of Vegetables. Beijing: China Agriculture Press: 43-44

Fang Y D. 2008.Research progress in the farmland nutrient balance of input/output. Journal of Shandong Agricultural University (Natural Science), 39: 492-494

Huang D F, Wang G, Li W H, et al. 2009. Present status, mechanisms, and control techniques of nitrogen and phosphorus non-point source pollution from vegetable fields. Chinese Journal of Applied Ecology, 20: 991-1001

Huang D M. 2001. Fertilizer Technical of Vegetables. Beijing: China Agriculture Press: 28

Huang Z P, Xu B, Zhang K Q, et al. 2007. Accumulation of heavy metals in the four years' continual swine manure-applied greenhouse soils. Transactions of the Chinese Society of Agricultural Engineering, 23: 239-244

Jin Z F, Wang F E, Chen Y X, et al. 2004. Nitrate pollution of groundwater in urban area. Acta Pedologica Sinica, 41: 252-258

Liu H X, Wang D L, Wang S Y, et al. 2002. Changes of crop yield and soil fertility under long-term application of fertilizer and recycled nutrients in manure on a black soil. III. Soil nutrient budget. Chinese Journal of Applied Ecology, 13: 1410-1412

Liu Z H, Jiang L H, Zhang W J, et al. 2008. Evolution of fertilization and variation of soil nutrient contents in greenhouse vegetable cultivation in Shandong. Acta Pedologica Sinica, 45: 296-303

Lu R K, Liu H X, Wen D Z, et al. 1996. Study on the cycling and balance of nutrients of agro-ecosystems in the typical area of China. IV. Evaluation method and principle of nutrients balance. Chinese Journal of Soil Science, 27: 197-199

Ma D W. 1997. Marrow Cultivation in Protected Fields. Beijing: Jindun Press: 20

Malhia S S, Nyborg M, Harapiak J T. 1998. Effects of long-term N fertilizer-induced acidification and liming on micronutrients in soil and in bromegrass hay. Soil and Tillage Research, 48: 91-100

Qiu X X, Huang D F, Cai S X, et al. 2004. Effect of applying fertilizer on nitrate accumulation in vegetables. Chinese Journal of Eco-Agriculture, 12: 111-114

Suo D R. 2008. Nutrient balance and long-term location test study of fertilizer use efficiency. Phosphate & Compound Fertilizer, 23: 65-69

Shen H, Zou G Y. 2004. Parameters selection for evaluation of vegetable soil quality and its gradation. Chinese Journal of Soil Science, 35: 553-557

Wang S, Li T X, Zhang X Z, et al. 2006. Study on changes of microbial characters in greenhouse soil. Journal of Soil and Water Conservation, 20: 82-86

Xiong G H, Lin X Y, Zhang Y S, et al. 2004. Effects of fertilization on nitrate accumulation in vegetable crops. Chinese Journal of Soil Science, 35: 217-221

Xu R K, Coventry D R. 2002. Soil acidification as influenced by some agricultural practices. Agro-Environmental Protection, 21: 385-388

Yu H Y, Li T X, Zhou J M. 2006. Salt in typical greenhouse soil profiles and its potential environmental effects. Acta Pedologica Sinica, 43: 571-576

Yu H Y, Li T X, Zhou J M. 2007. Salt accumulation, translocation and ion composition in greenhouse soil profiles. Plant Nutrition and Fertilizer Science, 13: 642-650

2.3 Potassium movement and transformation in an acid soil as affected by phosphorus

2.3.1 Introduction

The choice of fertilizer application methods for a particular nutrient depends upon the mobility of the nutrient in the soil. Phosphorus fertilizer applied to the soil typically moves only a short distance (Sharpley, 1986; Eghball and Sander, 1989). The movement of P in soil is primarily by diffusion (Barber et al., 1963). The rate and extent of phosphorus (P) movement into soil from fertilizers depends upon soil compaction, soil moisture and P sorption capacity (Benbi and Gilkes, 1987). Potassium (K) moves in soil through diffusion and mass flow, but diffusion is the most important mechanism involved in the movement of fertilizer K to absorbing roots. Potassium diffusion occurs in soil solution and is affected by several factors including volumetric moisture and temperature (Barber, 1985). Equilibrium exists between solution K and exchangeable K, and between exchangeable K and fixed K. Fixation and release is a reversible process that is dependent on the concentration of K ion on the clay surface, which in turn is dependent on the concentration of K ion in the soil solution (Foth, 1984). Clay minerals responsible for K fixation are 2 : 1 types such as vermiculite, illite, and montmorillonite (Dennis and Ellis, 1962; Page et al., 1967; Rich, 1968; Ross and Cline, 1984).

Phosphorus and K are important macronutrients, and are often co-applied in a band or on the soil surface directly. The interaction between P and K might influence their movement and bioavailability after co-application to the soil. The influences of K applied in combination with P on the behavior of P in soil have been investigated by various researchers (Bouldin et al., 1960; Isensee and Walsh, 1972; Ernani and Barber, 1991; Akinremi and Cho, 1993). However, there are fewer studies on K status as affected by P applied to soils. Earlier studies demonstrate that K can be retained when applied with phosphorus fertilizers in acid tropical and subtropical soils, possibly due to the increase in CEC of the soil as a result of the formation of aluminum-phosphates (Ayres and Hagihara, 1953; Perkins, 1958; Thorup and Mehlich, 1961). Zhou and Huang (1995) found that application of $NH_4H_2PO_4$ fertilizer enhanced K release from three Chinese soils, i.e. Oxisol, Alfisol and Entisol, thus increased their K-supplying rate. However, no reports have been published on the effect of monocalcium phosphate (MCP) on the behavior of K from fertilizer in soil.

The nature of the reactions of phosphate fertilizers with soil constituents may vary with the distance from the fertilizer granule when P and K are applied together, due to the change in phosphate concentration and pH. This change might cause substantial alterations of soil constituents and subsequent K release. However, no research is available on the effect of applied phosphorus on the movement and transformation of K in the chemical environment surrounding the fertilizer sources.

The objective of this study was to investigate the effects of monocalcium phosphate (MCP) applied in combination with KCl on the movement of K from the fertilizer sources and consequent changes in different K forms in an acid red soil, which is common in south of China.

2.3.2 Materials and Methods

An acid red soil (Haplic Acrisol, FAO Soil Classification System) was collected from the top 15 cm of a cultivated layer in the Ecological Experiment of Red Soil, the Chinese Academy of Sciences (28°15′30″N, 116°55′30″E), located in Yingtan, Jiangxi province, China. Some selected chemical and physical properties of the soil are listed in Table 1. The clay minerals are dominated by kaolinite and hydrous mica, and also contain a small amount of vermiculite.

Table 1 Selected physical and chemical properties of the soil used

Property	Value
Texture	Clay (18.1% sand, 33.4% silt, 48.5% clay)
pH	4.6 (1 : 2.5 soil to H_2O)
Organic matter[①] (g/kg)	10.8
Cation-exchange capacity (CEC)[②] (cmol/kg)	9.88
Available P[③] (mg/kg)	1.22
Available K[④] (mg/kg)	49.25
Nonexchangeable K[⑤] (mg/kg)	96.80
Oxalate-extracted Al_2O_3[⑥] (g/kg)	2.91
Oxalate-extracted Fe_2O_3[⑥] (g/kg)	1.43

Notes: ① Wet oxidation method (Lu, 1999). ② $BaCl_2$ method (Hendershot et al., 1993). ③ Bray-I extractant (Bray and Kurtz, 1945). ④ 1 mol/dm³ NH_4OAc extractant (Lu, 1999). ⑤ Difference between 1 mol/dm³ HNO_3 extractable K and 1 mol/dm³ NH_4OAC extractable K (Lu, 1999). ⑥ Extracted by 0.2 mol/dm³ $(NH_4)_2C_2O_4$ (pH 3.0) (Lu, 1999).

Prior to incubation, the soil was air-dried and sieved using 1 mm mesh. The containers used for soil incubations in this study were wax columns, 150 mm in height with a cylindrical cavity having an internal diameter of 50 mm. These were prepared using the method of Khasawneh and Soileau (1969). The materials used to make wax column were the molten mixture of 2 parts of paraffin wax and 1 part of petroleum jelly. The wax column was closed on the bottom with two pieces of filter paper (Whatman 40, Whatman Inc., Clifton, NJ). Soil (360 g) was packed into each column to obtain a final bulk density of 1.22 g/cm^3. To keep a constant bulk density, we filled and packed the soil in 3 cm intervals. More specifically, we calculated the weight of 3 cm depth of the soils based on the cross-sectional area of the wax column and the bulk density. The weighted soil was filled into the wax column and packed carefully to a depth of 3 cm. We brushed the soil surface to maintain a good contact with added soil and repeated the procedure. One filter paper 49 mm in diameter was placed on the soil surface to separate the soil from fertilizer. The wax blocks were next put on fine sand and wetted by capillary rise to a moisture content of 370 g/kg (field capacity, determined experimentally). The top and bottom of the packed wax blocks were covered with parafilm to prevent moisture loss (Fan and MacKenzie, 1993) after abandoning the filter papers covering the bottom and then allowed to equilibrate for 48 h at 25°C before P and K treatments were applied.

Fertilizer treatments were 0.5 g KCl alone or in combination with 0.98 g monocalcium phosphate (MCP). Reagent grade fertilizers were applied uniformly as finely ground materials (< 1mm) by spreading over the whole surface of the filter paper on the soil cylinder. Both K and P were at a rate of 0.26 g per column. This application is equivalent to the addition of 300 kg P_2O_5/ha or 160 kg K_2O/ha in a 5 cm wide band alongwith one side of crop rows spaced 45 cm apart and were intended to simulate conditions similar to those near fertilizer granules or banding. A control treatment without application of P and K was also carried out. Three replicates were used in a completely randomized design. Each treatment had six soil columns. Three were sampled after 7 d and three after 28 d incubation. The wax blocks were covered and then incubated vertically at 25°C. After 7 d and 28 d, the filter paper on the soil cylinder was taken off for K analysis to evaluate the K retained by filter paper. The wax columns were sectioned into 25 slices, 2 mm thick, and then 20 slices, 5 mm thick, using a sharp stainless knife. The wax column was put on a wooden base of a self-designed apparatus (Fig. 1) and pushed by a screw rod with a 2 mm screw pitch. Each 2 mm thick slice was precisely extruded each time by turning the screw rod one circle each time. While being cut, the wax columns were fixed by turning clamps and fixing nut tightly.

Soil samples from each slice were analyzed for moisture content (Lu, 1999) and chemical composition. The chemical analysis involved several steps of procedures similar to Hao et al. (2002). First, 1.37 g of moist soil (equivalent to 1 g oven-dry soil) from each slice was weighed into a 15 mL centrifuge tube, 5 mL deionized water was added and the mixture was shaken for 10 min and centrifuged at 10000 r/min to separate the supernatant. Second, an additional 5 mL deionized water was added to the residual soil samples in the centrifuge tubes and the supernatant was again extracted as described in the first step. The two extracts

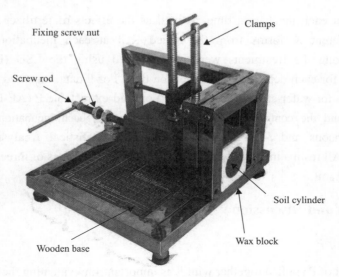

Fig.1 Diagram of sectioning apparatus

obtained from step 1 and step 2 were combined and termed as the water-extractable fraction. Third, the same sample was then successively extracted three times with 5 mL 1.0 mol/dm^3 HCl as described in step 1. The three acid extracts were combined for the acid-extractable fraction. The concentration of K in water extractant was determined using a flame photometer (Model 6410, Shanghai Analytical Instrument Company, China). The concentration of Al and Fe in water and acid extractant was determined using an ICP (Inductively Coupled Plasma) apparatus. Concentration of P in extracts was determined by the ascorbic acid-ammonium molybdate method of Murphy and Riley (1962). The entire filter papers were extracted only with 1.0 mol/dm^3 HCl as described above in step 3 and analyzed for K using the flame photometer. For P, water was used as the extractant to provide information on the mobile phase, while the acid extractant was used to estimate sorbed phase (Akinremi and Cho, 1993; Hao et al., 2002).

To determine exchangeable K and non-exchangeable K, another 1.37 g of moist soil was extracted with 1.0 mol/dm^3 NH$_4$OAc. The exchangeable K was taken as the difference between the K extracted by 1.0 mol/dm^3 NH$_4$OAc and that extracted by water. Then, the residual soil sample was extracted with boiling 1.0 mol/dm^3 HNO$_3$ for 10 minutes. The concentration of K was determined using the flame photometer (Lu, 1999) and used to determine the non-exchangeable K.

The pH measurement of each slice was made using a thin combined glass/calomel electrode (Model 206-C, Shanghai San-Xin Instrument Company, China). Deionized water was added to a 15 mL centrifuge tube containing 1.0 g of moist soil from each slice at a soil /water ratio of 1/2.5. Soil and water were thoroughly mixed with a vortex mixer and the pH was determined after stirring for 10 min (Lu, 1999).

The data for each layer were analyzed as a completely randomized design. One-way ANOVA (Analysis of Variance) was used to evaluate the effects of fertilizer application on the concentrations of K in different forms and pH changes at each distance from the fertilizer application site at each incubation time, as well as the effects of fertilizer treatment on the amounts of different K forms from the added KCl at each incubation time. Multiple comparisons among the treatments were characterized using the LSD (Least Significant Differences) test for each depth at each incubation time. For P data, unpaired t-test was used to analyze difference for water-extractable P (WE-P) or acid-extractable P (AE-P) between KCl + MCP treatment and the control treatment for each depth at each incubation time. ANOVA, multiple comparisons and *t*-test were performed using Statistical Analysis System (SAS Institute, 1996). All results in figures and tables were given as means of three replicates on the basis of oven-dry soil.

2.3.3 Results and Discussion

1. *Movement of P*

The movement of P applied together with K is important for evaluating the effects of P on K movement in soil columns. The amount of P in columns receiving MCP and KCl exceeding the values obtained for unfertilized soil in control was assumed to be from the fertilizer (Benbi and Gilkes, 1987; Fan and MacKenzie, 1993). Total P was calculated as the amount of WE-P plus that of AE-P. The amounts of P in different forms from fertilizer in each column were obtained by summing the quantities of P from MCP extracted from each slice of the column. Total P amount in each slice and the concentration distributions of WE-P and AE-P in soil columns receiving both MCP and KCl are presented in Table 2 and Table 3.

Table 2 Distribution of applied P with distance from fertilizer placement in the soil column treated with monocalcium phosphate and KCl after 7 days

Distance (mm)	WE-P[①] (mg/kg)		AE-P (mg/kg)		Total P[②] (mg)
	KCl + MCP	Control	KCl + MCP	Control	
0-2	601.7(54.8)[③]	0.2(0.0)	5530(510)	2.2(0.1)	29.4(2.7)
2-4	741.8(52.2)	0.2(0.1)	6102(419)	2.0(0.3)	32.8(2.0)
4-6	970.2(69.7)	0.2(0.1)	5203(705)	1.9(0.2)	29.6(3.7)
6-8	902.1(67.3)	0.2(0.0)	3595(420)	2.2(0.3)	21.6(2.3)
8-10	824.9(57.1)	0.2(0.1)	2708(262)	2.1(0.1)	17.0(1.5)
10-12	546.4(71.4)	0.2(0.0)	1990(346)	2.0(0.2)	12.2(2.0)
12-14	390.2(66.6)	0.2(0.0)	1655(46.6)	2.1(0.4)	9.8(0.5)
14-16	291.0(16.5)	0.2(0.1)	1291(170)	2.0(0.3)	7.6(0.9)
16-18	89.4(5.0)	0.2(0.1)	787.2(27.2)	2.1(0.2)	4.2(0.1)
18-20	16.1(3.7)	0.2(0.1)	367.1(55.1)	1.9(0.4)	1.8(0.3)
20-22	0.7(0.2)	0.2(0.0)	73.6(11.6)	2.1(0.2)	0.3(0.1)
22-24	0.2(0.0)	0.2(0.1)	15.0(1.8)	2.2(0.4)	0.1(0.0)
24-26	0.2(0.1)	0.2(0.1)	2.1(0.4)	2.0(0.4)	0.0(0.0)
26-28	0.2(0.0)	0.2(0.0)	2.7(0.5)	2.2(0.1)	0.0(0.0)
28-30	0.2(0.0)	0.2(0.1)	2.2(0.3)	2.1(0.2)	0.0(0.0)
Sum[④]					164.6

Notes: ① WE-P = water-extractable P; AE-P = acid-extractable P. ② Total P refers to the P from MCP moved into each slice of the column treated with KCl and MCP. ③ Values in parentheses are standard deviations (*n*=3). ④ Sum refers to all P from fertilizer moved into soil cylinder.

Table 3 Distribution of applied P with distance from fertilizer placement in the soil column treated with monocalcium phosphate and KCl after 28 days

Distance (mm)	WE-P[①] (mg/kg)		AE-P (mg/kg)		Total P[②] (mg)
	KCl + MCP	Control	KCl + MCP	Control	
0-2	262.5(48.1)[③]	0.2(0.0)	5083 (325)	2.1(0.1)	25.7(1.8)
2-4	213.3(19.0)	0.2(0.0)	5514(387)	2.0(0.5)	27.5(2.0)
4-6	202.0(25.6)	0.2(0.1)	4623(766)	2.1(0.2)	23.2(3.8)
6-8	196.7(15.0)	0.2(0.0)	4539(655)	2.2(0.3)	22.7(3.2)
8-10	199.0(37.2)	0.2(0.0)	3618(685)	2.0(0.1)	18.3(3.5)
10-12	167.6(10.1)	0.2(0.0)	3218(370)	2.1(0.2)	16.2(1.8)
12-14	141.1(3.0)	0.2(0.1)	2654(117)	1.9(0.4)	13.4(0.6)
14-16	130.2(13.4)	0.2(0.1)	2471(355)	1.9(0.3)	12.5(1.8)
16-18	58.1(0.9)	0.2(0.0)	1700(116)	2.0(0.3)	8.4(0.6)
18-20	29.3(5.3)	0.2(0.1)	1113(57.1)	1.9(0.5)	5.5(0.3)
20-22	11.5(2.9)	0.2(0.0)	816.7(58.6)	2.1(0.2)	4.0(0.2)
22-24	4.7(1.3)	0.2(0.1)	496.0(22.8)	2.1(0.4)	2.4(0.1)
24-26	0.9(0.2)	0.2(0.0)	182.7(29.7)	2.0(0.4)	0.9(0.1)
26-28	0.2(0.0)	0.2(0.0)	12.4(1.8)	2.2(0.1)	0.1(0.0)
28-30	0.2(0.0)	0.2(0.0)	2.2(0.1)	2.2(0.3)	0.0(0.0)
Sum[④]					178.4

Notes: ① WE-P = water-extractable P; AE-P = acid-extractable P. ② Total P refers to the P from MCP moved into each slice of the column treated with KCl and MCP. ③ Values in parentheses are standard deviations ($n=3$). ④ Sum refers to all P from fertilizer moved into soil cylinder.

The distance of P movement in the red soil was relatively small. After 7 d and 28 d, added P moved vertically to 24 and 28 mm, respectively. About 60% and 70% of added P were recovered in soil after 7 d and 28 d, respectively. This indicated that added P moved faster within the first week, but then slowed in the subsequent 3 weeks due to its strong affinity for soil particles and reactions with Al and Fe in soil. These measurements of P movement are comparable to those obtained by Fan and MacKenzie (1993) in a similar investigation. They observed that 77%, 81% and 95% of the applied P moved into a Canadian acid soil from triple superphosphate with a movement depth of 20 mm, 20 mm and 25 mm after 5 d, 10 d and 20 d, respectively. The main physical and chemical properties of this Canadian soil included 56.8% clay, pH 5.0 (1 : 2.5 soil/water suspension), 43 g/kg organic matter and 18 kg/ha P (Mehlich No.3 extractant). Soil moisture was adjusted to 250 g/kg before the application of triple superphosphate. In their study, the total P retained in soil was calculated by summing the amount of soil solution P and that of sulphuric acid soluble P. The differences between these studies in the amount and distance of P movement can be attributed to the different characteristics of soils such as P adsorption/fixation capacities, water content, and bulk density. The latter two affect the tortuosity and hence the diffusion.

2. Soil pH changes

Soil pH decreased as a result of KCl application (Fig. 2). However, the reduction was modified by added MCP. Compared with KCl alone, the addition of MCP with KCl increased soil pH at 0-2 mm distance from the fertilizer site by 0.30 and 0.45 units after 7 d and 28 d, respectively. This increase extended to 9 mm and 11 mm at 7 d and 28 d, respectively, which were within the distance of P movement from the fertilizer site. The reason why a pH difference between KCl and KCl + MCP treatments was not observed as deep as P movement was probably due to the buffer capacity of soil.

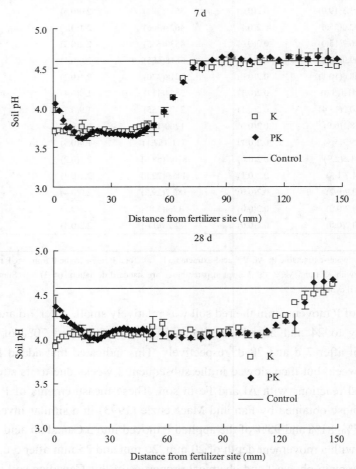

Fig. 2   Changes of pH in soil columns treated with KCl alone (K) or in combination with monocalcium phosphate (PK) at different incubation period

Vertical bars represent standard deviation (n=3)

Although KCl is neutral in water solution, applying KCl will reduce the pH of this red soil through cation exchange. In this red soil, a large portion of the acidic cations on the soil surface could be displaced by K ions resulting in a soil pH decrease (Wang et al., 2003). The significant change in soil pH close to fertilizer site after the addition of MCP with KCl probably was a result of Al-P interactions. Aluminum ions in soil solution could make soil pH decrease due to hydrolysis. As Al ions precipitated with P after addition of MCP, the amount of Al ions involved in hydrolysis was decreased. Therefore, the decrease of soil pH induced by KCl application was buffered to a certain extent. Another possible explanation is exchange of $H_2PO_4^-$ and OH^- on soil surfaces, which would slightly increase soil pH (Zhu, 1982).

3. Movement of K and water-extractable K

The concentrations of water-extractable K (WE-K) contained in soil sections with increasing distance from the fertilizer application site are shown in Fig. 3. The addition of MCP did not change the depths of K movement, which were 60 mm and after 7 d and 110 mm after 28 d of incubation.

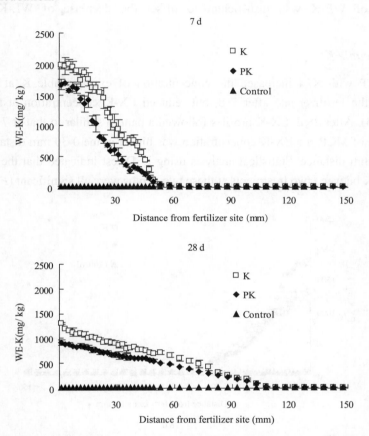

Fig. 3 Distributions of water-extractable K (WE-K) in soil columns treated with KCl alone (K) or in combination with monocalcium phosphate (PK) at different incubation period

Vertical bars represent standard deviation ($n=3$)

The addition of MCP with KCl decreased the concentration of WE-K at 0-60 mm distance from fertilizer site in columns after 7 d and 0-110 mm after 28 d. Statistical analysis using LSD test indicated that the differences in WE-K concentration at the same distance from fertilizer site were all significant ($P < 0.05$) at the same incubation period.

The WE-K concentration reduction in soil columns as affected by the addition of MCP was probably due to the reaction of K and P with Al and Fe. As the dissolution of MCP, H_3PO_4 was formed resulting in a solution pH of about 1.5 near the fertilizer grain (Lindsay and Stephenson, 1959; Sample et al., 1980). Other soil minerals in contact with the H_3PO_4 may be dissolved, increasing the concentration of cations such as Al and Fe near the fertilizer (Lehr et al., 1959; Lindsay and Stephenson, 1959; Low and Black, 1947). A part of K in soil solution would be precipitated with P and Al as the non-crystal analog of taranakite (Lindsay et al.,

1962). Precipitation of P, rather than adsorption processes would dominate the process due to the very high initial P concentration (van Riemsdijk et al., 1984). As a result, the concentration of WE-K could be reduced because of the formation of K-bearing precipitations. It is worth noting that some Ca ions from MCP could displace K ions on the sorption sites simultaneously, which would result in higher WE-K concentration in soil solution. Probably, this increase of WE-K was insufficient to offset the decrease of WE-K induced by precipitations.

4. *Exchangeable K*

Adding MCP with KCl increased the concentration of exchangeable K at the 0-18 mm distance from the fertilizer site after 7 d, but reduced EX-K concentration at the 28-40 mm distance (Fig. 4). After 28 d, EX-K profiles followed a pattern similar to that at 7 d. As affected by the addition of MCP, the EX-K concentration was higher at the 0-30 mm distance, but lower at the 50-105 mm distance. Statistical analysis using LSD test indicated that the differences in exchangeable K between two treatments at these two depths were all significant ($P<0.05$).

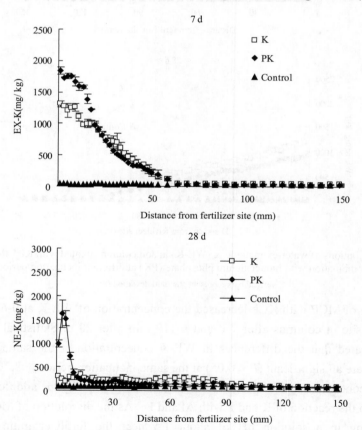

Fig. 4 Distributions of exchangeable K in soil columns treated with KCl alone (K) or in combination with monocalcium phosphate (PK) at different incubation period

Vertical bars represent standard deviation ($n=3$)

A possible explanation for the results is the reaction of P with Al. The application of KCl on the soil surface provided K ion in soil column. The K ion in soil solution displaced Al ion on the exchange sites. In the soil columns treated with KCl and MCP, the higher Al ion concentration in the vicinity of a high P concentration increased the interaction between the two ions. At the same time, this interaction promoted more Al ion to be exchanged by K ion, which led to higher EX-K concentrations. The phosphate ion is capable of neutralizing positive charges of Fe hydroxides in the clay fraction generating electronegative sites (Mekaru and Uehara, 1972), which could also explain why MCP increased K adsorption. Another possible explanation is the competition between Al (Fe) ion and K ion. When KCl alone was added in this acid soil containing kaolinte, soil pH was reduced below 4 close to the fertilizer site (Fig. 3). As a result, kaolinite was decomposed and thus more Al and Fe ions existed in solution (Lindsay, 1979). Those can compete with the K ion on the exchangeable sites. In contrast, when MCP was co-applied with KCl, the P precipitated with Al (Fe) and then less of these ions were left in solutions. In addition, the pH close to the fertilizer site was a bit higher after the addition of MCP than only KCl, which caused less dissolution of kaolinite. As a result, less Al and Fe ions competed with K ion on the sorption sites. Therefore, adding MCP with KCl increased EX-K concentration in soil close to fertilizer site.

5. *Nonexchangeable K*

The distributions of nonexchangeable K (NE-K) in soil column are presented in Fig. 5.

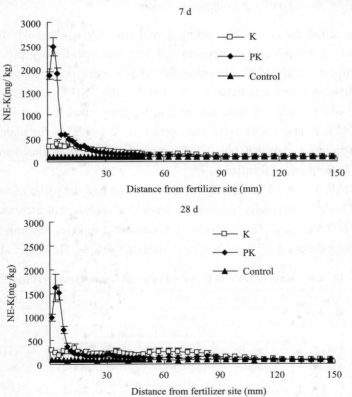

Fig. 5 Distributions of nonexchangeable K in soil columns treated with KCl alone (K) or in combination with monocalcium phosphate (PK) at different incubation period

Vertical bars represent standard deviation ($n=3$)

Since the K-fixation ability of red soil is relatively low (Xie et al., 2000), the concentration of NE-K increased only a little in the zone close to the fertilizer site after the addition of KCl. However, the addition of MCP with KCl greatly increased the NE-K concentration close to fertilizer site after both the 7 d and 28 d incubation. The maximum concentration of NE-K occurred at 2-6 mm from the surface of the soil column.

When MCP was applied with KCl, the much higher concentration of NE-K close to the fertilizer site was believed to be derived from newly formed precipitations containing K. After the reaction between MCP and soil, a certain amount of K would be precipitated with P and Al as the non-crystal analog of taranakite, but not yet converted to the acid-insoluble crystalline form (Hao et al., 2002). Zhou (1995) also found the concentration of 1 mol/dm^3 HNO$_3$ extractable K was greatly enhanced in Oxisol by the application of MCP. Taranakite has been identified in soils as a reaction product of phosphate fertilizer (Lindsay et al., 1962). As the concentration of K ion in the soil-fertilizer reaction zone decreases, taranakite becomes unstable and its solubility increases (Lindsay, 1979). The formation of taranakite in soil can result in the transformation of K from the readily available form to the slowly available form and thus affect the dynamics and bioavailability of K in soils to plants, especially in the immediate vicinity of phosphate fertilizer bands. This transformation may reduce the rate of K supply at the early stage of plant growth.

6. Amounts of K from fertilizer in different forms

Since the extraction method using boiling 1 mol/dm^3 HNO$_3$ obviously involved a rather drastic treatment of the mineral components of soil, and the freshly precipitated reaction products involving P, K, Al and Fe are vulnerable to acid extraction (Sample et al., 1979), the sum of K amounts in soil slices extracted by 1.0 mol/dm^3 NH$_4$OAc and 1 mol/dm^3 HNO$_3$ above that from unfertilized soil were assumed to represent total K from fertilizer applied in this study. The WE-K and EX-K were considered to be available for plant growth. In this experiment, the amounts of WE-K plus EX-K above that from unfertilized soil were assumed to the available K from added fertilizers.

The data in Table 4 showed that nearly all the K moved into the soil column was recovered. The addition of MCP significantly reduced the amount of WE-K, but increased the amount of EX-K or NE-K at 7 d or 28 d. The recovery of K was not complete. This might have been due to the incomplete extraction of K by boiling 1 mol/dm^3 HNO$_3$. However, above 93% of the

Table 4 Mass balance of K moved into soil column from fertilizer[1]

Treatment	K in filter paper	K added	WE-K	EX-K	NE-K	Available K	Total K
			(mg/column)				
			After 7 d incubation				
K	0.53 a[2]	261.75	147.15 a	87.77 b	20.22 b	234.92 a	255.14 a
PK	0.56 a	261.75	99.76 b	103.49 a	47.15 a	203.25 b	250.24 a
			After 28 d incubation				
K	0.54 a	261.75	135.41 a	93.75 b	27.04 b	229.16 a	255.83 a
PK	0.50 a	261.75	101.29 b	109.58 a	35.27 a	210.87 b	246.14 a

Notes: [1] WE-K, water-extractable K; EX-K, exchangeable K; NE-K, nonexchangeable K. [2] Data represent average of three columns. Means in the same column followed by different letters are significantly different at $P < 0.05$ by LSD.

added K was accounted for in all the columns. Addition of MCP with KCl significantly reduced the amount of available K maintaining in the soil column. The result may imply the addition of MCP could reduce the bioavailability of K after the fertilizer application.

7. *Acid-extractable Al and Fe*

The addition of MCP with KCl resulted in significant variations of Al and Fe concentrations extracted by 1 mol/dm^3 HCl. In this experiment, acid-extractable Al (AE-Al) and acid-extractable Fe (AE-Fe) is assumed to represent the fraction of Al and Fe that had been precipitated with P as the non-crystal analog of variscite and strengite, but not yet converted to the acid-insoluble crystalline forms which cannot be extracted by 1 mol/dm^3 HCl. Close to the fertilizer placement site, the concentrations of AE-Al and AE-Fe were higher and steadily decreased with increasing distance from the fertilizer site as affected by the addition of MCP with KCl (Fig. 6). The distribution of AE-Al and AE-Fe in soil columns treated with MCP and KCl followed a pattern similar to that of the acid extractable P in soil column, which indicated that the changes of AE-Al and AE-Fe were the results of added P and confirmed the formation of new non-crystal precipitations that could be extracted by 1 mol/dm^3 HCl.

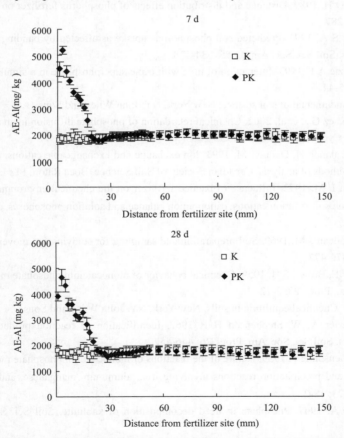

Fig. 6 Distributions of acid-extractable Al (AE-Al) and acid-extractable Fe (AE-Fe) in soil columns treated with KCl alone (K) or in combination with monocalcium phosphate (PK) at different incubation period

Vertical bars represent standard deviation (*n*=3)

References

Akinremi O O, Cho C M. 1993. Phosphorus diffusion retardation in a calcareous system by co-application of potassium chloride. Soil Sci. Soc. Am. J., 57: 845-850

Ayres A S, Hagihara H H. 1953. Effect of the anion on the sorption of potassium by some humic and hydrol humic latosols. Soil Sci., 75: 1-17

Barber S A, Walter J M, Vasey E H. 1963. Mechanisms for the movement of plant nutrients from the soil and fertilizer to the plant root. J. Agric. Food Chem., 11: 204-207

Barber S A. 1985. Potassium availability at the soil-root interface and factors influencing potassium uptake. In: Munson R D. Potassium in agriculture. ASA, CSSA, and SSSA, Madison, WI

Benbi D K, Gilkes R J. 1987. The movement into soil of P from superphosphate grains and its availability to plants. Fert. Res., 12: 21-36

Bouldin D R, Lehr J R, Sample E C. 1960. The effects of associated salts on transformations of monocalcium phosphate monohydrate at the site of application. Soil Sci. Soc. Am. Proc., 24: 464-468

Bray R H, Kurtz L T. 1945. Determination of total, organic and available forms of P in soils. Soil Sci., 59: 39-45

Dennis E J, Ellis R Jr. 1962. Potassium equilibrium and lattice changes in vermiculite. Soil Sci. Soc. Am. Proc., 26: 230-233

Eghball B, Sander D H. 1989. Distance and distribution effects of phosphorus fertilizer on corn. Soil Sci. Soc. Am. J., 53: 282-287

Ernani P R, Barber S A. 1991. Predicted soil phosphorous uptake as affected by banding potassium chloride with phosphorus. Soil Sci. Soc. Am. J., 55: 534-538

Fan M X, MacKenzie A F. 1993. Interaction of urea with triple superphosphate in a simulated fertilizer band. Fert. Res., 36: 35-44

Foth H D. 1984. Fundamental of soil science. New York, NY: John Wiley and Sons

Hao X, Cho C M, Racz G J, et al. 2002. Chemical retardation of phosphate diffusion in an acid soil as affected by liming. Nutr. Cycl. Agroecosys., 64: 213-224

Hendershot W H, Lalande H, Duquett M. 1993. Ion exchange and exchangeable cations. In: Cater M R. Soil sampling and methods of analysis. Canadian Society of Soil Science. Boca Raton, FL: Lewis Publishers

Isensee A R, Walsh L M. 1972. Influence of banded fertilizer on the chemical environment surrounding the band. II. Effect on soil-solution cations, cation-anion balance and solution phosphorus. J. Sci. Food. Agric., 23: 509-516

Khasawneh F E, Soileau J M. 1969. Soil preparation and sampling for studying ion movement. Soil Sci. Soc. Am. Proc., 33: 476-477

Lehr J R, Brown W E, Brown E H. 1959. Chemical behavior of monocalcium phosphate monohydrate in soils. Soil Sci. Soc. Am. Proc., 23: 3-12

Lindsay W L. 1979. Chemical equilibria in soils. New York, NY: John Wiley and Sons

Lindsay W L, Frazier A. W, Stephenson H F. 1962. Identification of reaction products from phosphate fertilizers in soils. Soil Sci. Soc. Am. Proc., 26: 446-452

Lindsay W L, Stephenson H F. 1959. Nature of the reactions of monocalcium phosphate monohydrate in soils: II. Dissolution and precipitation reactions involving iron, aluminum, manganese, and calcium. Soil Sci. Soc. Am. Proc., 23: 18-22

Low P F, Black C A. 1947. Phosphate induced decomposition of kaolinite. Soil Sci. Soc. Am. Proc., 12: 180-184

Lu R K. 1999. Analytical methods for soil and agro-chemistry. Beijing: China Agricultural Science and Technology Publishing House (in Chinese)

Mekaru T, Uehara G. 1972. Anion adsorption in ferruginous tropical soils. Soil Sci. Soc. Am. Proc., 36: 296-300

Murphy J, Riley J P. 1962. A modified single solution method for the determination of phosphorus in nature waters. Anal. Chim. Acta., 27: 31-36

Page A L, Burge W D, Ganje T J, et al. 1967. Potassium and ammonium fixation by vermiculite soils. Soil Sci. Soc. Am. Proc., 31: 337-341

Perkins A R. 1958. Effect of phosphate on the cation-exchange capacity of minerals and soils. Soil Sci. Soc. Am. Proc., 22: 509-511

Rich C I. 1968. Mineralogy of soil potassium. In: Kilmer V J. The role of potassium in agriculture. SSSA, Madison, WI

Ross G J, Cline R A. 1984. Potassium exchange characteristics in relation to mineralogical properties and potassium uptake by grapes of selected soils in Niagara Peninsula of Southern Ontario. Can. J. Soil Sci., 64: 87-98

SAS Institute. 1996. SAS User's guide software release 6.09. Cary, NC

Sample E C, Khasawneh F E, Hashimoto I. 1979. Reactions of ammonium ortho- and polyphosphate fertilizers in soil: III. Effects of associated cations. Soil Sci. Soc. Am. J., 43: 58-65

Sample E C, Soper R J, Racz G J. 1980. Reactions of phosphate fertilizers in soils. In: Khasawneh F E, Sample E C, Kamprath E J. The role of phosphorus in agriculture. ASA, CSSA, and SSSA, Madison, WI

Sharpley A N. 1986. Disposition of fertilizer phosphorus applied to winter wheat. Soil Sci. Am. J., 50: 953-958

Thorup R M, Mehlich A. 1961. Retention of potassium meta- and ortho-phosphates by soils and minerals. Soil Sci., 91: 38-43

van Riemsdijk W H, van der Linden A M A, Boumans L J M. 1984. Phosphate sorption by soils: III. The P diffusion-precipitation model tested for the three acid sandy soils. Soil Sci. Soc. Am. J., 48: 545-548

Wang H Y, Zhou J M, Chen X Q, et al. 2003. Interaction of NPK fertilizers during their transformation in soils: I. Dynamic changes of soil pH. Pedosphere, 13: 257-262

Xie J C, J M Zhou, Härdter R. 2000. Potassium in Chinese Agriculture. Nanjing: Hehai University Press (in Chinese)

Zhou J M. 1995. Kinetics and mechanisms of phosphate-induced potassium release from selected K-bearing minerals and soils. Canada: Saskatchewan, University of Saskatchewan, Ph D Thesis

Zhou J M, Huang P M. 1995. Kinetics of monoammonium phosphate-induced potassium release from selected soils. Can. J Soil Sci., 75: 197-203

Zhu Z X. 1982. Phosphorus in soil. In: Zhu Z X. Soil science. Beijing: Agricultural Publishing House (in Chinese)

2.4 Potash application patterns and soil potash fertility change

2.4.1 Introduction

It's not a new topic to study the application methods or patterns of potash and the change of soil potassium fertility. With the increase of demand for high-quantity and -quality agricultural products and more application of N and P fertilizers, it's still necessary to study how to use the limited potassium fertilizer effectively and how to meet the requirement for the normal growth of plants and maintain or improve soil K fertility rationally.

Many researchers have studied the effect of potash with different dressing methods on crop yield under rotation systems (Ma and Du, 1982; Bao and Xu, 1993; Suyamto, 1993; Janardan et al., 1994; Tao et al., 1994). In order to use potash most efficiently, some people compared the benefit from the intervallic application pattern (applying K only in one cropping season in a rotation system) with that from the successive application pattern (applying K in each of the cropping seasons) (Long et al., 1992; Hu et al., 1993; Chen, 1997; Chen and Zhou, 1999). Hu et al. (1993) found that when K was applied only to wheat, more benefit could be gained. But results of more benefit from splitting application of K to each crop were indicated from Zhejiang Province (The Science and Technology Bureau of Ministry of Agriculture, 1991). Long et al. (1992) also demonstrated that the application pattern of K used in each season is more beneficial. In most of the above studies the total annual amount of potash was not the same under different K application patterns in the experiments. From the five cropping seasons' data, we found that applying the total K in the rice season could be better than applying it in the wheat season or splitting it to each season (Chen and Zhou, 1999). But few of the studies revealed the effect of K fertilization patterns on K-supplying power. Based on the preliminary work (Chen and Zhou, 1999), we continued this long-term field experiments to compare the benefit and potash use efficiency of the different potash application patterns and to demonstrate the soil K-supplying power changes induced by these patterns.

2.4.2 Materials and Methods

1. Field experiment design

A rice-wheat rotation, which is a main cultivation system in Jiangsu Province and some other provinces of China, was conducted on the paddy soils in Siyang County and Liyang City of Jiangsu Province, respectively. In the experiments, rice was planted in the first, third, fifth, and seventh season, and wheat was in the second, fourth, sixth, and eighth season.

The field experiment in each place was designed with four treatments: ① no K was applied (K_0); ② half of the total annual amount of K was equally applied to each crop (K_1); ③ the total annual amount of K was applied only to rice but not to wheat ($K_{2(rice)}$); and ④ only to wheat ($K_{2(wheat)}$). Each treatment was replicated four times in a completely randomized block arrangement. Each plot has an area of 40 m^2. The annual amount of potash was the same, 216 kg K_2O/ha, for all K treatments. The application amount and methods of other fertilizers were the same as farmers used (Chen and Zhou, 1999).

2. Potash use efficiency calculation

The potash use efficiency of the first cropping year was calculated according to the following formula:

$$\text{K use efficiency (\%)} = [(T_1 + T_2) - (C_1 + C_2)] / (C_1 + C_2) \times 100$$

Where, T and C is the K removed with the harvest of crops in K treatments (K_1 or K_2) and CK treatments (K_0); 1 and 2 mean the first and second cropping season, respectively. The K use efficiency of the first 2 years was calculated as follows:

$$\text{K use efficiency (\%)} = [(T_1 + T_2 + T_3 + T_4) - (C_1 + C_2 + C_3 + C_4)] / (C_1 + C_2 + C_3 + C_4) \times 100$$

Where 3 and 4 mean the third and fourth cropping seasons, respectively. The K use efficiency of the first 3 and 4 years is obtained in a similar manner.

3. *Soil sampling and analysis*

Soils of 0-15 cm depth were taken from the field before the trials were conducted and after each cropping season was completed. The soil samples were air-dried and gently ground to pass through a 2 mm sieve in preparation for laboratory analysis. The basic fertility properties of the soil samples were analyzed using the systematic approach (PPIC, 1992) by the Sino-Canada Soil and Plant Test Laboratory in Beijing before the experiments (Table 1). Comparing with the critical values from Table 1, N and K were found to be deficient in both soils and B in the soil of Liyang and Zn in the soil of Siyang were rather deficient.

Table 1 Fertility conditions of the experimented soils[①]

Site	pH	OM (%)	N	P	K	S	B	Ca	Mg	Cu	Fe	Mn	Zn
								(µg/mL)					
Liyang	5.7	0.62	23.0	8.9	43.0	24.1	0.00	1623.2	245.4	4.8	168.0	31.9	2.3
Siyang	8.1	0.32	11.1	25.0	39.1	77.4	0.26	1863.7	151.9	2.1	31.5	9.9	1.6

Note: ① The critical values of N, P, K, S, Ca, Mg, Cu, Fe, Mn, Zn, and B in the systematic approach are 50, 12, 78.2, 12, 400.8, 121.5, 1, 10, 5, 2, and 0.2 µg/mL, respectively (PPIC, 1992).

4. *Soil K fertility analysis*

The exchangeable and nonexchangeable K of the samples were also determined with the conventional chemical analysis methods, i.e., exchangeable K was extracted with 1 mol/dm^3 neutral NH_4OAc for 30 min (Jones, 1973), and nonexchangeable K with 1 mol/dm^3 boiling HNO_3 for 10 min (Pratt, 1965). The potassium was measured using flame spectrophotometry.

2.4.3 Results and Discussion

1. *Effect of potash application patterns on crop yield*

The yields and yield increases of eight cropping seasons for different K treatments in the rice-wheat rotation experiments conducted on the soils in Siyang and Liyang were presented in Tables 2 and Table 3. The data show that the yields of treatment K_2 in each current season, i.e. the yields of crops in the seasons with the total annual amount of potash were always the highest, although with increasing number of cropping seasons, yields for the three K application patterns were not significantly different. The yield increases of crops in the treatments $K_{2(rice)}$ of the four rice seasons or treatments $K_{2(wheat)}$ of wheat seasons were increasing with the increase of cropping years, except for those in the wheat seasons on Liyang soil. These results indicate that the benefit from potash application was inclined to getting larger. It may imply that the potassium fertility of soils with no potash applied was decreasing as cropping seasons increased. During the wheat seasons in Liyang, especially the fourth and sixth cropping seasons, serious drought in the early growing stage of wheat and too much rain and impeded drainage in the latter stage may have induced a great response of wheat yield to potash. The yield increases in all treatments on Liyang soil were always larger than those in corresponding treatments on Siyang soil which might indicate that the soil K-supplying power on Liyang soil was less than that on Siyang soil.

Table 2 Yields and yield increases of eight cropping seasons in the rice-wheat rotation experiment conducted on Aquept in Siyang County

Treatment	First season Yield (10^3 kg/ha)	Increase (%)	Second season Yield (10^3 kg/ha)	Increase (%)	Third season Yield (10^3 kg/ha)	Increase (%)	Fourth season Yield (10^3 kg/ha)	Increase (%)	Fifth season Yield (10^3 kg/ha)	Increase (%)	Sixth season Yield (10^3 kg/ha)	Increase (%)	Seventh season Yield (10^3 kg/ha)	Increase (%)	Eighth season Yield (10^3 kg/ha)	Increase (%)
K_0	8.00 b[1]	-	7.42 c	-	6.60 c	-	4.48 c	-	7.46 b	-	5.94 b	-	8.33 d	-	4.66 b	-
K_1	8.70 a	8.8	8.01 b	7.9	7.46 ab	13.1	5.16 b	15.26	9.06 a	21.37	6.98 a	17.49	9.52 b	14.3	5.54 a	19.0
$K_{2(wheat)}$	8.00 b	-	8.29 a	11.7	7.06 b	7.0	5.34 a	19.39	8.92 a	19.59	7.14 a	20.17	8.90 c	6.8	5.57 a	19.7
$K_{2(rice)}$	8.97 a	12.2	7.85 b	5.7	7.57 a	14.7	5.30 a	18.47	9.08 a	21.83	6.85 a	15.44	10.1 a	21.0	5.53 a	18.8

Note: [1] Column values followed by the same letter are not significantly different at the 0.05 probability level.

Table 3 Yields and yield increases of eight cropping seasons in the rice-wheat rotation experiment conducted on Aqualf in Liyang City

Treatment	First season Yield (10^3 kg/ha)	Increase (%)	Second season Yield (10^3 kg/ha)	Increase (%)	Third season Yield (10^3 kg/ha)	Increase (%)	Fourth season Yield (10^3 kg/ha)	Increase (%)	Fifth season Yield (10^3 kg/ha)	Increase (%)	Sixth season Yield (10^3 kg/ha)	Increase (%)	Seventh season Yield (10^3 kg/ha)	Increase (%)	Eighth season Yield (10^3 kg/ha)	Increase (%)
K_0	7.38 b[1]	-	2.85 c	-	4.73 b	-	1.90 c	-	4.43 b	-	2.98 c	-	7.81 d	-	0.395 c	-
K_1	8.56 a	16.1	3.73 ab	30.7	6.75 a	42.8	3.88 b	104.48	6.58 a	48.43	4.42 a	48.16	9.49 b	21.6	2.04 a	415.8
$K_{2(wheat)}$	7.38 b	-	3.91 a	37.2	6.65 a	40.7	4.06 a	113.74	6.59 a	48.76	4.71 a	58.08	8.55 c	9.5	2.11 a	434.9
$K_{2(rice)}$	8.80 a	19.3	3.56 b	24.7	6.81 a	44.2	3.88 b	104.14	6.67 a	50.58	4.00 b	34.18	9.90 a	26.8	1.36 b	245.3

Note: [1] Column values followed by the same letter are not significantly different at the 0.05 probability level.

Table 4 presents the K use efficiency of the first cropping year, the first 2 years, the first 3 years, and the first 4 years. The data show that the K use efficiency tended to become larger with the increase of the cropping year. Such results are consistent with those of yield increase. In Table 4, the highest use efficiency of potash is observed in $K_{2(rice)}$ treatment and the lowest in the $K_{2(wheat)}$ treatment on both soils. This indicates that applying the same amount of potash only in rice season may provide more benefit than splitting it in both rice and wheat season, and much more than applying it only to wheat.

Table 4 The K use efficiency of different potash application patterns in the initial one, two, three, and four cropping years on Aquept and Aqualf

Soil	Treatment	The first year (%)	The initial 2 years (%)	The initial 3 years (%)	All 4 years (%)
Aquept	K_1	49.3 a	56.5 a	50.4 b	56.3 b
	$K_{2(wheat)}$	39.6 b	44.8 c	46.6 c	54.5 b
	$K_{2(rice)}$	33.9 c	52.3 b	55.9 a	59.2 a
Aqualf	K_1	49.9 b	69.0 b	74.3 b	74.8 b
	$K_{2(wheat)}$	23.7 c	47.2 c	57.1 c	62.0 c
	$K_{2(rice)}$	65.4 a	92.1 a	90.4 a	85.9 a

2. Effect of potash application patterns on soil K fertility

The variation of the contents of exchangeable and nonexchangeable K in the experiments after one to four cropping years are shown in Fig. 1, Fig. 4.

Fig. 1 shows that on Siyang soil, the variation of exchangeable K for the same treatments is not very distinct with the increase of cropping year. Applying potash in the current cropping season could correspondingly increase the exchangeable K, however, there's no significant difference found between the increases caused by the application of whole or half amount of K in the current season. But applying K only in the previous season, the exchangeable K is obviously less than that in the other two K treatments. On the Aqualf (Fig. 3), with an increase of cropping year, the exchangeable K in K_1 and $K_{2(rice)}$ treatments visually decreases, while that in the $K_{2(wheat)}$ treatment varied little. The differences among the three K treatments were distinct. With no K applied, the exchangeable K is significantly decreased in both soils. Such results may imply that K application could increase K-supplying power compared with the K_0 treatment, but the K application rate was not big enough to maintain the K fertility of the soils.

Fig. 2 illustrates that with the increase of cropping year, the nonexchangeable K of Aquepts slowly declined. When the total annual potash was applied only in the rice season, the nonexchangeable K was always relatively lower. Meanwhile, the gap of nonexchangeable K

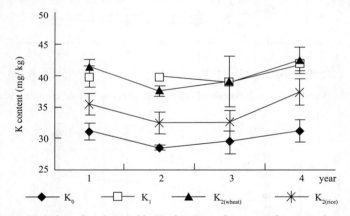

Fig. 1 Variation of exchangeable K of Aquept after one to four cropping years

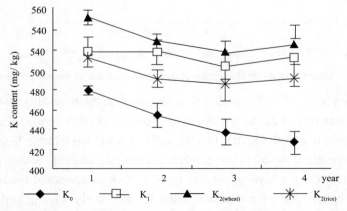

Fig. 2 Variation of nonexchangeable K of Aquept after one to four cropping years

between soils with or without K applied was widened, mainly caused by the relatively greater decrease of nonexchangeable K of soils in CK treatment. The nonexchangeable K values of soils in all treatments were lower than the original content of soil K indicating that some of the nonexchangeable K were released to maintain the exchangeable K during the growing seasons. The higher nonexchangeable K in the Siyang soil means a higher buffer capacity in K supply. For the Liyang soil, the nonexchangeable K of soils with K_1 and $K_{2(wheat)}$ treatments was slightly higher than the original soil, while in $K_{2(rice)}$ treatment, the nonexchangeable K was always lower than that in the other two K treatments indicating that potash application patterns influenced the K-supplying power. The nonexchangeable K in K_0 treatment was significantly lower than the original soil, demonstrating K exhaustion without K application in the rotation.

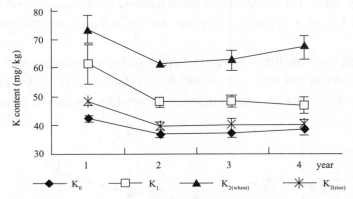

Fig. 3 Variation of exchangeable K of Aqualf after one to four cropping years

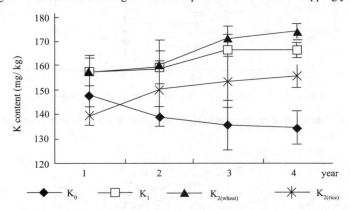

Fig. 4 Variation of nonexchangeable K of Aqualf after one to four cropping years

Based on these results of exchangeable K, it appears that the K fertility level of the Liyang Aqualf is higher than that of the Siyang Aquepts. However, a more serious K deficiency and a higher crop response to potash indicate that the soil in Liyang has a lower K-supplying power. When the exchangeable K in soil is low enough, the nonexchangeable K may play the leading role in K supplying. On the Siyang soil, the growth of crops promoted the release of soil K mainly supplied by the part of nonexchangeable K. But on the Liyang soil, the content of nonexchangeable K was too low to supply sufficient K to meet the requirement of crops for normal growth and the K needed had to be supplied, in part, by exchangeable K of soils and the potash applied. In the K$_{2(rice)}$ treatment, both the exchangeable and nonexchangeable K were lower than those in the other two K treatments on Siyang and Liyang soils. This is mainly due to the higher K use efficiency of rice. In rice seasons, more potassium can be removed by the crop than by the harvest of the wheat seasons. Such results are not contrary to those of crop yields.

References

Bao S D, Xu G H. 1993. A study of contemporary and residual effects of potassium applied to a wheat-rice rotation. J. Nanjing Agric. Univ., 16 (4): 43-48

Chen F. 1997. The effect of long-term K application on the characteristics of fixation and release of soil K. Wuhan, China: Huazhong Agricultural University, PhD thesis

Chen X Q, Zhou J M. 1999. Effect of potash application patterns on crop yields under different cultivation systems. Pedosphere, 9 (3), 219-226

Hu S N, He C F, Portch S. 1993. Most efficient potassium use in a wheat-rice rotation in Sichuan Province, China. In: Barrow, N J. Plant Nutrition from Genetic Engineering to Field Practice. Sydney: Kluwer Academic Publishers: 633-636

Janardan S, Sharma H L, Singh C M, et al. 1994. Direct, residual and cumulative effects of potassium fertilization in rice (*Oryza sativa*)-wheat (*Triticum aestivum*) cropping system. Indian J. Agron., 39, (3): 345-355

Jones J B. 1973. Soil testing in the United States. Commun. Soil Sci. Plant Anal., 4: 307

Long C F, Chen F, Chen X C, et al. 1992. The studies on potash allocation under different crop systems. In: The Science and Technology Bureau of Ministry of Agriculture, Institute of Soil and Fertilizer, Chinese Academy of Agricultural Sciences, Potash and Phosphate Institute of Canada. Studies on Soil Potassium and Potash. Beijing: Chinese Agricultural Science and Technology Press: 65-69

Ma M T, Du C L. 1982. The role of potassium fertilizer in boosting rice yield in red earth areas of central China. J. Soil Sci., (1): 5-11

PPIC (Potash and Phosphate Institute of Canada). 1992. The Systematic Approach for Soil Fertility. Beijing: Chinese Agricultural Science and Technology Press: 1-117

Pratt P F. 1965. Potassium. In: Black C A, et al. Methods of Soil Analysis, Part 2. Agronomy, vol. 9. Madison: American Society of Agronomy: 1023-1031

Suyamto S. 1993. Direct and residual effects of potassium fertilizer on rice-maize cropping rotation on Vertisol. Indones. J. Crop Sci., 8 (2): 29-38

Tao Q X, Liu G R, Li Z Z, et al. 1994. Study on the K fertilizer distribution for rape-rice-rice crop system in poor K paddy soil. Acta Agriculture Jiangxi, 6: 53-60 (Supplement)

The Science and Technology Bureau of Ministry of Agriculture. 1991. Potassium in Agriculture in South China. Beijing: Agriculture Press: 57-106

2.5 Identification of reaction products of phosphate fertilizers with soil using chemical and FTIR-PAS methods

2.5.1 Introduction

In order to improve the fertility of soil, phosphate fertilizers, including manures and sewage sludge, are extensively used in agronomic practice (Ajiboye et al., 2004; Hunger et al., 2004).

However, fertilizer P application is inefficient, as only a small fraction of the P supplied is normally taken up by plants, with the remainders going to the buildup of soil reserves. Initially, soil can retain most of the excess P through such processes as immobilization, adsorption and precipitation, but long-term over-application of fertilizer P on adsorption-saturated soils can lead to increased transfer of P from soil reserves to ground and/or surface water by soil erosion and/or leaching (Smith et al., 2001; McDowell and Sharpley, 2004). Such losses are of greatly environmental concern as they can cause eutrophication of rivers and lakes (Sharpley et al., 2003; Toor et al., 2005a). One strategy to mitigate the problem is to reduce fertilizer P application and reactivate the retained P in soils as a new source for plant uptake, which should be based on the understanding of P forms in soils, manures and sludge.

Chemical fractionation had been widely used to identify P forms in soils (Chang and Jackson, 1957; Hedley et al., 1982; Gu and Jiang, 1990). However, chemical fractionation is based on the assumption that chemical extractant selectively dissolve discrete groups of P compounds, and such operationally defined soil P fractions are subject to broad interpretations (Sui et al., 1999). In addition, it is usually laborious and time-consuming. Therefore, many newly developed analytical techniques are attempted to identify P forms in soils, manures and sludge, including X-ray diffraction (XRD) (Lindsay et al., 1962; Huang and Shenker, 2004; Shand et al., 2005), nuclear magnetic resonance (NMR) (Hunger et al., 2004; Turner, 2004; Turner and Leytem, 2004; Shand et al., 2005), X-ray absorption near-edge structure (XANES) spectroscopy (Sato et al., 2005; Toor et al., 2005b) and enzyme hydrolysis (He and Honeycutt, 2001; Turner et al., 2002).

XRD is a good tool for identification of crystalline phosphates, however, due to the low content and amorphism of phosphate compounds present in soil, its application is greatly restricted. NMR is an alternative tool for the identification of P forms in soils and manures, including solution-state ^{31}P NMR and solid-state ^{31}P NMR. Using solution-state ^{31}P NMR, Toor et al. (2003) found that unreactive P in leachate from grassland soils was mainly comprised of monoester and diester forms of organic P. However, in this method, sample preparation uses a chemical extractant, typically NaOH-EDTA, that dissolves P in minerals, thus only allowing for quantification of the total concentration of inorganic P present (Toor et al., 2005b) and does not provide direct information about the mineral phase of P (e.g., Al-P, Ca-P, Fe-P). In addition, hydrolysis of some labile organic P species into orthophosphate in the alkaline extract may lead to the underestimation of organic P pools in the manure (Turner, 2004; Turner and Leytem, 2004). Finally, the recovery of polyvalent cations in the extract used for NMR analysis does not provide a direct evidence of their association with organic or inorganic P pools (Ajiboye et al., 2007). Solid-state ^{31}P NMR can help identifying inorganic P forms in soils and manures without extraction and dissolution of P minerals. For example, McDowell et al. (2002) used solid-state ^{31}P NMR to identify monetite ($CaHPO_4$), DCPD ($CaHPO_4 \cdot 2H_2O$) and wavellite in soil. But, in soil, inorganic P usually associated with paramagnetic cations, such as Fe, Al and Mn, which limit the applicability of this method (Hunger et al., 2004; Turner and Leytem, 2004). XANES is frequently used to identify P forms in soils (Beauchemin et al., 2003) and manures (TOOR et al., 2005b) in recent years. Beauchemin et al. (2003) reported that calcium phosphates in the form of hydroxyapatite and octacalcium phosphate were present in all soils, ranging from 11% to 59% and 24% to 53% of total P, respectively. In another study, Toor et al. (2005b) found that all broiler litters were rich in dicalcium phosphate (65%-76%), followed by aqueous phosphate (13%-18%) and phytic acid (7%-20%), but no hydroxylapatite was observed. However, XANES data analysis base on fitting techniques, which is inherently restricted by: ① the data quality (Beauchemin et al., 2003); and ② how well the chosen set of standards actually represents real species in the samples of unknown composition (Beauchemin et al., 2002). Furthermore, it is not easy to apply for XANES beam time due to the lack of facilities worldwide. Enzyme hydrolysis is another method to identify organic P species in soils and manures, but the results vary greatly from different researchers due to the lack of universal experimental conditions (Luo et al., 2009).

Infrared spectroscopy is also introduced to help identifying P forms in soils and manures. For example, Shand et al. (2005) used infrared spectra (KBr disks) to help identifying brushite and struvite in freeze-dried sheep feces. Due to its applicability for highly absorbing solid samples without any special pre-treatment, Fourier transform infrared photoacoustic spectroscopy (FTIR-PAS) has been successfully used to analyze soil properties (Du et al., 2008a, 2008b, 2009). With respect to soil analysis, FTIR-PAS is advantageous compared to transmittance measurements that require time-consuming preparation of KBr disks, or the attenuated total reflection (ATR) configuration that requires a saturated soil paste and suffers from interferences associated with the presence of water (Linker et al., 2004). To our knowledge, FTIR-PAS has not been used to identify P forms in soil.

The object of the present work was to identify P forms in soil-fertilizer reaction zone using chemical fractionation and FTIR-PAS after application of several widely used phosphate fertilizers to a Mollisol.

2.5.2 Materials and Methods

1. *Materials*

The selected soil is a fine clay loam, classified as a Mollisol according to Soil Taxonomy (Soil Survey Staff, 1999), obtained from plowed layers of farmland (0-20 cm) in Gongzhuling, Jilin province, China (43°39′ N, 124°59′ E). After air-dried at room temperature, the soil was grinded to pass a 20 mesh screen. Some soil properties, shown in Table 1, were measured using the methods recommended by Lu (2000). In order to eliminate potential interferences from impurities, the fertilizers used, including monocalcium phosphate monohydrate (MCP), monoammonium (MAP) and diammonium phosphate (DAP) were of analytical reagent, and properties of their saturated solutions were listed in Table 2.

Table 1   Properties of the studied soil

pH (H_2O)	O.M. (%)	Total P (%)	Total free iron oxide (%)	Total free alumina (%)	Amorphous iron oxide (%)	Amorphous alumina (%)
5.78	2.68	0.024	0.82	0.16	0.21	0.19

Table 2 Properties of the saturated solutions of the selected phosphate fertilizers (Lindsay et al., 1962)

Fertilizer	Composition of saturated solution at 25°C			
	Solution symbol	pH	P (mol/L)	Accompanying cation (mol/L)
$Ca(H_2PO_4)_2 \cdot H_2O$	MCP	1.48 (MTPS)	3.98	Ca 1.44
$NH_4H_2PO_4$	MAP	3.47	2.87	NH4 2.87
$(NH_4)_2HPO_4$	DAP	7.98	3.82	NH4 7.64

2. Experimental design

The movement of fertilizer P in soil is very limited. In order to simulate the high concentration of fertilizer P in soil-fertilizer reaction zone, ratio of soil to P was set at 100∶1 following a preliminary experiment. In each cylinder (about 6.5 cm in diameter and 9.5 cm in depth), 150 g air-dried soil and proportional fertilizer were mixed homogeneously, and then watered to 60% of the maximum field water holding capacity by spraying. The cylinders were sealed with parafilm, in which a few punctures were made to inhibit water loss, but allow the air to come through to keep the system aerobic, and then incubated at 25℃. Water were added regularly according to the weights to maintain the water content constant. Each treatment was repeated in triplicate and blank test (without fertilizer added) was carried out at the same time.

3. Chemical fractionation

Fresh soil samples were taken out from the cylinders after incubation for 1, 10, 50 and 150 days, respectively. In order to avoid any potential interaction during the air drying process, part of the fresh samples were used directly to extract water soluble P (We-P) and then to extract dicalcium phosphate (Ca_2-P), octacalcium phosphate (Ca_8-P), aluminum phosphate (Al-P), iron phosphate (Fe-P), occluded phosphate (O-P) and apatite (Ca_{10}-P) according to Gu and Jiang's (1990) method. Concentration of P in each extract was measured by inductively coupled plasma-atomic emission spectroscopy (ICP-AES). After fractionation, the residues were all dried and contents of various P fractions were calculated. The left soil samples were air-dried and grinded for FTIR-PAS determination (referred to as unknown soil samples).

4. FTIR-PAS determination

The calibration soil samples were prepared by adding increased variscite ($AlPO_4 \cdot 2H_2O$, purchased from Sinopharm Chemical Reagent Co. Ltd.) to air-dried soil, making the contents ranged from 0.05% to 2% stepped by 0.05%. FTIR-PAS spectra of variscite, the calibration and unknown soil samples were obtained using a Nicolet 380 FT-IR spectrophotometer (Thermo electron corporation, USA) equipped with a photoacoustic cell (Model 300, MTEC, USA). After filling the stainless steel sample holder (5 mm in diameter, 3 mm in height) with soil, purging the cell with dry helium (10 mL/min) for 30 s, the scans were conducted in the wave-number ranged from 4000 cm^{-1} to 400 cm^{-1} with a resolution of 8 cm^{-1} and a mirror velocity of 0.3165 cm/s. The spectra were the average of 32 successive scans.

5. Statistical analysis

The raw spectra data were smoothed in MATLAB R2007b with Savitzky-Golay filter (first-order and 25-point window). The calibration soil samples were split randomly into two parts, and 75% were used as calibration set to establish partial least square (PLS) regression model, while the left 25% were used as validation set to test the solidity of the model. The correlation between the contents of variscite in the unknown soil samples predicted by PLS model and Al-P measured by chemical fractionation were analyzed statistically in origin 8.

2.5.3 Results and Discussion

1. Chemical fractionation

The distribution of various P fractions in soil-fertilizer reaction zone was shown in Table 3. In the work, we also determined the difference (referred to as total organic P) between concentrations of P in each extract measured with colorimetric method (referred to as total inorganic P) and ICP-AES (referred to as total P). The results were very low and sometimes even minus. That is to say the fertilizer P applied was mainly transformed into inorganic compounds, or kept their original form in soil-fertilizer reaction zone. After incubation for 1 day, more We-P can be extracted from MAP and DAP treatments than that from MCP treatment. With the increased incubation time, We-P decreased, and after 150 days, We-P in various treatments followed the order: DAP > MAP > MCP. Contents of Ca_2-P and Ca_8-P produced nearly unaffected by the incubation time. Al-P has gradually produced and increased with the increase of incubation time and after 150 days, following the order: MCP > MAP > DAP, just in reverse to that of We-P, O-P and Ca_{10}-P produced were undetectable or very low. Apparently, Al-P was the dominant form in MAP and MCP treatments, but in DAP treatment, it was We-P that took the dominant position. This may be due to the difference of fertilizer properties. The saturated solutions of MAP and MCP are acid, while it is alkaline of DAP (Table 2). After application to the soil, MAP and MCP can dissolve more soil minerals than DAP does, so that more Al would be released from MAP and MCP treatments, which could

Table 3 Contents of various P fractions in the selected soil measured by chemical fractionation (mean ± STD)

Fertilizer	Incubation time (d)	Contents of various P forms (%)						
		We-P	Ca_2-P	Ca_8-P	Al-P	Fe-P	O-P	Ca_{10}-P
MAP	1	0.76±0.05	0.11±0.01	0.07±0.004	0.18±0.02	0.04±0.006	-	-
	10	0.62±0.10	0.14±0.03	0.06±0.01	0.33±0.07	0.07±0.02	-	-
	50	0.38±0.09	0.11±0.01	0.10±0.02	0.43±0.01	0.13±0.02	-	-
	150	0.27±0.03	0.10±0.006	0.07±0.006	0.37±0.02	0.12±0.008	-	-
MCP	1	0.24±0.07	0.16±0.08	0.11±0.05	0.12±0.03	0.02±0.002	-	-
	10	0.27±0.03	0.13±0.007	0.11±0.004	0.36±0.03	0.06±0.01	-	-
	50	0.20±0.03	0.14±0.02	0.11±0.003	0.41±0.01	0.09±0.01	-	-
	150	0.18±0.02	0.13±0.007	0.15±0.04	0.50±0.009	0.13±0.002		-
DAP	1	0.72±0.09	0.11±0.01	0.08±0.007	0.063±0.007	0.05±0.02	-	-
	10	0.73±0.09	0.11±0.01	0.09±0.005	0.11±0.02	0.04±0.008	-	-
	50	0.51±0.04	0.11±0.09	0.08±0.008	0.14±0.01	0.08±0.02	-	-
	150	0.62±0.02	0.12±0.03	0.13±0.03	0.25±0.07	0.15±0.06	-	-

Note: "-" represents undetectable or below 0.02%.

precipitate more fertilizer P as Al-P. Also interesting is that though the content of iron oxide in the studied soil is higher than that of alumina, Fe-P produced is lower than Al-P. Huffman and Taylor (1963) suggested that in the presence of phosphate and at the same pH within a given soil, Al was more soluble than Fe, Al-P apparently had much less tendency to form protective coatings on the surface of the hydroxide than Fe-P, so that Al-P was easier produced than Fe-P.

2. *FTIR-PAS spectra of the samples*

FTIR-PAS spectra of variscite, the calibration and unknown soil samples were shown in Fig. 1. The characteristic FTIR-PAS spectra of variscite were found in several spectral regions, including 1120-940 cm^{-1}, 1670-1600 cm^{-1} and 3600-3000 cm^{-1}. The absorbance of inorganic compounds in mid-infrared are mainly caused by the vibration of anion crystal lattice, so that the absorption peak is nearly unaffected by cations (Chen, 1993). On the FTIR-PAS spectra band of variscite, the absorbance at 1120-940 cm^{-1} was mainly caused by PO_4^{3-}, while at 1670-1600 cm^{-1} and 3600-3000 cm^{-1}, it was the absorbance of crystal water in variscite (Chen, 1993), and the band of 800-500 cm^{-1} was "fingerprint region". In the calibration and unknown soil samples, the characteristic FTIR-PAS bands were also found in the same regions

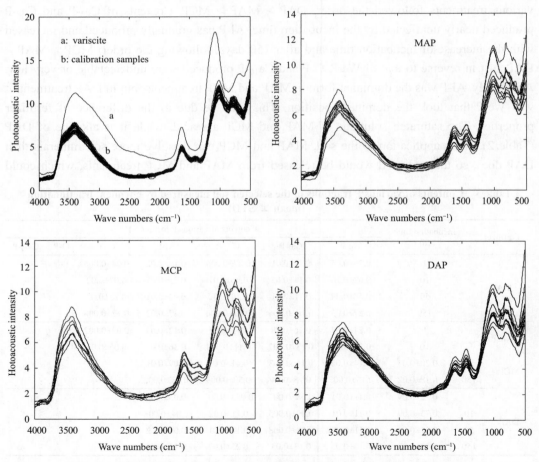

Fig. 1　FTIR-PAS spectra of variscite, the calibration and unknown soil samples

as mentioned above. However, these absorbance bands were attributed not only by phosphates, but also by the soil components. For example, soil water indicates a strong absorption in the region of 1600-1700 cm^{-1} and 2900-3600 cm^{-1} (Du et al., 2008b), which overlap the absorbance bands of crystal water in variscite. Therefore, multivariate calibration methods, such as PLS (Du et al., 2009), principal components analysis (PCA) and artificial neural networks (ANN) (Du et al., 2008a), are applied to extract the related information between soil properties and FTIR-PAS spectra. According to Du et al. (2009), the detected signal in photoacoustic spectroscopy is proportional to the sample concentration. Thus, PLS regression model, which assumes a linear relationship between predictor variables (e.g., spectral data) and the dependent variable of interest, is preferable for exploring the correlation between FTIR-PAS spectra data and soil properties.

3. *Establishment and optimization of PLS model*

PLS analysis can reduce the number of independent variables as spectra into a limited number of predictor variables as eigenspectra (PLS vectors), which are termed as PLS loadings or factors (Forouzangohar et al., 2008). In order to retain the ultra most spectra information, PLS model was established according to the raw spectra data ranging from 500 cm^{-1} to 3750 cm^{-1}. It was reported that the solidity of PLS model was heavily decided by the number of factors (Du et al., 2009). To evaluate the solidity of PLS model, statistics calculated included the correlation coefficient in calibration (R_{cal}) and validation (R_{val}), root mean square of standard error in cross-validation (RMSECV) and prediction (RMSEP). The desired PLS factor number is to produce lower RMSECV and RMSEP, but higher R_{cal} and R_{val}. The residual predictive deviation (RPD) was also calculated to confirm the predictive ability of the model. RPD is defined as the ratio of the standard deviation (SD) of the reference data to RMSECV (Fearn, 2002). If the standard error in cross-validation as measured by RMSECV is large compared with the spread in composition of that sample in the reference as measured by SD, and therefore produce a relatively small RPD, the calibration model is not considered to be robust. A RPD greater than three is considered to be appropriate for prediction purpose (Fearn, 2002). Table 4 reported the statistics calculated according to the number of PLS factors from 3 to 9. It is evident that 7 factors in the PLS model produce lowest RMSEP, and simultaneously, the correlation coefficients reach 1.000. The RPD value at 7 factors also indicates that the model has good ability of prediction. Therefore, 7 factors were chosen in the model for predicting the contents of variscite in the unknown soil samples.

Table 4 Calibration and validation statistics basing on the number of PLS factors

Factors	3	4	5	6	7	8	9
RMSECV (%)	0.016	0.007	0.004	0.003	0.001	0.000	0.000
RMSEP (%)	0.028	0.011	0.011	0.006	0.004	0.004	0.004
RPD	39.35	86.125	145.2	237.6	589.3	1619	3519
R_{cal}	0.990	0.998	0.999	1.000	1.000	1.000	1.000
R_{val}	0.987	0.998	0.997	0.999	1.000	1.000	1.000

4. Prediction of variscite in the unknown soil samples

In fact, the so called Al-P in chemical fractionation is just a broad interpretation, consisting of a group compounds that can be dissolved by special extractant, not only variscite. In order to testify the presence of variscite in Al-P fraction in the unknown soil samples, the correlations between the contents of variscite predicted by the PLS model established earlier and that of Al-P measured in chemical fractionation were analyzed. It was found that the contents of variscite predicted by the model in MCP and DAP treatments were significantly correlated to that of Al-P measured in chemical fractionation ($P<0.01$), but the correlation in MAP treatment was not very good ($P>0.05$) (Fig. 2). The results clearly demonstrate that in MCP and DAP treatments, variscite is present in Al-P fraction, but in MAP treatment, Al-P may consist some other P minerals. Due to the complexity of soil, it was not an easy work to distinguish the characteristic spectra of variscite from that of the unknown soil samples. Therefore, the predicted contents of variscite by the PLS model were higher than the measured Al-P in chemical fractionation. Nevertheless, FTIR-PAS still provides a satisfactory method to help explore the deep sight about the broad fractions of P interpreted in chemical fractionation.

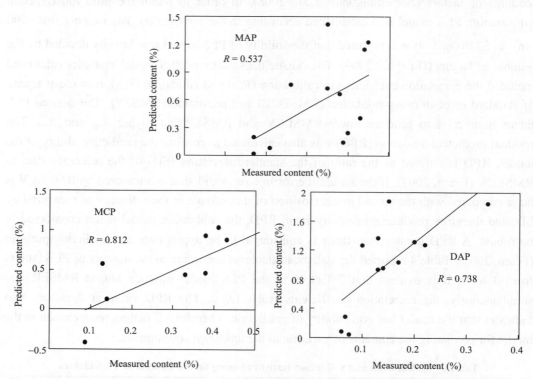

Fig. 2 Correlations between the contents of variscite predicted by PLS model and that of Al-P measured in chemical fractionation

References

Ajiboye B, Akinremi O O, Hu Y F, et al. 2007. Phosphorus speciation of sequential extracts of organic amendments using nuclear magnetic resonance and X-ray absorption near-edge structure spectroscopies. J. Environ. Qual., 36: 1563-1576

Ajiboye B, Akinremi O O, Racz G J. 2004. Laboratory characterization of phosphorus in fresh and oven-dried organic amendments. J. Environ. Qual., 33: 1062-1069

Beauchemin S, Hesterberg D, Chou J, et al. 2003. Speciation of phosphorus in phosphorus-enriched agriculture soils using X-ray absorption near-edge structure spectroscopy and chemical fractionation. J. Environ. Qual., 32: 1809-1819

Beauchemin S, Hesterberg D, Beauchemin M. 2002. Principal component analysis approach for modeling sulfur K-XANES spectra of humic acids. Soil Sci. Soc. Am. J., 66: 83-91

Chang S C, Jackson M L. 1957. Fractionation of soft phosphorus. Soil Sci., 84: 133-144

Chen Y K. 1993. Infrared Absorbance Spectroscopy and Its Applications. Shanghai: Shanghai Jiao Tong University Press

Du C, Linker R, Shaviv A. 2008a. Identification of agricultural Mediterranean soils using mid-infrared photoacoustic spectroscopy. Geoderma, 143: 85-90

Du C, Zhou J M, Wang H Y, et al. 2009. Determination of soil properties using Fourier transform mid-infrared photoacoustic spectroscopy. Vib. Spectrosc., 49: 32-37

Du C W, Zhou J M, Wang H Y, et al. 2008b. Study on the soil mid-infrared photoacoustic spectroscopy (in Chinese). Spectroscopy and Spectral Analysis, 28: 1246-1250

Fearn T. 2002. Assessing calibrations: SEP, RPD, RER and R^2. NIR News, 13: 12-14

Forouzangohar M, Kookana R S, Smernic R J, et al. 2008. Midinfrared spectroscopy and chemometrics to predict diuron sorption coefficients in soils. Environ. Sci. Technol., 42: 3283-3288

Gu Y C, Jiang B F. 1990. Methods of determination of inorganic phosphorus fractionation in calcareous soil (in Chinese). Soils, 22: 101-102, 110

He Z Q, Honeycutt C W. 2001. Enzymatic characterization of organic phosphorus in animal manure. J. Environ. Qual., 30: 1685-1692

Hedley M J, Stewart W B, Chauhan B S. 1982. Changes in inorganic and organic soil phosphorus fractions induced by cultivation practices and by laboratory incubations. Soil Sci. Soc. Am. J., 46: 970-976

Huang X L, Shenker M. 2004. Water-soluble and solid-state speciation of phosphorus in stabilized sewage sludge. J. Environ. Qual., 33: 1895-1903

Huffman E O, Taylor A W. 1963. The behavior of water-soluble phosphate in soils. Agriculture and Food Chemistry, 11: 182-187

Hunger S, Cho H, Sims J T, et al. 2004. Direct speciation of phosphorus in alum-amended poultry litter: solid-state ^{31}P NMR investigation. Environ. Sci. Technol., 38: 674-681

Lindsay W L, Fraziep A W, Stephenson H F. 1962. Identification of reaction products from phosphate fertilizers in soils. Soil Sci. Am. Proc., 26: 446-452

Linker R, Kenny A, Shaviv A, et al. 2004. Fourier transform infrared-attenuated total reflection nitrate determination of soil pastes using principal component regression, partial least squares, and cross-correlation. Appl. Spectrosc., 58: 516-520

Lu R K. 2000. Analytical Methods of Soil and the Agriculture Chemistry. Beijing: China Agricultural Science and Technology Press (in Chinese)

Luo C Y, Ji H J, Zhang R L, et al. 2009. Advance in the characterization of phosphorus in organic wastes. Chinese Journal of Soil Science, 40: 709-715 (in Chinese)

McDowell R W, Sharpley A N. 2004. Variation of phosphorus leached from Pennsylvanian soils amended with manures, composts or inorganic fertilizer. Agric. Ecosyst. Environ., 102: 17-27

McDowell R W, Condron L M, Mahieu N, et al. 2002. Analysis of potentially mobile phosphorus in arable soils using solid state nuclear magnetic resonance. J. Environ. Qual., 31: 450-456

Sato S, Solomon D, Hyland C, et al. 2005. Phosphorus speciation in manure and manure-amended soils using XANES spectroscopy. Environ. Sci. Technol., 39: 7485-7491

Shand C A, Coutts G, Hillier S, et al. 2005. Phosphorus composition of sheep feces and changes in the field determined by ^{31}P NMR spectroscopy and XRPD. Environ. Sci. Technol., 39: 9205-9210

Sharpley A N, Weld J L, Beegle D B, et al. 2003. Development of phosphorus indices for nutrient management planning strategies in the United States. J. Soil Water Conserv., 58: 137-152

Smith K A, Jackson D R, Withers P J A. 2001. Nutrient losses by surface runoff following the application of organic manures to arable land. 2. Phosphorus. Environ. Pollu., 112: 53-60

Soil Survey Staff. 1999. Soil taxonomy, a basic system of soil classification for making and interpreting soil surveys. USDA Agricultural Handbook, 2nd edn., Natural Resources Conservation Service, Vol. 436 Washington, D C: USDA

Sui Y, Thompson M L, Shang C. 1999. Fractionation of phosphorus in a Mollisol with biosolids. Soil Sci. Soc. Am. J., 63: 1174-1180

Toor G S, Condron L M, Cade-Menun B J, et al. 2005a. Preferential phosphorus leaching from an irrigated grassland soil. Eur. J. Soil Sci., 56: 155-167

Toor G S, Condron L M, Di H J, et al. 2003. Characterization of organic phosphorus in leachate from a grassland soil. Soil Biol. Biochem., 35: 1317-1323

Toor G S, Peak J D, Sims J T. 2005b. Phosphorus speciation in broiler litter and turkey manure produced from modified diets. J. Environ. Qual., 34: 687-697

Turner B L. 2004. Optimizing phosphorus characterization in animal manures by solution phosphorus-31 nuclear magnetic resonance spectroscopy. J. Environ. Qual., 33: 757-766

Turner B L, Leytem A B. 2004. Phosphorus compounds in sequential extracts of animal manures: chemical speciation and a novel fractionation procedure. Environ. Sci. Technol., 38: 6101-6108

Turner B L, McKelvie I D, Haygarth P M. 2002. Characterisation of water-extractable soil organic phosphorus by phosphatase hydrolysis. Soil Biol. Biochem., 34: 27-35

2.6 Influence of humic acid on interaction of ammonium and potassium ions on clay minerals

2.6.1 Introduction

Clay minerals and organic matter are important geochemical components that constitute the main body of soils (Huang et al., 1995a, 1995b). Clay minerals adsorb and fix nutrient ions from soil solutions and release them back into the soil solution for plant uptake. Some 2∶1 clay minerals such as illite and montmorillonite typically adsorb and fix potassium (K^+) and ammonium (NH_4^+), which has an ionic radius similar to K^+. Soil organic matter acts in a manner similar to clay minerals and plays a significant role in controlling the availability of nutrient ions in soils (Clapp et al., 2005). In field soils, organic matter often combines with clay minerals to form organo-mineral complexes. These mineral-bond organic substances may modify the inorganic surface properties, changing the nature and number of complexation sites for cations in the soil solution (Dumat et al., 2000; Arias et al., 2002).

Some researchers have investigated the influence of organic matter on the behavior of cations in soils by addition or removal of organic matter (Poonia and Niederbudde, 1990; Chung and Zasoski, 1994; Dumat et al., 1997; Wang and Huang, 2001; Fernando et al., 2005). By using hydrogen peroxide (H_2O_2) to remove soil organic matter, Poonia and Niederbudde (1990) found a dual role of natural organic matter in the exchange equilibria of potassium versus calcium, viz., increasing surface charge density (causing a decrease in K preference), and increasing the ratio of internal to external surface due to organo-mineral colloids (causing an increase in K preference). Chung and Zasoski (1994) used addition of organic matter to investigate the influence of organic matter on cation-exchange equilibria in gravelly loam, and found that organic matter significantly increased Ca^{2+}, NH_4^+, and K^+ retention, enhanced the preference for NH_4^+ over K^+, and significantly decreased the affinity for NH_4^+ compared to Ca^{2+}. Similar results were reported by Fernando et al. (2005), who also noted that high dissolved organic C contributed to NH_4^+ adsorption in soils containing liquid swine waste. However, the opposite effect of organic matter was found by Dumat et al. (1997), who showed that Cs adsorption was increased when the organic matter content was lower. Moreover, Wang and Huang (2001) found that organic matter increased the adsorption rate of K^+, and they speculated that organic matter might possess more accessible exchange sites compared to the clay minerals of the sampled soil, despite the fact that the affinity for K^+ by organic matter is smaller than the affinity for K^+ by layer silicate minerals (Salmon, 1964). However, Fernando et al.(2005) reported a slower NH_4^+ adsorption rate in soils exposed to organic substances. It is apparent from the investigations cited above that the role of organic matter in modifying the dynamics of soil cations is complex.

However, the method of removing organic matter by treatment with H_2O_2 has some side effects. Many researchers have reported that some clay minerals in soils, such as vermiculite, biotite, and montmorillonite, would be structurally altered by H_2O_2 treatment (van Langeveld et al., 1978; Amonette et al., 1985; Miles et al., 1985). Otherwise, organic matter would persist because of interaction with soil minerals that protect against oxidative treatment. Theng et al. (1992) and Righi et al. (1995) assumed that organic matter intercalated in expandable clay minerals is hardly decomposed by H_2O_2 due to limited accessibility to the reagent. In addition, soils are a relatively complex system and it is impossible to separate the effects of individual soil properties (Chung and Zasoski, 1994). These factors make it difficult to elucidate the effect of the organic matter alone.

In recent years, a new approach, using artificial organo-mineral complexes synthesized from pure organic components and clay minerals, has been developed (Wang and Xing, 2005). Many studies have used this new method successfully to examine the behavior of micronutrient ions, such as Cu^{2+}, Zn^{2+}, and others (Saada et al., 2003; Prado et al., 2006; Villaverde et al., 2009), but few studies have addressed the behavior of macronutrient ions on clay minerals in the presence of organic matter. Ammonium (NH_4^+) and K^+ are two important macronutrient cations in agricultural soils. They have similar ionic radii and compete with each other intensively for surface exchange or fixation sites (Nielsen, 1972; Bohn et al., 1985; Lumbanraja and Evangelou, 1990; Wang et al., 2010). Studying the interaction of NH_4^+ and K^+ on organo-mineral complexes will contribute to further understanding the behavior of NH_4^+ and K^+ in soils. Therefore, the objectives of this study were to evaluate the changes in major physicochemical properties of clay minerals after humic acid coating and measure the effects of these changes on the interaction of NH_4^+ and K^+ on clay minerals.

2.6.2 Materials and Methods

1. Clay minerals and humic acid

A kaolinite (Ka) KGa-1b, an illite (Il) IMt-1, and a montmorillonite (Mt) STx-1b were obtained from the Clay Minerals Society Source Clay Repository (Virginia, USA). The clay minerals were Ca-saturated by washing with 1 mol/L $CaCl_2$ (Wang and Huang, 2001) to maintain a homoionic surface. A humic acid (HA) 1S102H was purchased from the International Humic Substances Society (Minnesota, USA). Before use, HA was dissolved with NaOH (0.1 mol/L) into distilled water containing $HgCl_2$ (0.05 mmol/L) to prevent microbial degradation from significantly altering HA chemistry (Wolf et al., 1977). The HA solution was filtered through a 0.22 μm millipore filter before experiments to remove bacteria and any undissolved part. The basic characteristics of the clay minerals and HA are available on the websites of the Clay Minerals Society Source Clay Repository (http://www.clays.org/index.html) and International Humic Substances Society (http://www.humic substances.org/), respectively.

2. Organo-mineral complex preparation

Humic-clay complexes were prepared at an HA to clay ratio of 1 : 5 (w/w). The mixed suspensions were gently shaken on hematology mixers for 24 h to allow HA react with clay surfaces, and then centrifuged at 4224 × g for 20 min. The precipitates were freezedried and then washed repeatedly with deionized water to remove non-adsorbed HA. The washed precipitates were freeze-dried again, ground, and passed through a 0.15 mm sieve, and stored for subsequent uses (Wang and Xing, 2005).

3. Adsorption and fixation experiments

The following three cation solutions were prepared: 2 mmol/L KCl and 2 mmol/L NH_4Cl as a single system, and a mixture of 2 mmol/L KCl and 2 mmol/L NH_4Cl as a binary system. The pH of each solution was adjusted to 6.5 with 1 mmol/L NaOH and HCl. 10 mL of each cation solution was added to 0.2 g Ca-saturated clays or organo-mineral complexes that were weighed and put into 50 mL polypropylene centrifuge tubes. All treatments were replicated five times. The mixtures were shaken for 24 h on a shaker at 120 r/min (25℃) and then centrifuged for 20 min at 4224 × g. The supernatants were retained for analysis of NH_4^+ or K^+ content. The total adsorbed and fixed amounts of NH_4^+ or K^+ were calculated based on the difference between the initial and final NH_4^+ or K^+ concentrations.

Three replications of the solid residues were weighed to calculate the remaining solution. Then, 10 mL 0.5 mol/L CsCl was added, the mixed suspensions were shaken for another 12 h and centrifuged for 20 min at 4224 × g, and the suspension was retained to determine NH_4^+ and/or K^+ content. The amount of NH_4^+ or K^+ desorbed by CsCl was taken as the amount of NH_4^+ or K^+ adsorbed on clay minerals or organo-mineral complexes. The amount of fixed NH_4^+ or K^+ was calculated by subtracting the adsorbed amount from the total. The other two remaining replications of the solid residues were washed with deionized distilled water and centrifuged twice. After the suspension was discarded, the residues were air dried, ground, and passed through a 0.15 mm sieve, and stored for further analysis.

Different clay minerals or organo-mineral complexes were equilibrated with a series of five different 2 mmol/L Cl⁻ salt solutions containing various equivalent fractions of NH_4^+ and K^+ ranging from 0.1 to 0.9. The specific molar ratios of NH_4^+ /K^+ in the solution series were 0.9 ∶ 0.1, 0.7 ∶ 0.3, 0.5 ∶ 0.5, 0.3 ∶ 0.7, and 0.1 ∶ 0.9.

4. Measurements

The specific surface areas (SSA) of the clay minerals and organo-mineral complexes were measured using the N_2 adsorption-desorption technique on an ASAP 2010 (Micrometrics, USA) at liquid N_2 temperature. Infrared spectra were obtained on a Vector 27 Fourier transform infrared (FTIR) spectrometer (Bruker, Germany) equipped with deuterated triglycine detectors with a PA300 (MTEC, USA) photoacoustic cell. Each sample was placed in a columnar stainless sample cell with a Helium flow for 10 s, the unapodized resolution for the FTIR spectra was 2.0 cm^{-1}, and a total of 64 scans were collected for each spectrum. More detailed experimental was described elsewhere by Du and Zhou (2007) and Du et al. (2010). The cation exchange capacity (CEC) was determined using the ammonium acetate method (Lu, 1999). Organic carbon content was analyzed using a PE2400 elemental analyzer. NH_4^+ and K^+ contents were measured by indophenol blue colorimetry and emission spectrophotometry, respectively.

5. Calculations and statistical analysis

In the cation-exchange equilibria, an exchanger has the preference for one ion over another. Such preference is called selectivity. For an exchange system,

$$\nu A + \mu BX \Leftrightarrow \mu B + \nu AX \tag{1}$$

Where, ν and μ refer to the stoichiometric numbers of the reacting ions A and B, respectively; X denotes an exchanger phase with a negative charge. The values of the Vanselow selectivity coefficient for A and B ($K_{\nu, A \to B}$) were calculated from the isotherm as follows:

$$K_{\nu, A \to B} = \left(N_A^\nu / N_B^\mu\right)\left(a_B^\mu / a_A^\nu\right) = \left(N_A^\nu / N_B^\mu\right)\left(M_B^\mu \gamma_B^\mu / M_A^\nu \gamma_A^\nu\right) \tag{2}$$

Where, N_A^ν and N_B^ν denote the mole fractions of A and B in the exchange phase, respectively; a_A^ν, a_B^μ, M_A^ν, and M_B^μ denote the ion activity and molar concentration of A and B in the solution phase, respectively; and γ_A^ν and γ_B^μ refer to the activity coefficients of the cations A and B in the solution phase, respectively. Ion activity coefficient values were calculated using the extended Debye-Huckel equation.

All results in figures and tables were given as means of three replicates. Analysis of variance (ANOVA) was performed to evaluate differences among means with SAS 6.12.

Results were considered significantly different at $P<0.05$ unless noted otherwise. The error bars on the graphs represent one standard deviation above and below the mean. Multiple comparisons among the amounts of adsorption or fixation of NH_4^+ and K^+ between a clay mineral and the corresponding organomineral complex were characterized using the least significant difference (LSD) test.

2.6.3 Results

1. Changes in clay mineral properties after humic acid coating

The organic carbon (OC) content, CEC, and SSA of the clay minerals and organo-mineral complexes are listed in Table 1. Little OC, no more than 1.0 g/kg, was measured in montmorillonite (Mt) and kaolinite (Ka) but a higher OC content of 2.4 g/kg was found in illite (Il). After HA coating, the OC contents of Mt, Il, and Ka increased to 3.0, 3.9, and 1.9/g kg, respectively. Mt exhibited significantly greater CEC and SSA than the other two clays, but larger decreases in CEC (13%) and SSA (15%) were observed after HA coating. In contrast, slight changes in CEC and SSA were observed in Il and Ka after HA coating.

Table 1 Organic carbon (OC) content, cation exchange capacity (CEC), and specific surface area (SSA) of clay minerals and organo-mineral complexes

Clay mineral or organo-mineral complex[①]	OC content (g/kg)	CEC (cmol/kg)	SSA (m²/g)	CEC per unit area (10^3 cmol/m²)
Mt	0.8 ±0.2	76.93 ±3.44	83.79 ±0.22	0.92
MH	3.0 ±0.9	66.19 ±2.18	71.20 ±0.15	0.93
Il	2.4 ±0.1	25.09 ±0.82	20.16 ±0.25	1.24
It	3.9 ±0.6	23.81 ±0.90	17.90 ±0.13	1.32
Ka	0.4 ±0.0	3.87 ±0.05	10.05 ±0.08	0.39
KH	1.9 ±0.8	4.08 ±0.04	9.68 ±0.06	0.42

Note: ① Mt = montmorillonite; MH = humic-montmorillonite complex; Il = illite; IH = humic-illite complex; Ka = kaolinite; KH = humic kaolinite complex.

2. Effect of humic acid coating on NH_4^+ and K^+ adsorption on clay minerals

The influence of HA on NH_4^+ or K^+ adsorption on different clay minerals was as shown in Fig. 1. As expected, Mt exhibited a higher adsorption capacity for cations than the other two clay minerals, and the adsorption of NH_4^+ or K^+ differed significantly upon HA coating. Significant increases in NH_4^+ or K^+ adsorption were observed on organo-mineral complexes compared to clay minerals. Also, NH_4^+ adsorption was higher than K^+ adsorption on Ka and Mt, especially when the two clay minerals were coated by HA, whereas K^+ adsorption was higher on Il and humic-illite complex (IH). The effect of HA coating on NH_4^+ or K^+ adsorption was significantly and negatively correlated with the CEC of clay minerals (Fig. 2).

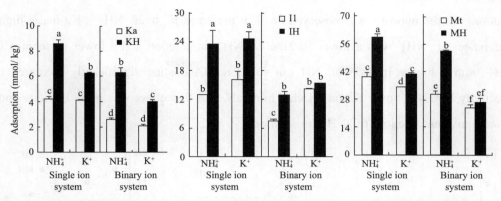

Fig. 1 NH_4^+ and K^+ adsorption on clay minerals and organo-mineral complexes in single and binary ion systems (NH_4^+ and/or K^+)

Values are means with standard deviations shown by vertical bars. Same letter(s) above the bars indicate no statistically significant differences ($P<0.05$)

Mt = montmorillonite; MH = humic-montmorillonite complex; Il = illite; IH = humic-illite complex; Ka = kaolinite; KH = humic-kaolinite complex

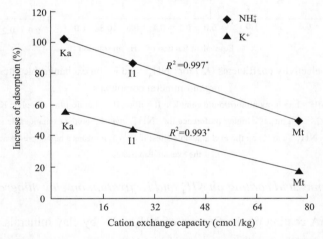

Fig. 2 Relationships between the cation exchange capacity of clay minerals and the percentage of NH_4^+ or K^+ adsorption increased by humic acid coating in a single ion system (NH_4^+ or K^+)

The asterisk indicates statistically significant differences ($P<0.05$)

Mt = montmorillonite; Il = illite; Ka = kaolinite

When NH_4^+ and K^+ were both present in solution, the adsorption of both cations decreased significantly on clay minerals and organo-mineral complexes (Fig. 1). The decrease in K^+ adsorption was greater than that of NH_4^+ adsorption on Mt and Ka, and this was more obvious on the relevant organo-mineral complexes. In contrast, a greater decrease in NH_4^+ adsorption than in K^+ adsorption was found on Il, and this did not occur on IH.

The differences in K_v for NH_4^+ over K^+ between clay minerals and organo-mineral complexes were compared as in Fig. 3. Vanselow selectivity coefficient values were not constant: they always decreased with an increasing equivalent fraction of NH_4^+ in solution. Mt exhibited a higher preference for NH_4^+ over K^+ at all the equivalent fractions of NH_4^+ in solution, but the opposite was observed on Il: it preferred K^+ over NH_4^+. Ka had a higher preference for NH_4^+ with a lower fraction of NH_4^+ in solution and a lower preference for NH_4^+ with a higher fraction of NH_4^+ in solution. HA coating significantly increased the selectivity for NH_4^+ over K^+ on all clay minerals, and this effect was more obvious under lower equivalent fractions of NH_4^+ in solution.

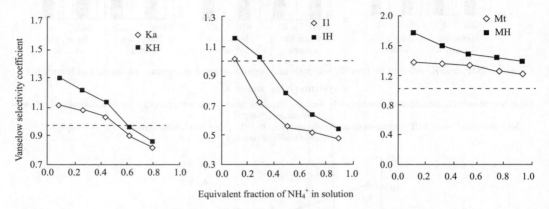

Fig. 3 Vanselow selectivity coefficients (K_v) for NH_4^+ and K^+ on exchangeable sites of clay minerals and organo-mineral complexes

Mt = montmorillonite; MH = humic-montmorillonite complex; Il = illite; IH = humic-illite complex; Ka = kaolinite; KH = humic kaolinite complex. If $K_v > 1$, it means a higher preference for NH_4^+ over K^+ on the exchangeable sites; if $K_v = 1$, it means a similar preference for NH_4^+ and K^+ on the exchangeable sites; if $K_v < 1$, it means a higher preference for K^+ over NH_4^+ on the exchangeable sites

3. Effect of humic acid coating on NH_4^+ and K^+ fixation on clay minerals

The effect of HA coating on the fixation of NH_4^+ or K^+ by clay minerals and organo-mineral complexes is shown in Fig. 4. As expected, NH_4^+ and K^+ fixation differed greatly, depending on the clay mineralogy. Very little NH_4^+ or K^+ was fixed on Ka; in contrast, much more NH_4^+ and K^+ fixation occurred on Il. HA coating increased NH_4^+ or K^+ fixation on Ka and Mt, and this effect was more pronounced for NH_4^+. HA coating inhibited NH_4^+ and K^+ fixation by Il.

In the binary ion system (NH_4^+ and K^+), varying degrees of decreases in NH_4^+ and K^+ fixation were observed for different clay minerals (Fig. 4). A significant decrease of K^+ fixation occurred on Mt and humicmontmorillonite complex (MH), but no statistically significant differences were observed in the changes of NH_4^+ or K^+ fixation on Ka and humic-kaolinite complex (KH). The presence of K^+ increased NH_4^+ fixation on Il and slightly decreased NH_4^+ fixation on IH, but the presence of NH_4^+ significantly suppressed K^+ fixation on Il and IH.

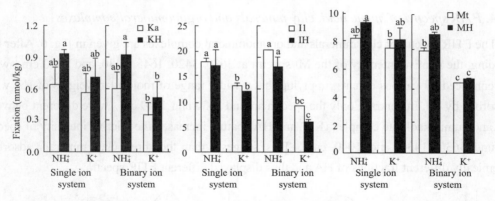

Fig. 4 NH_4^+ and K^+ fixation on clay minerals and organo-mineral complexes in single and binary ion systems (NH_4^+ and/or K^+)

Values are means with standard deviations shown by vertical bars. Same letter(s) above the bars indicate no statistically significant differences ($P<0.05$)

Mt = montmorillonite; MH = humic-montmorillonite complex; Il = illite; IH = humic-illite complex; Ka = kaolinite; KH = humic-kaolinite complex

The selectivity coefficients for NH_4^+ over K^+ on fixed sites of clay minerals and organo-mineral complexes were compared as shown in Fig. 5. Similar to the exchangeable sites, the selectivity coefficients for NH_4^+ and K^+ on fixed sites varied greatly with the equivalent fraction of NH_4^+ in solution. At a same fraction of NH_4^+ in solution, Il exhibited a higher preference for NH_4^+ than the other two clay minerals. HA coating increased the selectivity of NH_4^+ over K^+, and this was clearly observed for Ka and Il as the equivalent fractions of NH_4^+ increased in solution.

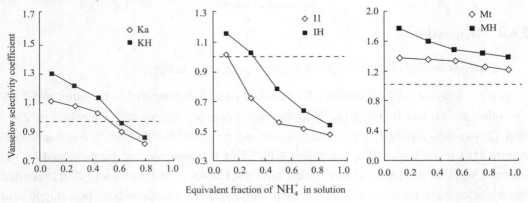

Fig. 5 Vanselow selectivity coefficients (K_v) for NH_4^+ and K^+ on fixed sites of clay minerals and organo-mineral complexes

Mt = montmorillonite; MH = humic-montmorillonite complex; Il = illite; IH = humic-illite complex; Ka = kaolinite; KH = humic-kaolinite complex. If $K_v > 1$, it means a higher preference for NH_4^+ over K^+ on the fixed sites; if $K_v = 1$, it means a similar preference for NH_4^+ and K^+ on the fixed sites; if $K_v < 1$, it means a higher preference for K^+ over NH_4^+ on the fixed sites

4. *FTIR spectra of humic acid, clay minerals and organomineral complexes*

The FTIR spectra of clay minerals and organomineral complexes are given in Fig. 6. After HA loading, the relative intensity of the Mt spectra at 3630, 3420, 1645, 1430, and 1040 cm^{-1} were all enhanced by various extents, implying that a wider range of molecular weights of HA were absorbed by Mt. In contrast, only the spectra around 3630 and 1040 cm^{-1} were different between Ka and humic-kaolinite complex (KH) and HA coating increased the relative intensity of spectra around 3630, 1430, and 1040 cm^{-1}. This suggests that the three clay minerals adsorbed chemically different fractions of HA and thus displayed different FTIR spectra.

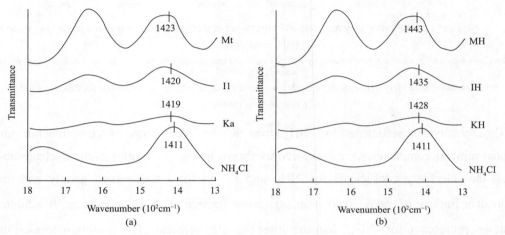

Fig. 6 Fourier transform infrared spectrometry (FTIR) spectra of clay minerals (a) and organo-mineral complexes (b) equilibrated for 24 h with 2 mmol/L NH$_4$Cl (pH 6.5) as well as the FTIR spectrum of NH$_4$Cl

Mt = montmorillonite; MH = humicmontmorillonite complex; Il = illite; IH = humic-illite complex; Ka = kaolinite; KH = humic-kaolinite complex

2.6.4 Discussion

1. *Changes in clay mineral properties after humic acid coating*

As a 1∶1 layered silicate mineral, Ka adsorbed more HA compared to the other two 2∶1 clay minerals, Mt and Il (Table 1). This observation is in accordance with the results of Zhou et al. (1994), who reported that Ka had a higher adsorption for HA than Mt in NaCl solutions, despite Mt's much larger SSA and higher CEC. This phenomenon can be explained by the selective sorption of HA by different clay surfaces. Chorover and Amistadi (2001) reported that Mt selectively adsorbed lower molecular weight organic compounds of peat humic acid (PHA) compared to goethite. However, Satterberg et al. (2003) found that Mt preferred the larger molecular weight component of PHA due to its higher CEC. Therefore, it may be true that the binding sites on Mt surfaces were occupied by HA with organic compounds having larger range of molecular weights, while Ka was predominantly covered with relatively larger molecules through specific sorption, which favored higher molecular weight polymers with a greater potential range of interactions with clay surfaces (Vermeer et al., 1998; Chorover and Amistadi, 2001; Feng et al., 2005). The specific adsorption of HA was also supported by the

FTIR spectra of clay minerals and organo-mineral complexes (Fig. 6) in this study. Different changes in FTIR spectra after HA loading clearly suggested that the three clay minerals adsorbed chemically different fractions of HA.

A significant decrease in SSA and CEC in Mt was observed after HA coating (Table 1). One possible explanation was that amorphous organic matter binds some clay mineral particles together through microaggregation, which would eventually reduce some of the available SSA (Feller et al., 1992), and consequently, some external exchangeable sites on the Mt surface would be covered or altered to form internal sites. A second explanation was that the clay micropores could be filled or clogged by organic matter, thus reducing the accessibility of layers to N_2 (Kaiser and Guggenberger, 2000). However, because of the apparent decrease in CEC in Mt, which chiefly reflected the properties of the external surfaces of clay minerals and HA or HA fractions, this mechanism should only play a minor role in reducing the SSA of Mt. In addition, small changes in SSA and CEC were found in Il, and a slight increase in CEC occurred in Ka, implying that HA coating had different effects on the properties of different clay minerals. This would contribute to the preferential sorption of HA and different binding mechanisms between clay minerals and HA.

2. *Role of humic acid coating in NH_4^+ and K^+ adsorption and fixation on clay minerals*

Humic acid coating-induced changes in SSA and the nature of adsorption sites resulted in changes in the behavior of NH_4^+ and K^+ on clay minerals. A significant increase in NH_4^+ and K^+ adsorption was observed for the organo-mineral complexes (Fig. 1). This suggests that HA provided some exchange sites that were more easily accessible to NH_4^+ and K^+. The same results were reported by Chung and Zasoski (1994), who found that the addition of organic matter to an Arbuckle gravelly loam significantly increased NH_4^+, K^+, and Ca^{2+} retention. However, the increased cation adsorption of organo-mineral complexes seemed at odds with the variation in CEC between clay minerals and organo-mineral complexes, especially for Mt and Il. A consideration of the nature of HA, its interaction with the clays, and the nature of the NH_4^+ and K^+ adsorption sites assumed to be involved may help to elucidate the mechanisms of HA effect on cation adsorption.

As far as we know, HA contains many highly reactive functional groups, such as carboxyl and phenolic hydroxyl. These acidic groups would contribute to cation exchange sites under moderate or alkaline conditions. Also, in contrast to the tortuous diffusion path for monovalent ions towards the exchange sites on the internal surfaces of these minerals (Rich, 1964), the ion exchange sites on the surface of organic matter are relatively easy to access because they are basically external sites (Stevenson, 1994). As mentioned above, organic matter can act as an aggregation agent for some clay particles and thus some surface exchangeable sites get blocked or become internal sites, which may induce a decrease in cation adsorption. The CEC values obtained were the result of repeated equilibration and under these ion saturation conditions, the adsorption retarding effect of HA coating would play an important role. In contrast, the NH_4^+ and K^+ adsorption occurred under one equilibration step with a lower ion concentration (2 mmol/L) where the HA coating had an adsorption-promoting effect, which was also confirmed by the negative relationship between the increased adsorption rates and the CEC values of the clay minerals(Fig. 2).

Compared with the effect of HA coating on NH_4^+ and K^+ adsorption, the effect of HA coating on NH_4^+ and K^+ fixation was more complex. A positive effect was observed for Ka and Mt, but the opposite was found on Il (Fig. 4). This suggests that the effect of HA coating on cation fixation depended on the nature of specific clay minerals. In contrast to Mt and Ka, Il contains more frayed edge sites through which and K^+ can be fixed in the Il interlayers. HA, anionic at pH 6.5, would be adsorbed on the anion exchange sites located on the crystal edges of Il. These sites are adjacent to, but distinct from, the frayed edge sites of Il, which are cation exchange sites. Therefore, the accessibility to the interlayer of Il would be impeded by the macromolecules of adsorbed HA, and in turn a decrease in NH_4^+ and K^+ fixation occurred on Il. The aggregation effect of HA as a cementing agent should be taken into account in explaining the fixation data in Mt and Ka. This effect promotes fixation because it affects the geometry of the microaggregates of clays in such a way as to increase the ratio of the internal to external surfaces (Williams et al., 1967). The lower the fixation capacity of a clay, the stronger the effect should be. Consequently, the steric impeding effect, caused by HA adsorption on anion exchange sites adjacent to the frayed edge sites, would be offset more by the fixation-promoting effect caused by the aggregation effect of the HA coating on clays having relatively lower fixation capacity such as Ka and Mt in this study.

3. *Role of humic acid coating in the interaction of NH_4^+ and K^+ on clay minerals*

Owing to their many shared physical and chemical properties, NH_4^+ and K^+ compete with each other intensively for surface exchange or fixation sites (Nielsen, 1972; Bohn et al., 1985; Lumbanraja and Evangelou, 1990; Wang et al., 2010). Here, the interaction of NH_4^+ and K^+ on clay minerals was mainly focused on the competition for adsorption and fixation of the two ions.

The adsorption of NH_4^+ was significantly increased by HA coating when NH_4^+ and K^+ were both present in solutions (Fig. 1). Consequently, the selectivity coefficients for NH_4^+ over K^+ on exchange sites were greatly increased by HA coating although selectivity coefficients varied with clay minerals and equivalent fractions of NH_4^+ in solution (Fig. 3). These results indicate that HA coating increased the competitiveness of NH_4^+ adsorption on clay minerals and enhanced the affinity of exchange sites of clay minerals for NH_4^+ over K^+. Chung and Zasoski (1994) also found that organic matter addition increased the selectivity for NH_4^+ over K^+ and Ca^{2+} over NH_4^+ in an Arbuckle gravelly loam. FTIR was employed to explore the reason why humic acid coating increased the competitiveness of NH_4^+ on clay minerals (Fig. 6(a)). Compared with the deformation band of free NH_4^+ at 1411 cm^{-1}, the NH_4^+ deformation bands at 1428 cm^{-1} in KH, 1435 cm^{-1} in IH, and 1443 cm^{-1} in MH seemed to be shifted to higher wavenumbers (Fig. 6(b)). This implies that complexation between the NH_4^+ and the anionic surface distorted the tetrahedral ion by weakening the N-H bond (Chappell and Evangelou, 2000). These shifts to higher wavenumbers were apparently the result of H-bonding between NH_4^+ and the clay surface (Nakamoto, 1986), an ability not shared by the K^+ ion. However, the shifts in deformation band of NH_4^+ were found not obvious in pure clay minerals. This suggests that humic acid coating contributed to the formation of hydrogen bonds between NH_4^+ and clay minerals. Therefore, in the organo-mineral complexes, except for electrostatic adsorption, H-bonds also played an important role in NH_4^+ adsorption, a process not shared by K^+ adsorption. As a result, higher selectivity coefficients were obtained in the organo-minerals complexes than in the clay minerals.

The selectivity coefficients for NH_4^+ over K^+ on the fixation sites of clay minerals were significantly increased by HA coating, especially at the lower equivalent fractions of NH_4^+ in solution (Fig. 5), implying that the competitiveness of NH_4^+ over K^+ for the fixation sites on clay minerals was enhanced by HA coating. It has long been known that soil organic matter can fix NH_4^+ through chemical mechanisms. Ammonium fixation by soil organic matter has been attributed to phenolic constituents associated with humic and fulvic acids. Ammonia is also known to react with reducing sugars to form brown-colored nitrogenous polymers (Stevenson, 1994). The formation of esters by reaction with carboxylic esters may also play a minor role (Burge and Broadbent, 1961). In the presence of HA, much more NH_4^+ would be fixed chemically compared with K^+, and thus a higher preference for NH_4^+ over K^+ was observed on fixed sites of the organo-mineral complexes.

References

Amonette J, Ismail F T, Scott A D. 1985. Oxidation of iron in biotite by different oxidizing solutions at room temperature. Soil Sci. Soc. Am. J., 49: 772-777

Arias M, Barral M T, Mejuto J C. 2002. Enhancement of copper and cadmium adsorption on kaolin by the presence of humic acids. Chemosphere, 48: 1081-1088

Bohn H L, McNeal B L, O'Connor G A. 1985. Soil Chemistry. 2nd Ed. New York: John Wiley and Sons Inc.

Burge W D, Broadbent F E. 1961. Fixation of ammonia by organic soils. Soil Sci. Soc. Am. Proc., 25: 199-204

Chappell M A, Evangelou V P. 2000. Influence of added K^+ on ammonium selectivity/mobility by soils with vermi cultic behavior. Soil Sci., 165: 858-868

Chorover J, Amistadi M K. 2001. Reaction of forest floor organic matter at goethite, birnessite and smectites surfaces. Geochim. Cosmochim. Acta, 65: 95-109

Chung J B, Zasoski R J. 1994. Ammonium-potassium and ammonium-calcium exchange equilibria in bulk and rhizosphere soil. Soil Sci. Soc. Am. J., 58: 1368-1375

Clapp C E, Hayes M H B, Simpson A J, et al. 2005. Chemistry of soil organic matter. In: Tabatabai M A, Sparks D L .Chemical Processes in Soils. Madison: Soil Sci. Soc. Am., Inc.

Du C W, Zhou J M. 2007. Prediction of soil available phosphorus using Fourier transform infrared-photoacoustic spectroscopy. Chinese J. Anal. Chem., 35: 119-122

Du C W, Zhou G Q, Zhou J M, et al. 2010. Characterization of animal manures using mid-infrared photoacoustic spectroscopy. Bioresource Technol., 101: 6273-6277

Dumat C, Cheshire M V, Fraser A R, et al. 1997. The effect of removal of soil organic matter and iron on the adsorption of radiocaesium. Eur. J. Soil Sci., 48: 675-683

Dumat C, Quiquampoix H, Staunton S. 2000. Adsorption of cesium by synthetic clay-organic matter complexes: effect of the nature of organic polymers. Environ. Sci. Technol., 34: 2985-2989

Feller C, Schouller E, Thomas F, et al. 1992. N2-BET specific surface areas of some low activity clay soils and their relationships with secondary constituents and organic matter contents. Soil Sci., 153: 293-299

Feng X J, Simpson A J, Simpson M J. 2005. Chemical and mineralogical controls on humic acid sorption to clay mineral surfaces. Org. Geochem., 36: 1553-1566

Fernando W A R N, Xia K, Rice C W. 2005. Sorption and desorption of ammonium from liquid swine waste in soils. Soil Sci. Soc. Am. J., 69: 1057-1065

Huang P M, Berthelin J, Bollag J M, et al. 1995a. Environmental Impact of Soil Component Interactions: Natural and Anthropogenic Organics, Vol. 1. Boca Raton: CRC Press/Lewis Publishers

Huang P M, Berthelin J, Bollag J M, et al. 1995b. Environmental Impact of Soil Component Interactions: Metals, Other Inorganics and Microbial Activities, Vol. 2. Boca Raton: CRC Press/Lewis Publishers

Kaiser K, Guggenberger G. 2000. The role of DOM sorption to mineral surfaces in the preservation of organic matter in soils. Org. Geochem., 31: 711-725

Lu R K. 1999. Analytic Methods for Soil and Agro-chemistry. Beijing: Chinese Agricultural Press (in Chinese)

Lumbanraja, J, Evangelou V P. 1990. Binary and ternary exchange behavior of potassium and ammonium on Kentucky subsoils. Soil Sci. Soc. Am. J., 54: 698-705

Miles N M, De Kimpe C R, Schnitzer M. 1985. Effect of soil organic matter oxidation on the collapse of vermiculite. Can. J. Soil Sci., 65: 593-597

Nakamoto K. 1986. Infrared and Raman Spectra of Inorganic and Coordination Compounds. New York: John Wiley & Sons, Inc.

Nielsen J D. 1972. Fixation and release of potassium and ammonium ions in Danish soils. Plant Soil, 36: 71-88

Poonia S R, Niederbudde E A. 1990. Exchange equilibria of potassium in soils: V. Effect of natural organic matter on K-Ca exchange. Geoderma, 47: 233-242

Prado A G S, Torres J D, Martins P C, et al. 2006. Studies on copper(II)- and zinc(II)-mixed ligand complexes of humic acid. J. Hazard. Mater., B136: 585-588

Rich C I. 1964. Effect of cation size and pH on potassium exchange in Nason soil. Soil Sci., 98: 100-106

Righi D, Dinel H, Schulten H R, et al. 1995. Characterization of clay-organic-matter complexes resistant to oxidation by peroxide. Eur. J. Soil Sci., 46: 423-429

Saada A, Breeze D, Crouzet C, et al. 2003. Adsorption of arsenic (V) on kaolinite and on kaolinite humic acid complexes: role of humic acid nitrogen groups. Chemosphere, 51: 757-763

Salmon R C. 1964. Cation exchange reactions. J. Soil Sci., 15: 273-283

Satterberg J, Arnarson T S, Lessard E J, et al. 2003. Sorption of organic matter from four phytoplankton species to montmorillonite, chlorite and kaolinite in seawater. Mar. Chem., 81: 11-18

Stevenson F J. 1994. Humus Chemistry, Genesis, Composition, Reactions. 2nd Ed. New York: John Wiley and Sons Inc.

Theng B K G, Tate K R, Becker-Heidmann P. 1992. Towards establishing the age, location, and identity of the inert soil organic matter of a spodosol. J. Plant Nutr. Soil.Sci., 155: 181-18

van Langeveld A D, van der Gaast S J, Eisma D. 1978. A comparison of the effectiveness of eight methods for the removal of organic matter from clay. Clay. Clay Miner, 26: 361-364

Vermeer A W P, van Riemsdijk W H, Koopal L K. 1998. Adsorption of humic acid to mineral particles. 1. Specific and electrostatic interactions. Langmuir., 14: 2810-2819

Villaverde P, Gondar D Antelo J,et al. 2009. Influence of pH on copper, lead and cadmium binding by an ombrotrophic peat. Eur. J. Soil Sci., 60: 377-385

Wang F L, Huang P M, 2001. Effects of organic matter on the rate of potassium adsorption by soils. Can. J. Soil Sci., 81: 325-330

Wang K J, Xing B S. 2005. Structural and sorption characteristics of adsorbed humic acid on clay minerals. J. Environ. Qual. , 34: 342-349

Wang H Y, Zhou J M, Du C W, et al. 2010. Potassium fractions in soils as affected by monocalcium phosphate, ammonium sulfate, and potassium chloride application. Pedosphere, 20: 368-377

Williams B G, Greenland D J, Quirk J P. 1967. The effect of polyvinyl alcohol on the nitrogen surface area and pore structure of soils. Aust. J. Soil Res., 5: 77-83

Wolf A, Bunzl K, Dietl F, et al. 1977. Effect of Ca^{2+}-ions on the absorption of Pb^{2+}, Cu^{2+}, Cd^{2+} and Zn^{2+} by humic substances. Chemosphere, 6: 207-213

Zhou J L, Rowland S, Mantoura R F C, et al. 1994. The formation of humic coatings on mineral particles under simulated estuarine conditions —a mechanistic study. Water Res., 28: 571-579

2.7 Minimum data set for assessing soil quality in farmland of Northeast China

2.7.1 Introduction

Soil is a natural resource essential for the existence of life on our planet. It provides services involving complex interactions among its biological, physical, and chemical properties (Karlen et al., 1997). Soil quality is defined as the capacity of soil to function within ecosystem boundaries to sustain biological productivity, maintain environmental quality and promote plant and animal health (Doran and Parkin, 1994). Soil quality assessment is best viewed as an integrative process of sustainable land management when used to evaluate environmental quality, food security, and economic issues (Larson and Pierce, 1994; Hussain et al., 1999). Soil assessment and monitoring rely on indicators that can integrate biological, chemical and physical attributes. Numerous soil quality evaluation methods have been developed since the USDA Soil Conservation Service released its land capability classification system in 1961 (Klingebiel and Montgomery, 1961), these methods include soil quality cards and test kits (Ditzler and Tugel, 2002), soil quality index (SQI) methods (Doran and Parkin, 1994; Andrews et al., 2002a), fuzzy association rules (Xue et al., 2010), dynamic soil quality models (Larson and Pierce, 1994), and the soil management assessment framework (Andrews et al., 2004; Masto et al., 2007; Karlen et al., 2008; Wienhold et al., 2009). Among these methods, the SQI approach is perhaps the most common (Andrews et al., 2002a) because of their simplicity and quantitative flexibility.

Soil quality indices are tools for adaptive soil resource management that can help farmers and their advisors determine soil health trends and thereby indicate whether one or more changes in practice are necessary (Karlen et al., 2001). Therefore, a universally accepted SQI should include a clear procedure for indicator selection, scoring function selection, and weighting, as well as a model that would aid in comparing soils from different regions.

The first step in soil quality assessment is selection of soil quality indicators, that is, the soil properties and processes that will provide a minimum data set (MDS) for evaluation. Soil quality indicators should also be linked to soil function (Doran and Parkin, 1994; Karlen et al., 1996). These functions may individually or collectively include serving as a medium for plant growth, an environmental filter, a buffer and transformer, and/or a habitat for biota (Bhardwaj et al., 2011). Several attributes were chosen as key indicators in many previous MDS methods. These included soil organic matter (SOM), available phosphorus (AvP), available potassium (AvK), cation exchange capacity (CEC), total content of nitrogen (TN), pH, soil texture, and bulk density (Bd) (Doran and Parkin, 1994; Karlen et al., 1997; Arshad and Martin, 2002; Qi et al., 2009). However, there is neither consensus on which soil properties should be used for soil quality monitoring nor on how the indicators should be interpreted (Schipper and Sparling, 2000). Andrews et al. (2002a) chose Na as indicator, which was seldom used in other papers. Indicators used by Qi et al. (2009) not only included common indicators, but also involved the thickness of obstacle horizons and drainage modulus which were not found in any other MDS methods. This lack of consensus can be partly attributed to the fact that soil quality is a complex concept and different site-specific soil conditions are required depending on the purpose of land use (de Lima et al., 2008).

Selection of soil quality indicators and the integration of such indicators into a single index could provide information about soil quality change and then guide land management decisions. Other studies employing similar methods have demonstrated the efficacy of SQI techniques in a variety of systems. Wander and Bollero (1999) used principal component analysis (PCA) to compare soil quality in tillage and no-till systems in Illinois. Andrews et al. (2002a) compared index methods composed of different indicator selection methods (expert opinion and PCA) with scoring functions (linear and non-linear) for vegetable production systems. Sharma et al. (2005) used PCA to establish a MDS for assessing soil quality affected by different management treatments.

The objectives of this study were: ① to establish a MDS of soil quality indicators; ② to evaluate soil quality of Hailun County, a typical agricultural county in Northeast China; and ③ to test whether this SQI was correlated with soybean yield.

2.7.2 Materials and Methods

1. Site description

This study was conducted in Hailun County, which is located in the middle part of Heilongjiang Province, Northeast China (46°58'-47°52' N, 126°14'-127°45' E). Hailun County covers a total terrestrial area of 4551 km^2 with a length 150 km from northeast to southwest and 78 km from north to south. The site has a continental monsoon climate with a long cold winter and short warm summer. The average annual temperature ranges from 1 to 2 ℃, with the highest temperatures at 37.7 ℃, and the lowest at −40.3 ℃. The annual precipitation (500 to 600 mm) is distributed mainly during the growing season from May to October. The general topography of the study area is mostly upland plains, with elevations ranging from 147 to 471 m. The primary soil type in this county, accounting for 63.4% of the area, is "black soil" (classified as Mollisols in the USA Soil Taxonomy system). This area is one of the four black soil lands in the world. Fields are cultivated with annual crops including soybean, maize, and rice from May to October. Continuous cultivation of soybean has become dominant in the region. Since the 1980s, crop land management was transitioned from collective farming to individual family farming, causing intensive agricultural production to flourish (Huang et al., 2006). These intensive production practices are characterized by increased fertilizer and pesticide applications, decreased dependence on organic amendments, and frequent tillage operations.

2. Soil sampling

A county-wide soil survey was conducted in 2008 with a total of 88 samples taken from agricultural lands based on spatial homogeneity, soil type, and land use. The latitude and longitude of each sample site were recorded using a GARMIN 60CSx global positioning system (GPS) handheld receiver. Interviews with landowners were also conducted at each sample site and information was collected on their agriculture management practices such as crop rotation, pesticide, fertilization, yield, irrigation, and drainage (Desbiez et al., 2004).

Soil samples were collected from tilled (0 to 20 cm), agricultural land after harvest in October 2008. Each sample was a composite of five sub-samples taken within a 350 m^2 area. Each subsample was collected in about 1000 cm^2 and weighed about 2 kg. The sub-samples were spread and mixed on a large clean plastic sheet, with clods thoroughly broken and vegetative material and stones removed to create a uniform mix. The mix was divided by coning and quartering until the weight of the composite sample was approximately 1 kg. Each sample was air-dried, placed in a plastic bag and stored until it could be analyzed. The air-dried samples were crushed and passed through a 2 mm (10 meshes) or 0.149 mm (100 meshes) sieve to determine its different attributes.

Soybean (*Glycine max* L. Merill) yields were determined in 2008 and 2009. Three of the five areas from which soil subsamples were collected were chosen as soybean sampling sites. Each soybean sample was collected within a 2 m^2 around the corresponding soil sampling site. The three soybean subsamples were collected, dried and weighed. Average yield of the three subsamples was calculated and used to estimate crop yield for the site. Ultimately, yield values used for correlation analysis were two-year yield averages for each site.

3. Laboratory analyses

The Bd of the soil was determined using the procedure of Veihmeyer and Hendrickson (1948), and the AvP was determined by extracting samples with 0.5 mol/L sodium bicarbonate ($NaHCO_3$) and subsequent colorimetric analysis (Olsen et al., 1954). Several other attributes were measured using methods described by Lu (2000), including soil pH in a 1 : 2.5

soil-water suspension; texture (clay, silt, and sand fractions) using the international pipette method; CEC by the ammonium chloride-ammonium acetate method; total phosphorus (TP) and total potassium (TK) after digesting soil samples with an acid mixture (HCl-HNO_3-HF-$HClO_4$); TN using the Kjeldahl digestion procedure; AvK by extracting with 1 mol/L ammonium acetate; available copper (AvCu), available zinc (AvZn), available manganese (AvMn) and available iron (AvFe) by extracting with 0.005 mol/L DTPA extraction; SOM using dichromate wet combustion; exchangeable calcium (exCa) and exchangeable magnesium (exMg) by extracting with 1 mol/L NH_4Ac; available boron (AvB) and available molybdenum (AvMo) by extracting with hot water and oxalate-ammonium oxalate, respectively. Concentrations of extracted Zn, Cu, Mn, Fe, Ca, Mg, B, Mo and digested P in solutions were determined using inductively coupled plasma atomic emission spectroscopy (ICP-AES), while K concentration was determined with a flame photometer.

4. Soil quality evaluation method

The development of a universal soil quality index should follow a logical path: ① choosing appropriate indicators for a MDS; ② transforming indicators to scores; and ③ combining the scores into an index (Andrews et al., 2002a). Each of these steps are discussed more completely below.

(1) Indicator selection

The first step was to select a MDS. The concept of having a MDS for soil quality assessment was widely accepted but initially the selection of MDS indicators relied primarily on expert opinion (Doran and Parkin, 1994; Larson and Pierce, 1994). However, selection of MDS variables can be simplified using statistical methods, such as PCA (Andrews and Carroll, 2001; Rezaei et al., 2005), factor analysis, and/or regression equations (Masto et al., 2008). We employed PCA as a data reduction tool to select the most appropriate indicators of site potential for the study area.

Principal components (PCs) are sets of indicators with high eigenvalues and factor loading. Eigenvalues less than one explains less variance than individual soil attribute. Therefore, only the PCs with eigenvalues ⩾1 and those that explained at least 5% of the data variation were considered potential members of the final MDSs. Communality describes the proportion of variance in each soil property explained by the PCA model. A high communality indicates that a high proportion of an indicator's variance can be explained by the factors (Brejda et al., 2000; Borůvka et al., 2007; Imaz et al., 2010). Highly weighted PCA indicators were defined as having absolute factor loading values ⩾0.50. Based on the MDS selection procedure described by Andrews and Carroll (2001), indicators with weighted absolute values within 10% of the highest indicator value for each PC were selected for the MDSs. Yemefack et al. (2006) stated that use of eigenvectors for variables in one PC does not provide information on the magnitude (norm) of the resulting vector for the variable in multi-dimensional space; either for the PCs or the original variables. Therefore, this approach might leave out some important indicators. The norm is the magnitude (length) of the vector representing the variable in the multi-dimensional space spanned by a set of PCs (Yemefack et al., 2006; Li et al., 2007), and was computed as follows:

$$N_{ik} = \sqrt{\sum_{1}^{k}(u_{ik}^2 \lambda_k)} \quad (1)$$

Where, N_{ik} is comprehensive loading of soil variable i on the first k PCs; λ_k is the eigenvalue of the PC; u_{ik} is the loading of soil variable i on PC k. Indicators receiving N_{ik} within 10% of the highest values were selected for the MDS.

When more than one variable were retained within a PC, multivariate correlation coefficients were employed to determine whether the variables could be considered redundant and thus can be eliminated from the MDS (Andrews et al., 2002a). If the highly weighted factors were not correlated, each was considered important and was retained in the MDS. Among the well-correlated variables, the variable within 10% of the highest norm was chosen for the MDS. Once all of the MDS indicators were chosen, a final check for correlations (between PC indicators) led to the selection of one replacement indicator for an indicator pair with correlation coefficients ⩾0.70 (Andrews et al., 2002a).

(2) *Indicator scoring*

Because of different indicator units, scoring functions (Andrews et al., 2002b) were used to score soil indicators. The linear and non-linear scoring functions were widely used for scoring (Diack and Stott, 2001; Liebig et al., 2001; Andrews et al., 2002b). In our study, the data of selected indicators were scored through the linear technique according to Diack and Stott (2001). Scores ranging from 0 to 1 were assigned to the indicators included in the MDS by applying either "More is better", or "Less is better" or "Optimum" function.

The equations of the score curves were as follows:

$$f(x) = \begin{cases} 0.1 & x \leqslant L \\ 0.9 \times \dfrac{x-L}{U-L} + 0.1 & L \leqslant x \leqslant U \\ 1 & x \geqslant U \end{cases} \quad (2)$$

$$f(x) = \begin{cases} 1 & x \leqslant L \\ 1 - 0.9 \times \dfrac{x-L}{U-L} & L \leqslant x \leqslant U \\ 0.1 & x \geqslant U \end{cases} \quad (3)$$

Where, $f(x)$ is the linear score; x is the soil property value; and L and U are the lower and upper threshold values, respectively. Eq. (2) was used for the "More is better" scoring function, whereas Eq. (3) was used for the "Less is better" function. For the "Optimum" function, indicators were scored as "More is better" for the increasing part and then scored as "Less is better" for the decreasing part. In this study, threshold values were obtained based on the relationship between the indicators and the soybean yields, and adjusted according to early literature (Ma et al., 2004).

(3) *Calculation and classification of SQI*

The weights of the integration could be accomplished in various ways. Regression coefficient, communality of the PCA, variation of PCA (Sharma et al., 2005; Masto et al.,

2008) and norm value were common weight methods, and thus were compared in this study. Regression was performed between the indicators in MDS and the soybean yield. The coefficient of each indicator was divided by the sum of the coefficient of indicators in MDS to determine the weight based on the regression. The weight derived from communality was calculated as the quotient of the communality divided by the sum of the communality of indicators in MDS. The weights based on norm values were calculated in similar way. Each PC explained a certain amount of the variation in the total data set (TDS). The percentage of the variation divided by the total percentage of variation explained by all PCs with eigenvectors >1 provided the weight based on variation of PCA. The weight of the TDS was calculated from the communality of the PCA. We then summed the weighted MDS variable scores for each observation. The final PCA-based soil quality equation was as follows:

$$SQI = \sum_{1}^{n} W_i S_i \quad (4)$$

Where, W_i is the PCA weighting and S_i is the indicator score. The SQI was then divided into five grades according to the classification criteria, i.e. values 10% more than or less than the average were regarded as Grade III. Other grades were divided according to the increments or decrements of 20% from the average. The quality of soil decreased as the grade increased. Thus, Grade I soil was considered suitable for plant growth, whereas Grade V soil had the most severe limitations for plant growth.

(4) *Evaluation of index methods*

To verify how well the MDS represented the TDS, the correlations between SQI calculated from TDS and MDS were analyzed. The correlation between SQIs and yield was analyzed to investigate whether the SQI based on MDS correlated with yield. Sensitivity (S) was calculated as follows:

$$S = SQI_{max}/SQI_{min} \quad (5)$$

Where, SQI_{max} and SQI_{min} are the maximum and minimum SQIs observed under each index procedure (Masto et al., 2008). To further describe relations between MDS and yield, multiple regressions using the PCs as independent variables and soybean yield as dependent variable were performed.

5. *Statistical analysis*

All statistical analyses (PCA, regression equations) were conducted with SPSS 17.0 software and Microsoft Excel. Correlation analysis was conducted to identify relationships between the measured parameters. The tests were performed at the 0.05 significance level or 0.01 significance level.

2.7.3 Results and Discussion

1. *Descriptive statistics of soil characteristics and yield*

Mean soil characteristic values and yields are summarized in Table 1. Silt was the dominant textural component, accounting for 535 g/kg, followed by 260 and 206 g/kg of clay and sand, respectively. Soil pH ranged from 5.9 to 7.6, with a mean and variation coefficient of 6.5 and 5.1, respectively. The soils were considered rich in organic matter, with mean SOM values of 48.0 g/kg. This was considerably higher than for soils in other areas of China, including Hainan, Anhui, and Sichuan Provinces where SOM values averaged 20, 23, and 12 g/kg, respectively (Cao and Zhou, 2008). However, the maximum (78.5 g/kg), minimum (22.5 g/kg) and mean SOM values in Hailun County confirmed that SOM had decreased significantly since the 1980s when SOM averaged 60.6 g/kg. The TN, TP, and TK of Hailun soils averaged at 2.2, 0.85, and 21.5 g/kg, respectively. AvP and AvK also had high values, averaging 37.7 and 159 mg/kg, respectively. For comparison, AvP in most soils in China usually ranges from 5 to 30 mg/kg (Zhu, 2010). The high nutrient and SOM levels confirmed that farmland in Hailun County was indeed quite fertile.

The average soybean yield for Hailun County was 2438 kg/ha with a range from 1608 to 3210 kg/ha. Land in the middle of the County generally had the highest yields, while yields in the North and South were lower. The highest average yield (2780 kg/ha) was recorded near Qianjin Town in the middle area, while the lowest average (2084 kg/ha) was recorded in Fengshan Town in south. Low yields were correlated with low TN and AvP and high sand content, presumably reflecting less plant available and water holding capacity.

Table 1 Descriptive statistics of soil characteristics in the site studied and the soybean yield for Hailun County

Property[①]	Mean	Median	Standard error	Minimum	Maximum	Coefficient of variation (%)
Sand (g/kg)	206	203	80	48	486	38.9
Silt (g/kg)	535	543	98	22	724	18.2
Clay (g/kg)	260	262	95	65	372	36.7
pH	6.5	6.5	0.3	5.9	7.6	5.1
CEC (cmol/kg)	32.8	32.6	3.2	27.1	43.0	9.7
SOM (g/kg)	48.0	47.4	10.0	22.5	78.5	20.9
Bd (g/cm^3)	1.04	1.02	0.23	0.81	1.27	22.4
TN (g/kg)	2.2	2.2	0.5	1.1	3.9	21.3
TP (g/kg)	0.85	0.85	0.25	0.41	1.96	28.9
TK (g/kg)	21.5	21.3	1.1	19.0	24.6	5.3
AvP (mg/kg)	37.7	32.5	19.9	10.2	105.0	52.7
AvK (mg/kg)	159	146	46	108	317	28.8

Continued

Property①	Mean	Median	Standard error	Minimum	Maximum	Coefficient of variation (%)
exCa (g/kg)	8.02	7.68	2.29	5.24	15.1	28.6
exMg (mg/kg)	944	932	200	406	1 500	21.2
AvFe (mg/kg)	81.8	78.7	32.8	21.1	206.0	40.1
AvMn (mg/kg)	42.7	44.8	12.9	10.6	69.7	30.4
AvCu (mg/kg)	1.82	1.79	0.28	1.14	2.69	15.4
AvZn (mg/kg)	1.70	1.44	0.87	0.37	3.96	50.8
AvB (mg/kg)	0.57	0.55	0.18	0.21	1.12	32.1
AvMo (mg/kg)	0.19	0.19	0.06	0.05	0.45	33.7
Soybean yield (kg/ha)	2 438	2 406	341	1 608	3 210	14.0

Notes: ①CEC = cation exchange capacity; SOM = soil organic matter; Bd = bulk density; TN = total nitrogen; TP = total phosphorus; TK = total potassium; AvP = available phosphorus; AvK = available potassium; exCa = exchangeable calcium; exMg = exchangeable magnesium; AvFe = available Iron; AvMn = available manganese; AvCu = available copper; AvZn = available zinc; AvB = available boron; AvMo = available molybdenum.

2. Indicator selection

Soil quality indicators can be classified as either inherent or dynamic. Inherent indicators reflect the soil forming factors of climate, parent material, time, topography, and biota. Dynamic indicators describe soil conditions due to recent land use or management decisions. Dynamic indicators are used to assess how soil management decision affects use-dependent soil properties (Wienhold et al., 2004). Our approach established a MDS from soil quality indicators that were able to mainly show effects as a result of agriculture management practices. Therefore, as many as 20 physical and chemical indicators were chosen for our TDS.

The 20 soil quality properties considered in the PCA were grouped into components. The first seven PCs had eigenvalues >1, each explaining at least 5% of the data variation and accounting for 74.9% of the total variance (Table 2). The first PC explained 22.3% of the variation, whereas PC2 and PC3 each explained more than 10% of the variation. The final four PCs accounted for 6.02 % to 7.81% of the variation.

Communalities for the soil properties indicated that the seven components explained > 90% of the variance in silt, clay, SOM and TN, as well as 80% of variance in CEC, AvK, exCa, and AvFe . The seven components explained variance ranging from 50% to 70% in other indicators, except for Bd, which exhibited generally weaker relationships with other characteristics. The results showed that the variability of most characteristics was well described.

The highly weighted loadings were ranked from high to low under each PC. Then, the norm values were calculated using Eq. (1). Within each PC, indicators receiving norm values within 10% of the highest value were selected for the MDS. The Pearson correlation test was used to examine the correlation among indicators to reduce redundancy (Table 3). Under PC1, CEC had the highest norm value at 1.91, SOM, TN, and exCa had values within 10% of this value, but CEC, SOM, TN, and exCa were significantly correlated with each other, thus, CEC was maintained in the MDS. Similarly, AvFe and AvCu were highly weighted under PC2, but were well correlated with each other. AvFe had a higher norm value and was thus retained. Under

Table 2 Results of principal component analysis (PCA) of soil quality indicators

Property[1]	PC1	PC2	PC3	PC4	PC5	PC6	PC7	Norm	Communality
Sand	0.204	-0.464	-0.255	-0.274	0.355	0.508[2]	0.030	1.25[4]	0.782
Silt	0.240	0.522	0.133	-0.534[2]	-0.476	-0.338	-0.007	1.39[3]	0.973
Clay	-0.425	-0.174	0.069	0.798[2]	0.215	-0.059	-0.018	1.40[4]	0.903
pH	0.433	-0.457	0.103	-0.056	0.004	0.299	0.181	1.26[3]	0.532
CEC	0.898[2]	0.031	-0.078	0.089	0.102	-0.112	-0.005	1.91[4]	0.844
SOM	0.737[2]	0.395	-0.346	0.177	0.012	0.004	0.313	1.81[3]	0.949
Bd	-0.124	0.262	-0.227	0.350	-0.229	-0.089	-0.304	0.87[3]	0.411
TN	0.771[2]	0.405	-0.245	0.140	0.095	-0.069	0.317	1.84[3]	0.953
TP	-0.077	-0.344	0.124	0.282	0.029	-0.470	0.560[2]	1.08[4]	0.754
TK	-0.500[2]	-0.146	0.420	-0.114	0.040	-0.361	0.205	1.35	0.635
AvP	-0.046	0.346	0.571[2]	0.044	-0.184	0.467	0.296	1.24[4]	0.789
AvK	0.193	0.014	0.776[2]	0.160	-0.196	0.264	0.198	1.34[3]	0.812
exCa	0.821[2]	-0.074	0.093	0.018	0.281	-0.281	-0.179	1.82[3]	0.879
exMg	0.394	0.016	0.592[2]	0.102	0.275	-0.100	-0.391	1.35[4]	0.755
AvFe	-0.226	0.712[2]	-0.243	0.043	0.348	0.168	0.251	1.42[4]	0.831
AvMn	-0.599[2]	0.388	-0.324	0.057	0.180	0.032	0.026	1.52	0.652
AvCu	0.098	0.514[2]	0.291	0.076	0.319	0.080	-0.337	1.12	0.586
AvZn	0.038	0.332	0.510[2]	-0.158	0.482	-0.114	0.100	1.15	0.652
AvB	0.308	0.237	0.061	0.453	-0.517[2]	0.189	-0.147	1.17[4]	0.684
AvMo	-0.628[2]	0.446	0.061	-0.092	0.100	0.029	0.047	1.53	0.618
Eigenvalue	4.46	2.67	2.33	1.56	1.46	1.30	1.20		
Percent	22.3	13.4	11.7	7.81	7.32	6.52	6.02		
Cumulative percent	22.3	35.7	47.3	55.1	62.4	68.9	74.9		

Notes: ①CEC = cation exchange capacity; SOM = soil organic matter; Bd = bulk density; TN = total nitrogen; TP = total phosphorus; TK = total potassium; AvP = available phosphorus; AvK = available potassium; exCa = exchangeable calcium; exMg = exchangeable magnesium; AvFe = available Iron; AvMn = available manganese; AvCu = available copper; AvZn = available zinc; AvB = available boron; AvMo = available molybdenum. ②Factor loadings considered highly weighted. ③Norms correspond to the indicators have norm in 10% of the highest norm. ④Norms correspond to the indicators included in the minimum data set.

Table 3 Correlation coefficients among soil indicators

Property[1]	Sand	Silt	Clay	CEC	SOM	TN	TP	TK	AvP	AvK	exCa	exMg	AvFe	AvMn	AvCu	AvZn	AvB
Silt	-0.438[2]	-															
Clay	-0.392[2]	-0.655[2]	-														
CEC	0.192	0.130	-0.293[2]	-													
SOM	0.046	0.208	-0.252[3]	0.712[2]	-												
TN	0.083	0.226[3]	-0.301[2]	0.771[2]	0.955[2]	-											
TP	-0.074	-0.156	0.222[3]	0.022	0.057	0.094	-										
TK	-0.106	-0.014	0.104	-0.391[2]	-0.474[2]	-0.404[2]	0.303[2]	-									
AvP	-0.085	0.108	-0.039	-0.129	0.011	0.038	-0.058	0.113	-								
AvK	-0.161	0.080	0.053	0.033	-0.031	-0.016	0.058	0.126	0.558[2]	-							
exCa	0.176	0.087	-0.240[3]	0.697[2]	0.542[2]	0.639[2]	0.084	-0.283[2]	-0.130	-0.056	-						
exMg	-0.026	0.032	-0.011	0.169	-0.056	0.000	-0.039	0.022	0.156	0.407[2]	0.168	-					
AvFe	-0.101	0.037	0.046	-0.206	0.200	0.153	-0.168	-0.006	0.149	-0.141	-0.318[2]	-0.073	-				
AvMn	-0.193	-0.054	0.217	-0.506[3]	-0.237[3]	-0.272[3]	-0.056	0.121	0.008	-0.292[2]	-0.480[2]	-0.223[3]	0.478[2]	-			
AvCu	-0.090	0.090	-0.020	0.155	-0.033	0.010	-0.220[3]	-0.021	0.168	0.160	0.013	0.244[3]	0.314[2]	0.060	-		
AvZn	-0.109	0.132	-0.048	-0.055	-0.066	0.020	0.054	0.206	0.253[3]	0.192	-0.025	0.327[2]	0.204	0.041	0.258[3]	-	
AvB	-0.195	0.113	0.050	0.246[3]	0.295[2]	0.234[2]	0.002	-0.278[3]	0.214	0.189	0.057	0.068	-0.069	-0.177	0.133	-0.105	-
AvMo	-0.291[2]	0.066	0.178	-0.525[2]	-0.213[3]	-0.226[2]	-0.003	0.197	0.218[3]	-0.106	-0.274[2]	-0.057	0.472[2]	0.408[2]	0.030	0.199	-0.095

Notes: ①CEC = cation exchange capacity; SOM = soil organic matter; TN = total nitrogen; TP = total phosphorus; TK = total potassium; AvP = available phosphorus; AvK = available potassium; exCa = exchangeable calcium; exMg = exchangeable magnesium; AvFe = available Iron; AvMn = available manganese; AvCu = available copper; AvZn = available zinc; AvB = available boron; AvMo = available molybdenum. ②、③ Significant at $P < 0.05$ and $P < 0.01$, respectively.

PC3, Avk was well correlated with AvP and exMg, but AvP and exMg were not well correlated with each other, so both AvP and exMg were kept in the MDS under PC3. Silt was correlated well with clay, and clay had a higher norm value. Thus, clay was retained under PC4. PC5, PC6, and PC7 had one highly weighted variable each, which were all retained. Thus, AvB, sand, and TP were kept under PC5, PC6, and PC7, respectively. Finally, no coefficients (Table 3) between PC indicators exceeded 0.7, so no more variables were excluded. The pH was eliminated because the range of values for the property within the study area was insufficient to result in substantial differences in plant growth as the coefficient of variation was 5.06%, which was less than the threshold of sensitivity at 10% (Xu et al., 2005). Similar results were reported by Andrews et al. (2003). Bd with a low communality indicated that much of this indicator's variance remained unexplained. The indicator was thus excluded. The variables finally retained in MDS were sand, clay, CEC, TP, AvP, exMg, AvFe, and AvB.

Several indicators used with the MDS method could be found in previously created MDS indicator groups (Doran and Parkin, 1994; Karlen et al., 1997; Andrews et al., 2002b; Qi et al., 2009). This is not unusual since AvP constituted nucleic acids, nucleoprotein, enzymes, and other important organic compounds within soybean (Li, 2002). Phosphorus deficiency therefore has a significant influence on the formation of phosphorus metabolites, which could have serious implications on soybean growth and thereby reduce soybean grain yield (Fatima et al., 2006). CEC influences the habitat for soil organisms and loss of surface CEC could lead to increased leaching losses of nutrients throughout the profile (Wang and Gong, 1998; Page-Dumroese et al., 2000). Inclusion of micronutrients, such as AvFe, helped farmers and scientists understand the importance of a more balanced view of soil nutrition not solely focused on N, P, and K.

3. *Indicator scoring*

The variables were transformed using linear scoring functions. After deciding the shape of the anticipated response ("more is better", "less is better" or "optimum"), the limits or threshold values were assigned for each indicator. The "more is better" function was applied to most of the variables based on their roles in nutrient availability. Soil pH was evaluated using an "optimum" curve, but applying a "more is better" relationship when pH values were less than 7 or "less is better" for higher values. A "less is better" curve was applied to Bd and sand content because high Bd values have an inhibitory effect on plant root growth and soil organisms (Andrews et al., 2004). The scores were then calculated by using Eq. (2) and Eq.(3).

4. *Calculation and classification of SQI*

The indicators in the TDS or MDS were weighted based on results of the mathematical statistics (Table 4). Indicators had weights relatively homogeneous based on communality, silt, clay, SOM, and TN had higher weights, whereas Bd had the lowest weight in the TDS. The indicators had weights from 0.1 to 0.14 in the MDS. The weights in the MDS based on variation had a large range, with the largest value being 0.258 and the lowest being 0.075. Also, since the indicators were classified into different PCs, the amount (%) of variation in the

TDS explained by different PCs varied. CEC was chosen in PC1 and this PC explained the largest amount of variation, but TP was under PC7 and explained the smallest amount of variation. The weight of indicators in the MDS based on norm value had similar trend to the weight on variation. The weight of indicators in the MDS based on regression had highest weight in AvB (0.397), which was much higher than that of others and was six times more than the lowest value of CEC (0.063). The enormous difference was explained by the fact that different indicators had different effects on soybean yield. Surprisingly, AvB played the most important role in the yield.

Table 4 Weight value of each soil quality indicator in the total data set (TDS) and minimum data set (MDS) indicator methods

Property[①]	TDS	MDS			
	Communality	Communality	Variation	Norm	Regression
Sand	0.052	0.118	0.075	0.115	0.110
Silt	0.065				
Clay	0.060	0.132	0.090	0.130	0.084
Ph	0.035				
CEC	0.056	0.129	0.258	0.177	0.063
SOM	0.063				
Bd	0.027				
TN	0.064				
TP	0.05	0.147	0.07	0.100	0.118
TK	0.042				
AvP	0.053	0.144	0.135	0.114	0.069
AvK	0.054				
exCa	0.059				
exMg	0.05	0.095	0.135	0.125	0.07
AvFe	0.055	0.109	0.154	0.131	0.088
AvMn	0.043				
AvCu	0.039				
AvZn	0.043				
AvB	0.046	0.126	0.085	0.108	0.397
AvMo	0.041				

Note: ① CEC = cation exchange capacity; SOM = soil organic matter; Bd = bulk density; TN = total nitrogen; TP = total phosphorus; TK = total potassium; AvP = available phosphorus; AvK = available potassium; exCa = exchangeable calcium; exMg = exchangeable magnesium; AvFe = available Iron; AvMn = available manganese; AvCu = available copper; AvZn = available zinc; AvB = available boron; AvMo = available molybdenum.

The SQIs were calculated using Eq. (4) and then ranked with the grades of each method shown in Table 5, which showed that classification criteria for SQI based on TDS (SQI-TDS), SQI based on MDS (SQI-MDS) calculated from norm (SQI-MDS$_{norm}$), SQI-MDS calculated from communality (SQI-MDS$_{communality}$), and SQI-MDS calculated from regression (SQI-MDS$_{regression}$) were almost the same. However, the threshold value for the same grade of SQI-MDS calculated from variation (SQI-MDS$_{variation}$) was lower than that of the others. Comparing results in Table 5, soil sites within Grades I and V in the TDS method were fewer than that in the MDS methods, but had a larger proportion of Grade Ⅲ except for SQI-MDS$_{regression}$. Although threshold values for corresponding grades of SQI-MDS$_{variation}$ method were lower than that of others, the trend was consistent. Grade Ⅲ was the dominant grade in the first four methods. The proportion of different grades in the SQI-MDS$_{regression}$ significantly differed from the other methods. Grades Ⅱ and Ⅲ were the dominant grades and had similar proportions, accounting for 55.5%. Grade Ⅰ accounted for 12% and had the largest proportion in corresponding grade in all methods. Though the grades of different methods were not consistent, the distribution of the same grade of SQI was similar. High quality lands were primarily distributed in middle area of the county and Hailun farm in northeast. Soil quality in the southwest area was relatively low.

Table 5 The criteria and grades of soil quality indices (SQI) in different methods

Index[①]		Grade				
		Ⅰ	Ⅱ	Ⅲ	Ⅳ	Ⅴ
TDS	Range[②]	<0.331	0.331-0.425	0.425-0.520	0.520-0.614	>0.614
	n[③]	2	23	40	18	4
	Percent[④] (%)	2.3	26.4	46.0	20.7	4.6
MDS$_{norm}$	Range	<0.330	0.330-0.425	0.425-0.519	0.519-0.614	>0.614
	n	6	21	32	22	7
	Percent (%)	6.8	23.9	36.3	25.0	8.0
MDS$_{communality}$	Range	<0.330	0.330-0.425	0.425-0.519	0.519-0.613	>0.613
	n	9	18	32	21	8
	Percent (%)	10.2	20.5	36.3	23.9	9.1
MDS$_{variation}$	Range	<0.320	0.320-0.411	0.411-0.503	0.503-0.594	>0.594
	n	6	21	35	18	8
	Percent (%)	6.8	23.9	39.8	20.4	9.1
MDS$_{regression}$	Range	<0.331	0.331-0.425	0.425-0.520	0.520-0.614	>0.614
	n	11	25	23	15	14
	Percent (%)	12.5	28.4	26.1	17.1	15.9

Notes: ①TDS = soil quality index (SQI) based on total data set (TDS); MDS$_{norm}$ = SQI based on minimum data set (MDS) calculated from norm; MDS$_{communality}$ = SQI-MDS calculated from communality; MDS$_{variation}$ = SQI-MDS calculated from variation; MDS$_{regression}$ = SQI-MDS calculated from regression.②Range of value of SQI in the corresponding grade.③Number of sample sites. ④Percentage of the sample sites of the corresponding grade in the total sample sites.

5. Evaluation of index methods

As is shown in Fig. 1, SQI-TDS had significant correlations with SQI-MDS$_{variation}$ (R^2 = 0.726), SQI-MDS$_{norm}$ (R^2 = 0.706), SQI-MDS$_{communality}$ (R^2 = 0.663) and SQI-MDS$_{regression}$ (R^2 = 0.415), respectively. The correlation coefficient between SQI-TDS and SQI-MDS$_{regression}$ was the lowest because the weight of the other three MDS and TDS methods were all based on the statistics of PCA, whereas the SQI-MDS$_{regression}$ was based on regression.

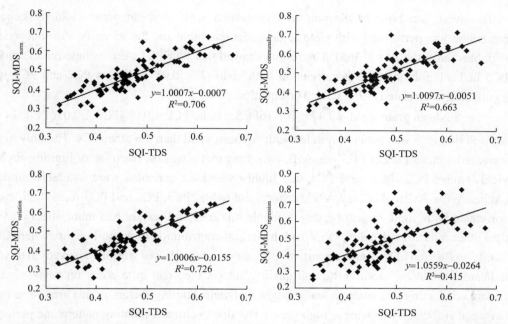

Fig. 1 The linear relationships between SQI-TDS (soil quality index (SQI) based on total data set (TDS)) and the values of SQI-MDS (SQI based on minimum data set (MDS)) calculated from norm (SQI-MDS$_{norm}$), communality (SQI-MDS$_{communality}$), variation (SQI-MDS$_{variation}$) and regression (SQI-MDS$_{regression}$)

Correlations between SQIs and soybean grain yield were evaluated. Among all methods, the method based on MDS$_{regression}$ had the maximum correlation coefficient, followed by MDS$_{communality}$, MDS$_{norm}$, MDS$_{variation}$, and TDS (Table 6). Sensitivity was calculated using Eq. (5). The sensitivities of different index methods diminished in the order MDS$_{regression}$ > MDS$_{norm}$ > MDS$_{variation}$ > MDS$_{communality}$ > TDS (Table 6). Thus, SQI-MDS$_{regression}$ was more sensitive and correlated better with soybean grain yield. Although the classification criteria of SQI-TDS, SQI-MDS$_{norm}$, SQI-MDS$_{communality}$ and SQI-MDS$_{regression}$ were almost the same, SQI-TDS had a narrower range and thus had the lowest sensitivity, whereas the SQI had a larger range and the highest sensitivity. Usually, more indicators represent soil quality more comprehensively. However, the method based on TDS showed the least sensitivity and worst correlation with soybean grain yield in this study. Indicators not included in the MDS could not indicate soil quality correctly as some indicators even resulted in reverse results.

Table 6 Evaluation of the index and indicator scoring methods applied for developing soil quality indices

Index[1]	Range	Sensitivity	r[2]
TDS	0.327-0.668	2.04	0.391[3]
MDS$_{norm}$	0.264-0.672	2.54	0.478[3]
MDS$_{communality}$	0.287-0.702	2.45	0.511[3]
MDS$_{variation}$	0.254-0.637	2.51	0.423[3]
MDS$_{regression}$	0.237-0.785	3.31	0.658[3]

Notes: [1]TDS = soil quality index (SQI) based on total data set (TDS); MDS$_{norm}$ = SQI based on minimum data set (MDS) calculated from norm; MDS$_{communality}$ = SQI based on MDS calculated from communality; MDS$_{variation}$ = SQI based on MDS calculated from variation; MDS$_{regression}$ = SQI based on MDS calculated from regression. [2]Correlation coefficient between SQI and soybean grain yield. [3] Significant at $P < 0.01$.

To investigate how MDS indicators correlated with soybean grain yield, a stepwise regression was performed with yield as dependent variable and the seven PCs as independent variables (adjusted $R^2 = 0.365$). The results showed that four PCs were included in the model. PC5 had a highly significant correlation with yield ($P < 0.01$), whereas PC2 and PC3 were significantly correlated at $P < 0.05$. The equation

$$\text{Soybean grain yield} = 2.471 - 0.116\text{PC5} + 0.085\text{PC2} + 0.084\text{PC3} + 0.076\text{PC4} \quad (6)$$

showed that PC5 was better correlated with soybean yield than the other PCs. The only highly weighted indicator under PC5 was AvB, indicating that it was the main factor limiting soybean yield. Under PC2, PC3, and PC4, the highly weighted variables were available nutrients (AvFe, AvCu, AvZn, AvK and AvP), texture, and exMg. PC5, PC2 and PC3 were significantly correlated with yield, indicating that available nutrients and texture had more effect on yield than other indicators. However PC1 had the largest contribution to the soil quality where CEC, SOM, exCa and TN were important indicators. Overall, our results showed that some indicators might be important to soil quality, but yet they had little effect on soybean yield. Before conducting this study, it was thought that yield-limiting factors in this area were being managed sufficiently to achieve high yields. But due to climate, planting pattern and perhaps a long-time misunderstanding, the limiting factors of our study have became more and more important to the soybean yield of this area.

Boron is an essential trace element for plant growth and has a significant influence on soybean production development, metabolism, and resistance (Goldbach, 1997). Lack of boron inhibits cell elongation and division, hinders root elongation, and affects the absorption of other nutrient elements (Stangoulis et al., 2000). Extremely low boron content can reduce soybean yield (Liu and Yang, 2000). The average content of AvB in Hailun County was 0.57 mg/kg (Table 1), which was higher than the critical value of AvB (0.5 mg/kg) in China used to decide whether boron fertilizer is needed. The black soil of this area was traditionally considered fertile, so fertilizer containing boron or other micronutrients was seldom used by farmers. This study indicated that AvB was a key factor to determine soybean yield in Hailun County and that boron fertilizer was urgently needed in this area to improve soybean yield.

An appropriate amount of phosphorous is also indispensable for soybean growth and yield formation (Tsvetkova and Georgiev, 2003). The soil of this area was one of the most fertile soils of China, and soil phosphorus content was generally significantly higher than that found in other regions. Soil phosphorus content in this region was also much higher than the value 20 years ago because of the high amount of ammonium phosphate that has been used for many years. Significant accumulation of soil phosphorus in Northeast China might lead to water pollution. More phosphorus was not recommended to prevent unintended environmental problems. However, the results of this study showed that soil phosphorus content had remarkable effect on soybean yield. More phosphorus was apparently needed because low soil temperatures can prevent phosphorus movement from bulk soil to the soybean roots. How much phosphate fertilizer should be applied to soil of this area is still a question need to be answered in future.

Potassium is also one of the most important nutrient elements for plant growth and is particularly important for soybean. Appropriate potassium fertilizer use could improve the photosynthesis rate of soybeans (Zheng, 2001). The soils of Hailun County had high AvK content with an average value of 159 mg/kg (Table 1). The soil of Northeast China was supposed to be rich in potassium and could meet the needs of plants, but K fertilizer has been routinely applied in this region for many years. The results of this study showed that soil AvK was correlated with soybean yield and that sites with AvK content exceeding 200 mg/kg had the highest yields. This could be due to the higher potassium supply improving soybean resistance to many kinds of stress including flooding, wind, cold, insects and diseases that affect soybean yield in this region. Thus it was necessary to apply some potassium fertilizer to maintain both higher soil potassium fertility and higher soybean yield.

Magnesium, which is also an essential crop nutrient, also has a significant effect on soybean growth and physiological metabolism. Soybean yield has increased significantly with the application of increasing amounts of NPK fertilizer, but the original soil nutrient balance is changing and thus causing a gradual decrease in soil magnesium content. Therefore, application of magnesium fertilizer to maintain soil magnesium element balance was an important measure to maintain soybean yield (Wu, 1998).

As a whole, this study showed that even though Hailun County soils were fertile and some indicators were much higher than normal criterion, more nutrients were needed to keep soybean yields high and stable because of the cold and dry weather in this region. The results were also considered reasonable and believable because they were based on an analysis of both soil quality indicators and soybean yield. The study also demonstrated that a MDS combined with regression analysis could not only appropriately assess soil quality but also be used as a powerful tool for diagnosing soil nutrient needs at the regional level.

References

Andrews S S, Carroll C R. 2001. Designing a soil quality assessment tool for sustainable agroecosystem management. Ecol. Appl., 11: 1573-1585

Andrews S S, Flora C B, Mitchell J P, et al. 2003. Growers' perceptions and acceptance of soil quality indices. Geoderma., 114: 187-213

Andrews S S, Karlen D L, Cambardella C A. 2004. The soil management assessment framework: a quantitative soil quality evaluation method. Soil Sci. Soc. Am. J., 68: 1945-1962

Andrews S S, Karlen D L, Mitchell J P. 2002a. A comparison of soil quality indexing methods for vegetable production systems in northern california. Agr. Ecosyst. Environ., 90: 25-45

Andrews S S, Mitchell J P, Mancinelli R, et al. 2002b. On-farm assessment of soil quality in California's central valley. Agron. J., 94: 12-23

Arshad M A, Martin S. 2002. Identifying critical limits for soil quality indicators in agro-ecosystems. Agr. Ecosyst. Environ., 88: 153-160

Bhardwaj A K, Jasrotia P, Hamilton S K, et al. 2011. Ecological management of intensively cropped agro-ecosystems improves soil quality with sustained productivity. Agr. Ecosyst. Environ., 140: 419-429

Borůvka L, Mládková L, Penížek V, et al. 2007. Forest soil acidification assessment using principal component analysis and geostatistics. Geoderma., 140: 374-382

Brejda J J, Karlen D L, Smith J L, et al. 2000. Identification of regional soil quality factors and indicators: II. Northern Mississippi loess hills and Palouse prairie. Soil Sci. Soc. Am. J., 64: 2125-2135

Cao Z H, Zhou J M. 2008. Soil Quality of China. Beijing: Science Press (in Chinese)

de Lima A C R, Hoogmoed W, Brussaard L. 2008. Soil quality assessment in rice production systems: establishing a minimum data set. J. Environ. Qual., 37: 623-630

Desbiez A, Matthews R, Tripathi B, et al. 2004. Perceptions and assessment of soil fertility by farmers in the mid-hills of Nepal. Agr. Ecosyst. Environ., 103: 191-206

Diack M, Stott D E. 2001. Development of a soil quality index for the chalmers silty clay loam from the midwest USA. In: Stott D E, Mohtar R H, Steinhardt G C. The Global Farm. Selected papers from the 10th International Soil Conservation Meeting held on May 24-29, 1999 at Purdue University and the USDA-ARS National Soil Erosion Research Laboratory. USA: 550-555

Ditzler C A, Tugel A J. 2002. Soil quality field tools of USDA-NRCS soil quality institute. Agron. J.,94: 33-38

Doran J W, Parkin T B. 1994. Defining and assessing soil quality. In: Doran J W, Coleman D C, Bezdicek D F, et al. Defining Soil Quality for a Sustainable Environment. SSSA Special Publication 35. Madison: Soil Science Society of America Inc.: 3-21

Fatima Z, Zia M, Chaudhary M F. 2006. Effect of rhizobium strains and phosphorus on growth of soybean (*glycine max*) and survival of rhizobium and P solubilizing bacteria. Pak. J. Bot., 38: 459-464

Goldbach H E 1997. A critical review on current hypotheses concerning the role of boron in higher plants: suggestions for further research and methodological requirements. J. Trace Microprobe T., 15: 51-91

Huang B, Shi X Z, Yu D S, et al. 2006. Environmental assessment of small-scale vegetable farming systems in peri-urban areas of the Yangtze River Delta Region, China. Agr. Ecosyst. Environ., 112: 391-402

Hussain I, Olson K R, Wander M M, et al. 1999. Adaptation of soil quality indices and application to three tillage systems in southern illinois. Soil Till. Res., 50: 237-249

Imaz M J, Virto I, Bescansa P, et al. 2010. Soil quality indicator response to tillage and residue management on semi-arid Mediterranean cropland. Soil Till. Res., 107: 17-25

Karlen D L, Andrews S S, Doran J W. 2001. Soil quality: current concepts and applications. Adv. Agron., 74: 1-40

Karlen D L, Mausbach M J, Doran J W,et al. 1997. Soil quality: a concept, definition, and framework for evaluation. Soil Sci. Soc. Am. J., 61: 4-10

Karlen D L, Parkin T B, Eash N S. 1996. Use of soil quality indicators to evaluate conservation reserve program sites in Iowa. In: Doran J W, Jones A J. Methods for Assessing Soil Quality. Madison: Soil Science Society of America Inc.: 345-355

Karlen D L, Tomer M D, Neppel J,et al. 2008. A preliminary watershed scale soil quality assessment in north central Iowa, USA. Soil Till. Res., 99: 291-299

Klingebiel A A, Montgomery P H. 1961. Land Capability Classification. United States Department of Agriculture, Washington, DC.

Larson W E, Pierce F J. 1994. The dynamics of soil quality as a measure of sustainable management. In: Doran J W, Coleman D C, Bezdicek D F, et al. Defining Soil Quality for A Sustainable Environment. Madison: Soil Science Society of America Inc.: 37-51

Li G L, Chen J, Sun Z Y, et al. 2007. Establishing a minimum dataset for soil quality assessment based on soil properties and land-use changes. Acta Ecol. Sin., 27: 2715-2724 (in Chinese)

Li X M. 2002. Study the rate of soybean absorbing the phosphorus in different soil types. Soybean Sci., 21: 75-77 (in Chinese)

Liebig M A, Varvel G, Doran J. 2001. A simple performance-based index for assessing multiple agroecosystem functions. Agron. J., 93: 313-318

Liu P, Yang Y. 2000. Effects of molybdenum and boron on membrane lipid peroxidation and endogenous protective systems of soybean leaves. Acta Bot. Sin., 42: 461-466 (in Chinese)

Lu R K. 2000. Soil Analytical Methods of Agronomic Chemical. Beijing: China Agricultural Science and

Technology Press (in Chinese)

Ma Q, Yu W T, Zhao S H,et al. 2004. Comprehensive evaluation of cultivated black soil fertility. Chin. J. Appl. Ecol., 15: 1916-1920 (in Chinese)

Masto R E, Chhonkar P K, Singh D, et al. 2007. Soil quality response to long-term nutrient and crop management on a semi-arid inceptisol. Agric. Ecosyst. Environ., 118: 130-142

Masto R E, Chhonkar P K, Singh D,et al. 2008. Alternative soil quality indices for evaluating the effect of intensive cropping, fertilisation and manuring for 31 years in the semi-arid soils of india. Environ. Monit. Assess., 136: 419-435

Olsen S R, Cole C V, Watanabe F S, et al. 1954. Estimation of Available Phosphorous in Soils by Extraction with Sodium Bicarbonate. USDA Circular 939. U.S. Government Ptinting Office, Washington, D.C.

Page-Dumroese D, Jurgensen M, Elliot W, et al. 2000. Soil quality standards and guidelines for forest sustainability in northwestern North America. Forest Ecol. Manag., 138: 445-462

Qi Y B, Darilek J L, Huang B, et al. 2009. Evaluating soil quality indices in an agricultural region of Jiangsu Province, China. Geoderma., 149: 325-334

Rezaei S A, Gilkes R J, Andrews S S, et al. 2005. Soil quality assessment in semiarid rangeland in Iran. Soil Use Manage., 21: 402-409

Schipper L A, Sparling G P. 2000. Performance of soil condition indicators across taxonomic groups and land uses. Soil Sci. Soc. Am. J., 64: 300-311

Sharma K L, Mandal U K, Srinivas K, et al. 2005. Long-term soil management effects on crop yields and soil quality in a dryland Alfisol. Soil Till. Res., 83: 246-259

Stangoulis J C R, Webb M J, Graham R D. 2000. Boron efficiency in oilseed rape: II. Development of a rapid lab-based screening technique. Plant Soil, 225: 253-261

Tsvetkova G E, Georgiev G I. 2003. Effect of phosphorus nutrition on the nodulation, nitrogen fixation and nutrient-use efficiency of bradyrhizobium japonicum-soybean (*glycine max* L. Merr.) symbiosis. Bulg. J. Plant Physiol, Special Issue: 331-335

Veihmeyer F J, Hendrickson A H. 1948. Soil density and root penetration. Soil Sci., 65: 487-493

Wander M M, Bollero G A. 1999. Soil quality assessment of tillage impacts in illinois. Soil Sci. Soc. Am. J., 63: 961-971

Wang X J, Gong, Z T. 1998. Assessment and analysis of soil quality changes after eleven years of reclamation in subtropical China. Geoderma., 81: 339-355

Wienhold B J, Andrews S S, Karlen D L. 2004. Soil quality: a review of the science and experiences in the USA. Environ. Geochem. Health., 26: 89-95

Wienhold B J, Karlen D L, Andrews S S,et al. 2009. Protocol for indicator scoring in the soil management assessment framework. Renew. Agr. Food Syst. 24: 260-266

Wu Y. 1998. Function of magnesiun in soybean nutrition. Soybean Sci., 17: 162-165 (in Chinese)

Xu M X, Liu G B, Zhao Y G. 2005. Assessment indicators of soil quality in hilly Loess Plateau. Chin. J. Appl. Ecol., 16: 1843-1848 (in Chinese)

Xue Y J, Liu S G, Hu Y M, et al. 2010. Soil quality assessment using weighted fuzzy association rules. Pedosphere, 20: 334-341

Yemefack M, Jetten V G, Rossiter D G. 2006. Developing a minimum data set for characterizing soil dynamics in shifting cultivation systems. Soil Till. Res., 86: 84-98

Zheng S Q. 2001. Effect of potassium on the physiology, yield and quality of soybean. Heilongjiang Agr. Sci., 4: 25-27 (in Chinese)

Zhu B S. 2010. Study on cultivated soil quality evaluation in Hailun of Heilongjiang Province. Xi'an: Southwest University Master Thesis (in Chinese)

2.8 Contributions of greenhouse soil nutrients accumulation to the formation of the secondary salinization

2.8.1 Introduction

Greenhouse agriculture provides vegetables, fruit and flowers to consumers even in the seasons that are not appropriate for their growth. It is also an important economic source for lower-income Chinese farmers. Although Chinese greenhouse agriculture can be dated back to 2000 years ago, real development was from the 1970s (Sun, 1999). Because greenhouse agriculture can play an important role in solving the contradiction between limited agricultural land and a large population in China, the greenhouse agriculture has been developed well in China, especially in recent 20 years with about 2.1 million hm^2 greenhouse vegetable land in 2004 according to 2004 China agriculture statistical report (Ministry of Agriculture, PRC, 2005). China has become the biggest greenhouse agriculture country in the whole world.

Greenhouse agriculture relies heavy on high technology and maintenance cost investment, which enables vegetables to grow normally when the climate condition is too harsh for the vegetable. The high investment and maintenance cost of the greenhouse result in the urgent demand of the higher yield and more expensive products than those from normal agricultural production. Therefore, it becomes more desirable for farmer to aggressively enhance the productivity to offset the high cost of the greenhouse agriculture. Those measures include applying more fertilizers, and planting unseasonal vegetables for the sake of profit.

The great changes of fertilization in greenhouse agricultural production can be characterized as ① the lack of appropriate education and knowledge in farmers resulting in excessive fertilization, and often imbalanced or sometimes subjective;② the plastic or glass cover of the greenhouse prevents rain water from leaching the ground soil.

Greenhouse agriculture implies high human intervention, which can affect the environment seriously (Masui et al., 1976; Merkle and Dunkle, 1944). Some apparent problems in the greenhouse agriculture in China include the frequent occurrences of diseases, the accumulation and change of water soluble salt in soil, the decrease of soil productivity, continuous cropping obstacles and high dependence on chemicals such as fertilizers, pesticides and hormones, (Xue et al., 1994; Tong and Chen, 1991; Li et al., 2001; Jiang et al., 2005). These problems not only affect the yield and quality of vegetables, but also seriously influence the environment quality and human health. Fortunately, much attention has been given to these problems and some people (Sun et al., 2005) have conducted the research on amending salinized soil.

Primary objectives of this research were: ① to use Yixing City, one of the typical Chinese greenhouse agriculture cultivation regions, as an example to investigate the status of some available nutrients in soil including micronutrients; ② to study the contribution of soil nutrients accumulation to the formation of the soil secondary salinization; ③ to provide advice on the revitalization of the secondary salinized soils and introduce measure to prevent the greenhouse soils from the formation of secondary salinization in the development of

Chinese greenhouse agriculture.

2.8.2 Materials and Methods

The sampling sites were located at Yixing City, Jiangsu Province in eastern China (31°14′ N, 119°53′ E). Mean annual temperature in this area is 15.7℃, average precipitation is about 1177 mm y/a.

Composite soil samples (7 subsamples) were collected from depths of 0-20 cm, 20-30 cm, 30-40 cm, and 40-50 cm, respectively (the depth of the underground water is about 60-80 cm) from 8 greenhouse agricultural fields and 4 neighboring open agricultural fields. The size of every greenhouse is 36 m × 6 m = 216 m^2. These greenhouses were established in 2001 and have been used to plant vegetables since August 2001. They were commonly managed by migrate workers from neighbor regions.

Soil samples were collected in December of 2005. Parts of the fresh moist soil samples were directly used to analyze soil NO_3^--N. All other soil samples were air-dried, ground with a wooden roller to pass a 1 mm sieve, homogenized and stored in plastic bags at room temperature in the dark before analysis.

Soil available nitrogen (NO_3^--N) was extracted using 2 mol/L KCl according to the method of Brenner (1965) and analyzed by continuous flow analysis (Soil Science Society of China, 2000). Soil available phosphorous (P) was extracted using 0.5 mol/L $NaHCO_3$ (pH 8.5) according to the method of Olsen et. al. (1954) and the P concentrations in the extracts were determined using the molybdenum-blue method (Soil Science Society of China, 2000). Soil rapidly and slowly available potassium (K) were respectively extracted by 1 mol/L NH_4OAc (pH 7) and 1 mol/L boiling HNO_3, and K concentrations in the extracts were analyzed by a flame photometer (Wood and de Turk, 1940; Soil Science Society of China, 2000).

For soil available calcium (Ca) and magnesium (Mg), the soils were rinsed by 1 mol/L NH_4OAc (pH 7) and the supernatant was retained to determine soil available Ca and Mg. The concentrations of Ca and Mg in the supernatant extracts were analyzed by inductively coupled plasma atomic emission spectroscopy (ICP-AES) (Soil Science Society of China, 2000). Soil available sulfur (S) was extracted by $Ca(H_2PO_4)_2$-HoAc and analyzed by spectrophotometer (Kilmer and Nearpass, 1960).

Soil available iron (Fe), manganese (Mn), copper (Cu) and zinc (Zn) were extracted by $[(HOCOCH_2)NCH_2 \cdot CH_2] \cdot NCH_2COOH$-$(HOCH_2CH_2)_3N$-$CaCL_2$ (DTPA-TEA-$CaCL_2$) and analyzed by ICP-AES according to the method of Tian and Chen (1993). Soil available boron (B) was extracted using hot water (Berger and Truog, 1944) and analyzed by ICP-AES (Soil Science Society of China, 2000); soil available molybdenum (Mo) was extracted by oxalic acid-ammonium oxalate (pH 3.3) (Agricultrual Chemistry Committee of China, 1983) and analyzed by ICP-AES.

Soil pH was measured in a 1∶2.5 soil/water suspension (Soil Science Society of China, 2000). Total water soluble salt and electrical conductivity (EC) of the soil were measured in a ratio of 1∶5 soil/water extract, the salt content in the extract was calculated from the sum of the determined ions content (Nanjing Agricultural University, 1996). Considering the soil nutrients accumulation condition, nitrate and micro elemental cations were also calculated. For the ionic composition of the salt, the Ca, Mg, K, Na, Fe, Mn, Cu, Zn, P and S were determined by ICP-AES. Since most of P and S in the extract are in the form of PO_4^{3-} and SO_4^{2-} respectively, P and S concentration in the extract were assumed to represent PO_4^{3-} and SO_4^{2-} concentration and the results were given as PO_4^{3-} and SO_4^{2-} concentration, respectively. The NO_3^- N in the extract was analyzed by continuous flow analysis, the K and Na by flame photometer, the CO_3^{2-} and HCO_3^- by the method of potentiometric titration and the Cl^- by the method of $AgNO_3$ titration (Soil Science Society of China, 2000).

Chicken manure of 2400 kg, 30 kg potassium sulfate, 10 kg urea and 50 kg fused calcium-magnesium phosphate were applied in one greenhouse every year. Phosphorous fertilizer, potash and chicken manure were used as basal fertilizers, and the urea was applied at seedling stage of the vegetable. Cucumber (*Cucumis sativus* L.) was planted in the greenhouse from August to November, and tomato (*Lycopersicon esculentum* Mill.) from December to next July.

Statistical analysis was performed with the SAS 8.2. Paired *t*-test was used to compare difference between the greenhouse soil and the open field soil.

2.8.3 Results and Discussion

1. Soil nutrients accumulation

In Chinese traditional cultivation management, fertilizer N, P and K were applied most intensively, resulting in the accumulation of a large amount of N, P and K nutrient ions in these greenhouse soils. The analytical result (Table 1) indicated that N, P and K had been significantly ($P < 0.05$) accumulated seriously in the top soil horizons (0-20 cm and 20-30 cm) in contrast to the open-field soils. After the cultivation of more than 4 years, the accumulations of N, P and K in the top horizon of greenhouse soil were averaged at 13.0, 13.0 and 1.7 fold greater than the open-field soils, and at 7.5, 8.8 and 0.5 fold greater than the open-field soils in the 20-30 cm soil, respectively. Phosphorus ($P < 0.01$) and K ($P < 0.05$) was significantly accumulated and averagely at 0.5 and 0.1 fold greater than the open-field soils in the 30-40 cm soil horizon, respectively. However, N wasn't significantly ($P > 0.05$) accumulated in this horizon. For the soil horizon of 40-50 cm, N, P and K were not significantly ($P > 0.05$) accumulated.

For the middle elements, the direct applications of Ca, Mg and S fertilizers in greenhouse cultivation practice were rare, but they were brought to the soils as the companion ions of P and K when the potassium sulfate and fused calcium magnesium phosphate were applied to the soils to meet with the growth need of the vegetable for the K and P, especially in the region of Yixing city with more serious remnant of S, Ca and Mg than the other regions resulted from frequent application of potassium sulfate and fused calcium magnesium phosphate. The analytical result (Table 2) showed that they were all significantly ($P < 0.01$) accumulated in the 0-20 cm soil horizons and averagely 0.6, 0.8 and 38.9 fold greater than the open-field soils,

Table 1 Content of soil available macro elements in different horizons of soils from greenhouses and open-fields

Nutrients	Soil Depth(cm)	No	Content(mg/kg)			T	Std Dev	CV(%)
			Minimum	Maximum	Mean			
N	0-20	GHS	117.8	382.5	228.2	5.12**	116.1	50.9
		OS	8.6	32.1	16.3		10.7	65.6
	20-30	GHS	101.8	188.3	147.2	6.11**	40.8	27.7
		OS	10.3	34	17.2		11.3	65.4
	30-40	GHS	7.4	14.1	10.6	-0.55ns	2.3	21.4
		OS	6.5	30.2	13.7		11.1	80.6
	40-50	GHS	6.0	11.3	9.1	-0.45ns	1.6	18.1
		OS	5.1	21.1	10.7		7.2	67.3
P	0-20	GHS	98.6	132.1	126.5	22.77**	13.3	10.5
		OS	8.0	8.9	9.0		0.4	4.1
	20-30	GHS	36.0	55.5	44.0	12.78**	5.9	13.5
		OS	2.9	6.3	4.5		1.8	38.9
	30-40	GHS	6.6	10.1	8.2	3.92**	1.4	16.7
		OS	4.6	5.1	5.4		0.5	10.2
	40-50	GHS	5.3	8.3	7.0	1.81ns	1.2	16.9
		OS	4.4	7.7	5.6		1.5	27.6
K	0-20	GHS	134.3	255.2	196.1	4.16**	47.9	24.4
		OS	18.0	135.2	72.8		49.6	68.0
	20-30	GHS	55.4	129.9	97.4	2.87*	22.1	22.7
		OS	49.3	75.3	63.3		10.8	17.1
	30-40	GHS	46.9	74.7	59.4	0.58ns	9.2	15.4
		OS	49.9	64.5	56.4		6.0	10.7
	40-50	GHS	30.3	51.0	42.5	0.57ns	6.9	16.3
		OS	31.3	45.5	40.1		6.8	16.9

Notes: ①GHS: greenhouse soils; OS: open soils; Std Dev: standard deviation; CV: coefficient of variance. ②Significance level of grouped comparison (*t*-test) of greenhouse soils vs. open-field soils was: * at $P < 0.05$; ** at $P < 0.01$; ns = non significant.

respectively. For 20-30 cm soils, they were significantly ($P < 0.01$) accumulated apart from Ca possibly due to the higher background content, averagely 0.1, 0.4 and 5.4 fold greater than the open-field soils, respectively. They were significantly ($P < 0.01$) accumulated in the 30-40 cm soil horizon and averagely 0.2, 0.2 and 2.1 fold greater than the open-field soils respectively; for the 40-50cm soil horizon, they were not accumulated significantly ($P > 0.05$) apart from S, which indicated S was accumulated more seriously than Ca and Mg.

Although microelement fertilizers were not commonly applied directly to greenhouse soils, the application of organic manure supplemented amble microelements. The analytical results (Table 3) indicated that magnesium and zinc were significantly ($P < 0.05$) accumulated in the 0-20cm soils, while iron and copper had the trend of decreasing but out of significant level. They were significantly ($P < 0.01$) accumulated in the 20-30 cm soils, probably due to the deep application of organic manure. For the 30-40 cm and 40-50 cm soil horizons, Cu and Zn were significantly ($P < 0.05$) accumulated but Fe and Mn.

Table 2 Content of soil available middle elements in different horizons of soils from greenhouses and open-fields

Nutrients	Soil Depth(cm)	No	Content(mg/kg)			T	Std Dev	CV(%)
			Minimum	Maximum	Mean			
Ca	0-20	GHS	808.8	1250.0	1059.9	4.52**	166.9	15.7
		OS	611.3	737.9	663.6		56.6	8.5
	20-30	GHS	821.1	1089.3	902.9	1.18ns	82.7	9.2
		OS	740.5	986.1	835.0		117.1	14.0
	30-40	GHS	896.4	1111.1	978.6	3.99**	67.4	6.9
		OS	699.3	880.3	803.6		80.4	10.0
	40-50	GHS	599.9	917.3	804.4	1.28ns	103.9	12.9
		OS	689.6	783.6	733.3		45.0	6.1
Mg	0-20	GHS	76.0	123.5	99.2	4.65**	17.2	17.4
		OS	47.5	69.1	55.4		9.8	17.7
	20-30	GHS	83.5	128.3	94.8	4.36**	15.4	16.3
		OS	63.0	70.8	68.2		3.6	5.3
	30-40	GHS	93.9	107.3	99.9	4.24**	5.3	5.3
		OS	80.2	92.3	86.3		4.9	5.7
	40-50	GHS	69.5	103.3	90.9	0.77ns	11.9	13.1
		OS	80.4	92.0	86.0		5.9	6.9
S	0-20	GHS	70.9	765.4	419.4	4.83**	239.4	57.1
		OS	3.8	15.3	10.5		5.0	47.3
	20-30	GHS	67.4	140.5	94.2	6.56**	22.9	24.4
		OS	8.4	27.4	14.7		8.8	59.9
	30-40	GHS	110.5	160.5	131.0	8.21**	18.0	13.8
		OS	26.4	65.6	42.2		16.7	39.6
	40-50	GHS	57.2	152.3	99.7	4.04**	29.4	29.5
		OS	20.2	49.7	36.4		12.5	34.4

Notes: ①GHS: greenhouse soils; OS: open soils; Std Dev: standard deviation; CV: coefficient of variance. ②Significance level of grouped comparison (t-test) of greenhouse soils vs. open-field soils was: * at $P < 0.05$; ** at $P < 0.01$; ns = non significant.

Table 3 Content of soil available micro elements in different horizons of soils from greenhouses and open-fields

Nutrients	Soil Depth(cm)	No	Content(mg/kg)			T	Std Dev	CV(%)
			Minimum	Maximum	Mean			
Fe	0-20	GHS	80.2	115.5	98.4	-2.78*	13.5	13.7
		OS	104.7	140.6	122.4		15.4	12.6
	20-30	GHS	67.4	107.2	89.6	4.96**	15.8	17.7
		OS	38.6	54.8	47.8		6.9	14.4
	30-40	GHS	10.9	29.9	16.7	1.63ns	5.9	35.4
		OS	8.7	15.8	11.4		3.2	27.6
	40-50	GHS	6.1	15.5	9.8	0.76ns	2.8	28.3
		OS	6.6	11.6	8.6		2.2	26.2

Continued

Nutrients	Soil Depth(cm)	No	Content(mg/kg)			T	Std Dev	CV(%)
			Minimum	Maximum	Mean			
Mn	0-20	GHS	26.6	85.3	53.1	4.40**	16.8	31.6
		OS	21.6	30.6	25.7		3.7	14.4
	20-30	GHS	29.0	49.1	37.9	5.33**	7.0	18.5
		OS	22.8	26.4	23.9		1.7	7.1
	30-40	GHS	14.6	23.1	19.4	1.06ns	3.1	16.0
		OS	13.7	19.8	17.5		2.8	16.1
	40-50	GHS	10.5	18.4	15.1	0.22ns	2.8	18.6
		OS	9.4	18.1	14.7		3.7	25.4
Cu	0-20	GHS	2.4	3.2	2.9	−1.87ns	0.3	9.0
		OS	2.9	3.6	3.2		0.3	10.3
	20-30	GHS	2.5	3.3	2.9	3.78**	0.3	10.8
		OS	2.0	2.4	2.3		0.2	7.7
	30-40	GHS	1.0	1.7	1.4	3.97**	0.2	16.4
		OS	0.7	1.0	0.9		0.2	17.3
	40-50	GHS	0.3	0.8	0.6	0.24ns	0.2	31.4
		OS	0.4	0.7	0.6		0.1	22.6
Zn	0-20	GHS	6.6	11.7	8.7	3.00*	1.9	22.2
		OS	4.2	7.3	5.4		1.4	27.1
	20-30	GHS	3.1	10.5	5.6	3.76**	2.7	48.8
		OS	1.9	2.0	2.0		0.1	3.0
	30-40	GHS	2.7	13.9	6.6	2.78*	4.5	68.0
		OS	1.7	2.5	2.2		0.3	16.0
	40-50	GHS	3.9	32.4	12.5	2.58*	10.9	87.0
		OS	0.9	4.4	2.4		1.5	62.5

Notes: ① GHS: greenhouse soils; OS: open soils; Std Dev: standard deviation; CV: coefficient of variance.② Significance level of grouped comparison (t-test) of greenhouse soils vs. open-field soils was: * at $P < 0.05$; ** at $P < 0.01$; ns = non significant

2. Water soluble salt and its composition

The accumulation of the nutrient ions results in the significant ($P < 0.01$) increase of the soil water soluble salt content and electric conductivity (EC) comparing to the open-field soils, Table 4 and Table 5 showed that in 0-20 cm soil horizons of the greenhouse and open-field water soluble salt content were averagely 3.43 g/kg and 0.33 g/kg and EC 0.89 mS/cm and 0.12 mS/cm, respectively. In the 20-30 cm soil horizons, the water soluble salt and EC in the soils of greenhouse were also greater than the open-field soils.

Table 4 The soil water soluble salt content in 0-20 cm and 20-30 cm horizons of soils from greenhouses and open-fields

Soil Depth(cm)	No	Salt content(g/kg)			T	Std Dev	CV(%)
		Minimum	Maximum	Mean			
0-20	GHSs	1.59	5.61	3.43	6.15**	1.42	41.42
	OSs	0.25	0.41	0.33		0.07	21.19
20-30	GHSs	1.18	1.95	1.47	7.37**	0.27	18.40
	OSs	0.31	0.58	0.40		0.12	29.88

Notes: ①GHS: greenhouse soils; OS: open soils; Std Dev: standard deviation; CV: coefficient of variance. ② Significance level of grouped comparison (*t*-test) of greenhouse soils vs. open-field soils was: ** at $P < 0.01$.

Table 5 EC in 0-20 cm and 20-30 cm horizons of soils from greenhouses and open-fields

Soil Depth(cm)	No	EC(mS/cm)			T	Std Dev	CV(%)
		Minimum	Maximum	Mean			
0-20	GHSs	0.43	1.43	0.89	6.10**	0.35	39.37
	OSs	0.10	0.16	0.12		0.03	24.39
20-30	GHSs	0.38	0.62	0.48	7.06**	0.09	18.84
	OSs	0.11	0.21	0.14		0.05	34.90

Notes: ①GHS: greenhouse soils; OS: open soils; Std Dev: standard deviation; CV: coefficient of variance. ②Significance level of grouped comparison (*t*-test) of greenhouse soils vs. open-field soils was: ** at $P < 0.01$.

The analytical results from Fig. 1 showed that in the 0-20 cm soil horizons, anions content was higher content than cations, making up about 74.68% of the total salt ions. Among total salt ions, sulphate (SO_4^{2-}) had the highest content accounting for about 46.64% of the anions and the second and third ones are nitrate (NO_3^-) and chloride (Cl^-), respectively. Among the cations, Ca^{2+} had the highest content and comprised about 70.62% of cations or 17.88% of total salt ions. In contrast, Mg^{2+} made up only about 9.79% of the cations. Other ions such as K^+, Na^+, PO_4^{3-}, CO^{2-} and micro elemental cations accounted for a very small part of the total

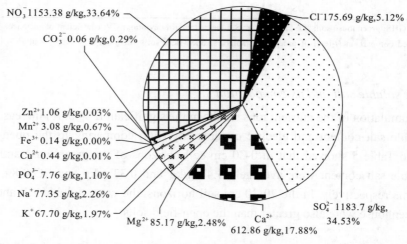

Fig. 1 The ionic composition and content of the water soluble salt in 0-20 cm greenhouse soil horizons

salt ions and contributed less to the formation of the soil secondary salinization.

For the soil horizon of 20-30 cm, the composition of water soluble anions followed a pattern similar to that of 0-20 cm (Fig. 2). Three main anions, i.e., SO_4^{2-}, NO_3^- and Cl^- approximately accounted for 70% of water soluble salt. For cations, Ca^{2+} content still was the highest one. However, Na^+ content ranked second in place of Mg^{2+} in contrast to the 0-20 cm soil horizon.

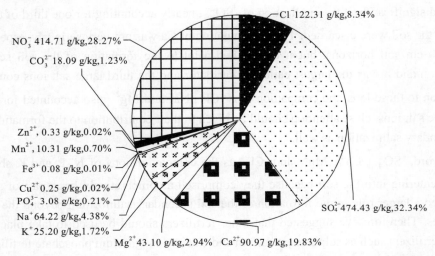

Fig. 2 The ionic composition and content of the water soluble salt in 20-30 cm greenhouse soil horizons

3. Discussions and conclusions

After converting crop fields into greenhouse agricultural fields, most nutrients had been accumulated significantly in the soils, especially in the shallow horizons. The accumulation of the nutrients resulted in the significant increase of the soil water soluble salt and EC and subsequently the formation of the greenhouse soil secondary salinization. The nutrient ions accounted for the most of the salt ions, but accumulation of different nutrient ions contributed differently to the formation of the soil secondary salinization. Among the nutrients, the accumulations of macro elements such as N, P and K were the most serious as compared with the corresponding open-field soils, but only nitrate contributed greatly to the formation the soil secondarily salinization and comprised approximately one-third of the total water soluble salt ions, which indicated nitrate was one of the largest and most important sources of the greenhouse soil salt ions. Even though P and K were accumulated seriously, they comprised a very small part of the soil salt ions and contributed a little to the formation of greenhouse soil secondary salinization. For P, it was probably because of the special chemical property and behavior in the soils. For K, it was probably because of the special behavior in the soils and large amount of need from the growth of the vegetable vegetables (Scaife and Yosef, 1995; Kilmer and Younts, 1968).

For the middle elements, Ca, Mg and S were all accumulated significantly, but sulphate contributed the most to the formation of the soil secondary salinization and was one of the most important sources of the soil salt ions, which accounted for the largest part of the salt ions. The significant accumulation of soil S and the larger contribution to the formation of greenhouse soil secondary salinization should be given close attention for the region where potassium sulfate is the major K source for application. In order to defer the forming course of the soil secondary salinization, SO_4^{2-}, as the companion ions of K^+ should be avoided being applied into the field with K^+, potassium nitrate was advised to take place of the potassium sulphate. From the analytical results, it was supposed that the soil salt content could be decreased significantly if the application of SO_4^{2-}, nearly accounting for one third of total soil salt, into the soil were discarded. The accumulation of Ca was not very seriously, especially in the 20-30 cm soil horizons, but it contributed a lot to the formation of the soil secondary salinization and it was the largest cations contributor and the third large salt ions contributor. In addition to these ions, the companion ions Cl^-, Na^+ and Mg^{2+} also accounted for a larger part of the salt ions, close attention should be given to their contribution to the formation of the soil secondary salinization as well.

In a word, SO_4^{2-}, Ca^{2+}, Mg^{2+} and Cl^- as the companion ions of N, P, and K should be avoided entering into the soil because they comprised a comparatively large part of salt ions and played a larger role in the formation of the soil secondary salinization beyond the need of vegetables. Therefore, we suggested that other fertilizers should be used to take place of the present fertilizers such as sulphate potassium and calcium-magnesium phosphate fertilizer.

References

Agricultural Chemistry Committee of China. 1983. Conventional Methods of Soil and Agricultural Chemistry Analysis. Beijing: Science Press (in Chinese)

Berger K C, Truog E. 1944. Boron tests and determination for soil and plants. Soil Sci., 57: 25-36

Bremner J M. 1965. Total nitrogen, inorganic forms of nitrogen, organic forms of nitrogen, nitrogen availability indexes. In : Blacketal C A. Methods of Soil Analysis. Part 2. Agronomy, 9: 1149-1255, 1324-1348. Madison, Wis: Am. Soc. of Agron., Inc.

Jiang Y, Zhang Y G, Liang W. 2005. Influence of greenhouse vegetable cultivation on composition of soil exchangeable base cations. Journal of Soil and Water Conservation, 19(6): 78-80 (in Chinese)

Kilmer V J, Nearpass D C. 1960. The determination of available sulfur in soils. Soil Sci. Soc. Am. Proc., 24(5): 337-340

Kilmer V J, Younts S E. 1968. The Role of Potassium in Agriculture. Madison: American Society of Agronomy: 509

Li W Q, Zhang M, van der Zee S. 2001. Salt contents in soils under plastic greenhouse gardening in China. Pedosphere, 11(4): 359-367

Lv F T, Si D X. 2004. Study on soil salinity accumulating and ion constitution changing of sunlight greenhouse. Soils, 36 (2): 208-210 (in Chinese)

Masul M, Nukaya A, Ishida A. 1976. Salt content of well water of greenhouse growers in Shizuoka Prefecture. Bulletin of the Faculty Agriculture Shizuoka University, 25: 15-22

Merkle F G, Dunkle E C. 1944. Soluble salt content of greenhouse soils a diagnostic aid. J. Am. Soc. Agron., 36(1): 10-19

Ministry of Agriculture, PRC. 2005. China agriculture statistical report. Beijing: China Agriculture Press

Nanjing Agricultural University.1996. Soil Agricultural Chemistry Analysis. Beijing: Agriculture Press (in

Chinese)

Olsen S R, Cole C V, Watanabe F S, et al. 1954. Estimation of available phosphorus in soils by extraction with sodium bicarbonate. US Dept Agric Circ., (939): 1-19

Scaife A, Yosef B. 1995. Fertilization for High Yield and Quality Vegetables. International Potash Institute, Basel: 104

Soil Science Society of China. 2000. The Analysis Method of Soil Agricultural Chemistry. Beijing: Agricultural Science and Technology Press (in Chinese)

Sun Z F. 1999. Preliminary investigation on present situation and essential problems of horticulture engineering in China. In: Cao Z H, Zhou J M. Techniques on Greenhouse Agriculture. Beijing: China Agricultural Science and Technology Press

Sun Z Q, Zhao W X, Zhang W B. 2005. Effects of different nitrogen fertilization patterns on the quality of the lettuce and soil environment. Transactions of the CSAE, 21(Supp): 159-161 (in Chinese)

Tian X Y, Chen C Z. 1993. The research on the method of analyzing 27 elements in soils by using ICP-AES. Chinese journal of soil science, 24 (4): 188-190 (in Chinese)

Tong Y W, Chen Y F. 1991. The study of secondary salinization in greenhouse soils and its control. Journal of Horticulture, 18: 159-162 (in Chinese)

Wood L K, de Turk E E. 1940. The adsorption of potassium in soil in non-replicable forms. Soil Sci. Soc. Am. Proc., 5: 152-161

Xue J C, Bi D Y, Li J J, et al. 1994. The soil factors of vegetable physiologic disorder in the protected land production and the controlling measures. Soils and Fertilizers, (1): 4-9 (in Chinese)

Zhang Y S, Ni W Z. 1997. The state of the available nutrients of the vegetable garden soil in the suburb of Hangzhou and the fertilizing countermeasures. In: Xie J C, Chen J X. Vegetable Garden Soil Fertility and Rational Fertilization of Vegetable. Nanjing: Hehai University Press: 43-49

3 Nutrient migration and cycling in environment

3.1 Phosphorus fractions in sediment profiles and their potential contributions to eutrophication in Dianchi Lake

3.1.1 Introduction

The importance of lake sediments as a nutrient sink and a nutrient source has long been recognized. Lake sediments play an important role in the phosphorus metabolism in lakes. The impact depends on the tendencies to retain and release phosphorus under certain limnological conditions (Pettersson, 1998). The main sources of phosphorus in lakes include external loading, such as rainfall, runoff, soil leaching, industrial and municipal effluents, and internal loading (Kaiserli et al., 2002). The internal loading will often determine the eutrophication status of the lake and the time lag for recovery after reduction of the external loading. During recovery following a reduction in external phosphorus loading, excessive internal loading from a phosphorus-rich sediment may prevent or delay improvement in lake water quality for several years (Boström et al., 1985; Marsden, 1989). The phosphorus concentration in sediments depends on the lake trophic status, sediment composition, sedimentation rate, physico-chemical conditions and the extent of diagenetic processes (Gonsiorczyk et al., 1998; Frankowski et al., 2002). Zhou et al. (2001), Kaiserli et al. (2002) reported that the total concentration of phosphorus in sediments can not predict the potential ecological danger and the fraction of available phosphorus is an important parameter for predicting future internal P-loading. It is very important to definitude P forms in sediments, which helps to understand P dynamics and release trends into lake water (Andrieux-loyer and Aminot, 2001). Fractionation schemes as a method to characterize phosphorus binding to a variety of organic and inorganic sediment components have been widely used (Hieltjes and Lijklema, 1980; Psenner et al., 1988; Hupfer et al., 1995). According to Hupfer et al. (1995), phosphorus in lake sediments is divided into NH_4Cl-P, BD-P, NaOH-P, HCl-P and organic P (Org-P). NH_4Cl-P represents the loosely adsorbed P, which may contain porewater P, P released from $CaCO_3$ associated phosphorus or leached P from decaying cells of bacterial biomass in deposited phytodetrital aggregates. BD-P represents the redox-sensitive P forms, mainly bound to Fe-hydroxides and Mn compounds. The NaOH-P represents P bound to metal oxide, mainly of Fe and Al, which is exchangeable again with OH^- and inorganic P compounds soluble in bases. The HCl-P represents P forms which are sensitive to low pH, assumed to consist mainly of apatite P, P bound to carbonates and traces of hydrolysable organic P. Rydin (2000) reported that mobile phosphorus comprises BD-P and NaOH-P, and mobilizations of the two fractions are the most important mechanism of P release under high pH values and anaerobic conditions.

Dianchi Lake is in the middle of the Yungui Plateau, Southwestern China. The lake has a surface area of 294 km^2 and is divided into two parts, Cao-hai and Wai-hai, covering 2.7% and 97.3% of the total area, respectively. The mean depth is 4.3 m and the hydrological retention time is 2-3 years. The water temperature ranges from 9.8℃ to 24.5℃, with an annual average of about 16.0℃, and the lake water is alkaline (Jin et al., 1990; Whitmore et al., 1997). In autumn and winter, wind stirring ensures that the lake water does not stratify. Dianchi Lake is adjacent to Kunming City, the capital of Yunnan Province; therefore large quantities of industrial wastewater and municipal sewage have been discharged into the lake. In the Dianchi Catchment, phosphate rock and phosphatic chemical enterprises exist in the southern area, and industrial estates, farmlands and human habitation surround the remaining lakeshore. Consequently, the catchment runoff drains a significant amount of nitrogen and phosphorus nutrients into the Dianchi Lake. At present, the lake water is heavily polluted and suffers from eutrophication (Wu et al., 1992; Li et al., 2003). Total phosphorus (TP) concentrations in the two areas are 0.28 mg/L and 1.06 mg/L, and total nitrogen concentrations reach 1.06 mg/L and 11.89 mg/L in Wai-hai and Cao-hai, respectively, showing an increasing trend with time (Zhe, 2002). Eutrophication has become more and more serious in the lake in the recent 20 years, and algal blooms occur frequently from April to November each year (Wu et al., 1992; Li et al., 2003). The control of water pollution in Dianchi Lake began in the 1980s, with strategies mainly including the control of external load, recovery technology of large aquatic plants (such as Eichhornia crassipes Solms), mechanical harvesting of algae, the establishment of swamp, removal of surface sediments, etc. (Chen et al., 2002) In spite of the large investment (about US $4.8 million) and efforts, the water quality has not improved markedly.

External phosphorus load is important in water quality, but the internal load must also be taken into account. P is the most critical nutrient limiting lake productivity in Dianchi Lake, but few results have been reported concerning the contribution of sediment P to water quality. The objective of this study was to investigate the trophic status of lake sediments, spatial and vertical distribution of various P-binding forms, and to evaluate their potential contributions to the overlying water in Dianchi Lake.

3.1.2 Materials and Methods

1. *Study sites*

In the summer of 2002, 118 sampling sites were selected throughout the Dianchi Lake to investigate the total phosphorus concentration in the sediments, and the sediment cores were sliced into 0-5 cm, 5-10 cm, and 10-20 cm layers. Based on the data obtained from the 118 sites, six sampling sites in the east, west, south, north and the center of the main part of the lake, Wai-hai, were selected in December 2002 for the present study (Fig. 1), to define the variation with depth and speciation of different P forms. Site 1 (S1) was in the south of the lake, near the Kunyang Phosphate Fertilizer Factory; site 2 (S2) located in the Chaihe Delta area; site 3 (S3) was on the north of S2, far from the lakeshore; site 4 (S4) was in the center of the whole lake; site 5 (S5) was near Dounan Town, a district famous for flower production;

site 6 (S6) was around Haigeng, in northern lake, which was close to Kunming City. Some general characteristics of the 6 sampling sites are summarized in Table 1.

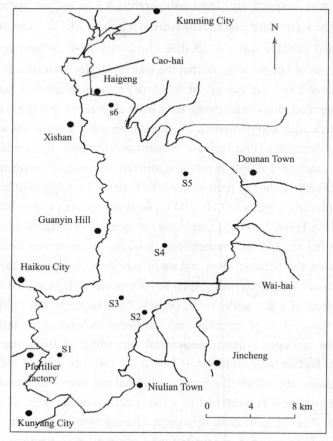

Fig. 1 Distribution of sampling sites in Dianchi Lake

Table 1 General characteristics at the 6 sampling sites selected in Dianchi Lake

Parameters	S1	S2	S3	S4	S5	S6
Water depth (m)	4.6	4.1	5.8	6.5	5.8	4.2
Secchi disc (m)	0.54	0.48	0.50	0.49	0.55	0.41
Water temperature (℃)	10.5	11.0	9.8	9.8	10.0	10.1
Sediment depth (m)	0.40	0.30	0.35	0.40	0.35	0.40
Bottom water						
DO (mg/L)	8.72	9.12	8.84	8.56	8.80	9.24
pH	8.48	8.45	8.11	8.40	8.23	8.13
TP (mg/L)	0.19	0.23	0.16	0.20	0.24	0.30
SRP (mg/L)	0.026	0.054	0.049	0.046	0.038	0.075

2. Sediment and water sampling

Water samples in the bottom layers were collected with a water sampler (made in the Institute of Hydrobiology, Chinese Academy of Sciences). After being transported to the laboratory, total phosphorus, soluble reactive phosphorus (SRP), pH and dissolved oxygen (DO) were determined immediately.

The sediment cores were sampled using a UWITEC gravity corer (Austria). If possible, the sediment cores were sliced into 0-2 cm, 2-4 cm, 4-6 cm, 6-8 cm, 8-10 cm, 10-15 cm, 15-20 cm, 20-25 cm, 25-30 cm, 30-35 cm and 35-40 cm layers, each of which was analyzed separately. Moisture content was determined immediately. Sediment samples were kept at 4 ℃ until analysis. Different P fractions were analyzed immediately by sequential extraction, and the remaining portion were dried at room temperature and ground (0.149 mm) for total P and organic P analysis.

The porewater samples at different depths were obtained by centrifugation (4000 r/min for 30 min), followed by filtration through a 0.45 μm pore-size phosphorus-free membrane.

3. Chemical analysis

Total P concentration in the water was determined by the molybdenum blue method after digestion with H_2SO_4 and HNO_3. SRP was analyzed by the molybdenum blue method after lake water was filtrated through a 0.45 μm phosphorus-free membrane (NEPB, 1998). Dissolved oxygen was analyzed with the Winkler titration method (Parson et al., 1984).

The moisture content was determined by drying at 105 ℃ for 2 h. Total P content in sediments was measured by the molybdenum blue method after digestion with $HClO_4$ and HF. Organic P was analyzed using the method outlined by Ruttenberg (1992); it was determined by the difference between the extraction with 0.5 M H_2SO_4 (16 h) with and without combusted (550 ℃, 1 h) sediment.

Phosphorus fractions: the sequential extraction scheme of Hupfer et al. (1995) with slightly modification by Fu et al. (2000) was used for P fractionation. Fresh sediment was subjected to sequential chemical extraction with 1.0 md/dm³ NH_4Cl (0.5 h), 0.11 md/dm³ $NaHCO_3/Na_2S_2O_4$ (40 ℃, 1 h), 1.0 md/dm³ NaOH (16 h) and 0.5 md/dm³ HCl (16 h). The ratio of sediments (DW) to water volume was 1 : 20. The extracts were centrifuged and the supernatants were filtered through a 0.45 μm pore-size phosphorus-free membrane. The SRP concentration in each fraction was determined as described above.

3.1.3 Results

1. Phosphorus concentrations in porewater and bottom water

Total P concentrations in the bottom water were higher than 0.15 mg/l (Table 1), with the highest at S6 (0.30 mg/L). SRP concentrations in bottom water were low, ranging from 0.026 mg/L to 0.075 mg/L. Porewater profiles at the 6 sites showed that SRP concentration was relatively low in the oxidized surface layers. Below the interface of water and sediment, it increased with sediment depth, and then declined slightly in deeper layers (30-40 cm) (Table 2). Among the 6 sites selected, the highest SPR concentrations in porewater and water were both found at S6. The SPR concentrations were generally higher in the surface porewater than in the bottom water, and there was a concentration gradient at the sediment-water interface (Table 1 and Table 2).

Table 2 Soluble reactive phosphorus concentrations in porewater at the 6 sampling sites selected in Dianchi Lake (Unit: mg/L)

Sites	Depth (cm)										
	0-2	2-4	4-6	6-8	8-10	10-15	15-20	20-25	25-30	30-35	35-40
S1	0.039d[①]	0.048d	0.059cd	0.066d	0.064e	0.058d	0.067c	0.061d	0.058d	0.054c	0.054c
S2	0.056b	0.058c	0.064c	0.072c	0.075d	0.072c	0.065c	0.071c	0.067c	-[②]	-
S3	0.058b	0.078b	0.079b	0.087b	0.088c	0.081b	0.083b	0.079b	0.080b	0.077b	-
S4	0.060b	0.079b	0.085b	0.091b	0.095b	0.098a	0.084b	0.081b	0.087b	0.075b	0.078b
S5	0.047c	0.049d	0.054d	0.058e	0.057f	0.059d	0.052d	0.049e	0.047e	0.052c	-
S6	0.084a	0.089a	0.097a	0.107a	0.114a	0.098a	0.107a	0.109a	0.098a	0.084a	0.083a

Notes: ①The same letter means not significant at $P<0.05$ level according to Duncan's MRT. ②Data were not measured.

2. Total phosphorus concentration in lake sediments

A survey of the 118 sampling sites showed that total P concentrations in the surface sediments (0-5 cm) changed in the range of 0.44-6.66 g/kg, with a mean value of 2.27 g/kg. TP concentrations in surface sediments of 70% samples were higher than 2.0 g/kg. In 5-10 cm sediment layers, TP concentrations of 58% samples ranged from 1.0 to 2.0 g/kg. For the 10-20 cm layers, TP concentrations of 69% samples changed in the range of 1.0-2.0 g/kg. On the whole, total P concentrations in surface sediments in Dianchi Lake were high, which indicated that the lake sediments had a great potential to supply P to the overlying water (Marsden, 1989).

The total P concentrations in the sediment profiles from the 6 sampling sites ranged from 1.13 g/kg to 3.45 g/kg (Fig. 2). Comparison of the 6 sites, the order of TP concentrations in the top 10 cm layers was S1>S6>S4>S3>S5>S2. At S1, in the phosphate rock area, TP contents (at the 2.71-3.46 g/kg range) were much higher than at the other sites in the whole profile. At

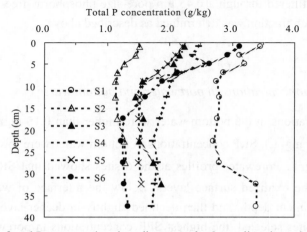

Fig. 2 Depth profiles of total P concentrations in sediments at the 6 sampling sites selected in Dianchi Lake

S6, close to Kunming City, TP concentrations in the top 10 cm layers were the second highest. And, TP concentrations in the central lake (at S4) were higher than near the lakeshore (at S2, S3 and S5).

Fig. 2 demonstrates that the changing tendency of the TP concentration in the sediment profiles were different at the 6 sites. At S1, TP concentrations decreased with sediment depth in the surface 0-10 cm layers (from 3.45 g/kg to 2.75 g/kg), changed slightly in 10-30 cm, and then increased drastically in the deeper layers. At S6, TP concentrations decreased rapidly with depth in 0-20 cm sediment layers, from 3.13 g/kg at 0-2 cm to 1.27 g/kg at a 15-20 cm depth, and then changed slightly in 20-40 cm layers. Generally, the TP concentrations at the 6 sites were higher in surface sediments than in deep sediments and had an increasing trend upward toward the sediment surface in the top 10 cm layers. This indicated that the P load aggravated with time in Dianchi Lake and the human activities in the catchment area intensified in recent years. Comparison of the 6 sites, TP concentration gradient was the greatest at S6 (from 3.13 g/kg at 0-2 cm to 1.88 g/kg at a 8-10 cm depth), which showed that the impact of city sewage on lake trophic status became more and more serious in recent years.

3. *Distribution of different P forms in sediment profiles of Dianchi Lake*

Fig. 3 shows the concentrations of different P forms in the surface sediments (0-2 cm). NH_4Cl-P, a seasonally variable pool of P compounds dissolved in the porewater, was very low (ranged at 0.26-1.84 mg/kg) at the 6 sites, accounting for less than 1‰ of the total phosphorus. In the oxidized surface layers, the proportions of BD-P to TP varied from 1.7% to 4.3%, and the mean BD-P concentration was 0.08 g/kg. The BD-P concentration at S6 was highest (0.15 g/kg, 4.3% of the total). NaOH-P accounted for 19.9%-41.9% of TP, and the highest was also found at S6. HCl-P comprised 18.2%-46.2% of TP, with the highest value being found at S1. Org-P proportions in the surface sediments were high, ranging from 29.5% to 51.5%.

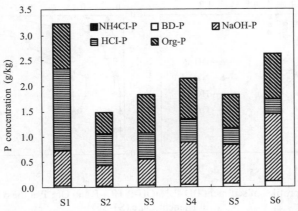

Fig. 3   Concentrations of different P forms in surface sediments (0-2 cm) at the 6 sampling sites selected in Dianchi Lake

Profile concentrations of various P forms in the sediments from the 6 sampling sites are shown in Fig. 4. NH_4Cl-P in the sediments was too low to express in the figure. The concentrations of various P forms at S4, S5 and S6 followed the order of NaOH-P, Org-P>HCl-P>BD-P>NH$_4$Cl-P in the whole profile. At S1 and S2 (in the southern lake), HCl-P comprised a larger proportion, and the rank order was HCl-P>NaOH-P, Org-P >BD-P> NH$_4$Cl-P in the top 15-cm sediment layers. Different forms of P had an increasing trend upward toward the sediment surface (Fig. 4).

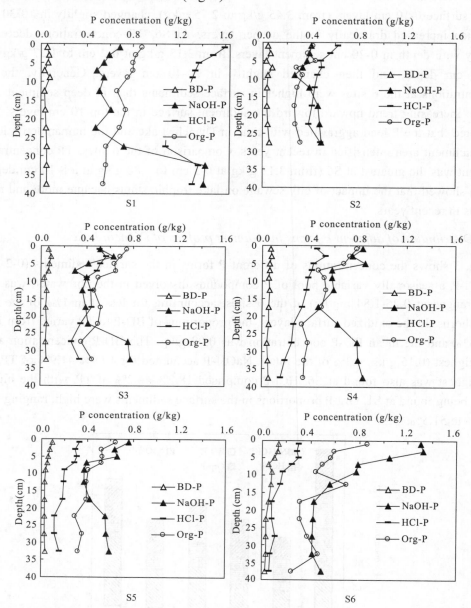

Fig. 4 Concentration profiles of sediment BD-P, NaOH-P, HCl-P and Org-P at the 6 sampling sites selected in Dianchi Lake (S1-S6)

BD-P, the redox-sensitive P forms, has been known to be a source for internal P loading (Rydin, 2000). BD-P concentrations declined rapidly with depth in 0-10 cm layers and then changed slightly (Fig. 4). At S6, BD-P concentration was relatively high and declined from

0.14 g/kg at 0-2 cm to 0.08 g/kg at a 10-15 cm depth. At S4, it changed little throughout the sediment profile.

NaOH-P represents P bound to metal oxide, mainly of Fe and Al. Concentration profiles of NaOH-P at the 6 sites showed that it first declined rapidly with depth in 0-10 cm layers (0-20 cm layers at S6), and then changed slightly (at S2, S5 and S6) or increased greatly at depths greater than 25 cm (at S1, S3 and S4). The proportions of NaOH-P to TP were higher in the northern lake sediments.

HCl-P concentrations decreased with depth in the 0-20 cm sediment layers at the 6 sampling sites, and then changed slightly or increased at depths greater than 30 cm (at S1 and S3). In the southern lake sediments (at S1 and S2), HCl-P concentrations were high, whose proportion to the total P was more than 35%. The HCl-P was relatively stable under alkaline conditions.

Organic P in the sediment probably comes from in-washed organic P and from autochthonous sources, such as biological material generated within the lake itself. The Org-P proportions at 6 sites were high and the proportion to the TP ranged between 20.6% and 30.4% (Fig. 4). Except for the low concentration at S2, there was no great difference among the other sampling sites. Org-P concentration profiles were S-shaped, and the general trend was that concentrations in surface sediments were higher than that in deeper sediments. At S6, Org-P concentration decreased from 0.87 g/kg (0-2 cm) to 0.25 g/kg (35-40 cm).

4. Potential contribution of sediment P fractions to eutrophication in Dianchi Lake

Mobile phosphorus in sediments can be used to estimate the potential release of sediment P to the overlying water (Rydin, 2000). At the study sites except S6, the mobile P concentrations (the sum of BD-P and NaOH-P) first declined with sediment depth in the top 10-cm sediment layers, changed slightly in 10-25 cm, and then increased in deeper layers (Fig. 4). At S6, it decreased sharply with depth (from 1.44 g/kg at 0-2 cm to 0.46 g/kg at a 20-25 cm depth), and then changed slightly below 25 cm depth. Comparison of the different sampling sites, mobile P concentration in the surface 0-10 cm was highest at S6, S4 was the second, and was lowest at S2. The mobile P concentrations were large and increased rapidly toward the surface in the top 10 cm sediments, which is in accordance with more and more serious eutrophication of the lake in recent 20 years.

3.1.4 Discussion

1. Spatial distribution of total phosphorus in lake sediments

In general, phosphorus concentration in sediments depends on the lake trophic status, sediment composition, sedimentation rate, physico-chemical conditions and the extent of diagenetic processes. However, the primary influence factor is different in different lakes. In Dianchi Lake, sedimentation rate was slightly fewer in central lake than near the lakeshore, and it had no great difference in the whole lake. The sedimentation rate was not the primary factor influencing the differences of P concentration among different areas in the lake. Total P concentration in the sediments had great differences among different sites in the Dianchi Lake (Fig. 2), which mainly resulted from different geographical locations, anthropogenic P sources, sediment types, and so on (Gonsiorczyk et al., 1998; Frankowski et al., 2002). Surface total phosphorus concentrations were greatest at sites S1 and S6, probably for the following reasons: in the southern Dianchi Lake, phosphate rock and phosphatic chemical enterprises exist; whereas in the northern area, industrial estates, farmlands and human habitation surround the lakeshore. At S1, in the phosphate rock area and near the phosphate fertilizer factory, TP concentrations in the whole profile were much higher than at the other sites, which may be attributed to the sedimentation of phosphate rock particles and the drainage from the nearby phosphate fertilizer factory. S6 was close to Kunming City, where large quantities of city sewage containing phosphorus were drained into the lake; therefore, the TP concentrations in sediments at S6 were also very high. TP concentrations in the central lake sediments (at S4) were higher than near the lakeshore (at S2 and S5), the possible reason being that the sediments in the center were mainly clay, whose grain size was much finer than near the shore (Andrieux-loyer and Aminot, 2001). On the east lakeshore, dense rivers lead to the appearance of deltas (such as Panlong River Delta and Chaihe Delta). At S2, in the Chaihe Delta, TP concentrations were the lowest.

2. *Vertical distribution of total phosphorus in lake sediments*

The vertical P distribution in sediment profiles is the net result of the history of P sedimentation as well as many transformation processes (Hupfer et al., 1995). In Dianchi Lake, sediment profiles at different locations showed that TP concentrations were much higher in surface sediments (0-10 cm) than in deep layers (except at S1) (Fig. 2). At S1, high TP concentrations in deep layers possibly resulted from the laggard mining technology, the sedimentation of phosphate rock particle and the serious soil erosion at that time. High TP concentrations in 0-10 cm layers at all the sites suggest that the P inputs increased in recent years. The rapid development of industry, agriculture, urbanization and population in Dianchi Catchment lead to the excessive P inputs into the lake. As the P loading increase, so does autochthonous sedimentation and the P concentration in the surface sediments (Hupfer et al., 1995; Søndergaard et al., 1996; Whitmore et al., 1997). Yu et al. (1996) reported that the average sedimentary rate in the Dianchi Lake was about 0.28 cm/year. In addition, Zhang et al. (2005), using ^{137}Cs dating method, found that the sediment depth, which deposited in 1963, ranged from 11 cm to 14 cm at different sediment cores in the Dianchi Lake. Hence, it can be inferred that high total P concentrations in surface 0-10 cm was a result of increasing P inputs in the recent 30-40 years.

Total P concentrations had an increasing trend upward toward the sediment surface and the concentration gradients were very large in the top 10 cm layers at all of 6 sites (Fig. 2), which indicated the phosphorus inputs had great differences at various periods. In Dianchi Lake, the sedimentation rate had no great difference among various years after 1963 (Zhang et al., 2005), and the sedimentation rate was not the primary factor determining the variation with depth of TP in 0-10 cm layers. Large concentration gradients of total P in the top 10 cm layers were closely related to the increasing anthropogenic P inputs in recent 40 years (Zhang et al., 1996; Chen et al., 2002). In the 1950s the water in Dianchi Lake was very clear and became turbid in the 1970s. Since the early 1980s, the water quality has deteriorated and the lake has been classified as hypertrophic (Jin et al., 1990; Whitmore et al., 1997). The deterioration of water quality had close connection with socio-economic development in the Dianchi Catchment in recent years. The anthropogenic P inputs increased rapidly during the periods of industrial operation and urban development. In the southern lake the first phosphate fertilizer factory was built in 1954. During the 1980s a great number of phosphate industries and chemical enterprises were established. At S1, high TP concentration in surface sediments was likely due to the over-exploitation of phosphorite and the increasing P inputs from the nearby phosphatic chemical enterprises in recent years. Since the early 1980s, social economy in Kunming City has come into a period of rapid development. In addition, the speed of population growth and urban expansion of the city was rapid. For example, the human population of the city was approximately 0.85 million in 1957, 1.26 million in 1980, and increased to 1.73 million in 1998. The area of Kunming City zone was 48.6 km^2 in 1959, 70 km^2 in 1980, and up to 132 km^2 in 2003 (Shen, 2003). As a result of urban development, spreading industrialization and population growth, the increasing industrial and domestic effluent containing phosphorus have been drained into the Dianchi Lake. In Kunming City, the first wastewater treatment plant was built in 1991, and large quantities of urban wastewater were directly drained into the lake before this time. Up to now six sewage treatment plants have been built, but only 60% of urban wastewater was treated (Chen et al., 2002). At S6, close to Kunming City, high TP concentration in surface sediments was attributed to the increasing anthropogenic inputs of phosphorus from the city. On the other hand, agricultural production in the catchment area has transferred from conventional organic agriculture to fertilizer agriculture in recent 20 years. Due to the increasing runoff losses and untreated rural domestic discharge from the catchments, the non-point pollution currently has an increasing tendency with time. In conclusion, spreading industrialization, rapid urbanization, intensive agriculture and population growth resulted in a rapid increase in anthropogenic inputs of phosphorus. Large TP concentration gradients in surface 0-10 cm was a result of rapid socio-economic development and frequent human activities in Dianchi Catchment during the last 40 years. Comparison of the 6 different locations, the worst water quality and the greatest difference of TP concentration in the sediment profile were both found at S6 (Table 1, Fig. 2), which indicated that the rapid urbanization in recent years accelerated the pollution of the lake. The discharge of industrial wastewater and municipal sewage from Kunming City was the predominant reason of the eutrophication in Dianchi Lake (Whitmore et al., 1997; Chen et al., 2002).

3. P fractions in lake sediments

P fractions in lake sediments depend on the origin of sediments, land-use types, biological activity, physico-chemical conditions, and so on (Lopez and Morgui, 1993; Ulrich, 1997). Due to the change of biological environment and physico-chemical conditions, the P fractions in sediments had great differences among different districts. In the Dianchi Lake, HCl-P comprised a large proportion in the TP in southern sediments; whereas NaOH-P concentration was high in the north. From the south to the north of the lake, HCl-P concentration in the sediments declined and NaOH-P concentration increased. Except for the low concentration at S2, Org-P proportions had no great difference among the other sampling sites (Fig. 4).

A great amount of phosphate rock of early Cambrian Age exists in the south of Dianchi Catchment area. The essential mineral is collophane, whose main component is supermicroscopic low-carbonate fluorapatite aggregates. The aggregates have the general formula: $Ca_{10}[(PO_4)_{6-Z}(CO_3)_X]$ (F, Cl, OH, I)$_{Z+X}$, where the concentrations of Cl, OH and I are few and the X value is very small. In the southern lake (at S1 and S2), HCl-P comprised a larger proportion in the TP (Fig. 4). The phosphate rock here was enriched in calcium; therefore, in the early diagenesis P was deposited in the form of Ca-P compound. In addition, high HCl-P depended on the co-precipitation of phosphate with calcite. In the Dianchi Catchment, the oldest fertilizer factory was built in 1954, which has a history of 50 years. Therefore, high HCl-P concentration in surface sediments partly related to the separation of gypsum and superphosphate in phosphate fertilizer manufacturing. In the northern lake, high HCl-P in surface sediments was likely due to the increasing inputs of organic matter and external P load (Søndergaard et al., 1996).

Hisashi reported that the sum of Al-P and Fe-P content in lake sediments was one of the pollution parameters (Hisashi, 1983), and the NaOH-P concentration in the sediments partly related to the pollution degree. The highest NaOH-P concentration at S6 (Fig. 4) showed that site 6 was polluted heavily, which corresponds with the results obtained above (Table 1). High NaOH-P concentration in surface sediments at S6 also indicated that the rapid urbanization accelerated the deterioration of water quality. On the other hand, Dianchi Lake is a tectonic lake and the soil-forming rocks are mainly basalt, limestone and sandstone. Red earth and paddy soil are the prevailing soils in the surrounding lakeshore. Hence, the sediments in Dianchi Lake are enriched in iron and alum. As the P inputs into the lake increase, so does the Fe-P and Al-P concentration (NaOH-P) in surface sediments (Søndergaard et al., 1996). At 3 of the 6 sampling site (S1, S3, S4), NaOH-P concentration increased greatly in deeper layers, especially at S1. TP, NaOH-P and HCl-P concentrations increased greatly at depths greater than 35 cm at S1, especially NaOH-P. These conditions may relate to the origin of sediments, the early sedimentary environment and physico-chemical conditions of the time, which needs further investigation. At S6, high NaOH-P concentration in surface sediments mainly related to the increasing P inputs in recent years. In Kunming City, the first wastewater treatment plant was built in 1990; therefore, high NaOH-P concentration at S6 had no close connection with the treatment of wastewater.

Organic P is quantitatively one of the most important phosphate phases buried in the sediment and thus directly affects the availability levels of dissolved phosphorus for primary production (Edlund and Carman, 2001). The mobilization of recently sedimentary labile Org-P seems to be the driving force of the P release (Gonsiorczyk et al., 1998). The Org-P proportions at six sites were high and the proportion to the TP varied between 20% and 30% (Fig. 4). Hydrolysis of this fraction contributes to the pool of inorganic P, which may be released to the overlying water.

4. *Release potential of sediment P*

Mobile phosphorus in sediments can be used to estimate the potential release of sediment P to the overlying water (Rydin, 2000). The mobile P concentrations in the sediments were very high and decreased rapidly with depth in the top 10-cm layers in Danchi Lake (Fig. 4). This suggests that P release potential in the surface sediments from Dianchi Lake was very large. The northern industry estates and the southern phosphate diggings were the two areas rich in phosphorus in sediments in Dianchi Lake. However, in southern lake sediments, HCl-P was the most prevailing P form, which is relatively stable in the alkaline water of Dianchi Lake where pH values range from 8.2 to 9.5 (Jin et al., 1990; Zhe, 2002). Hence, the sediments P in the northern lake had a greater release potential than those in the southern lake.

The vertical distribution of porewater SRP was strongly influenced by the decomposition of organic matter, reduction of iron oxides and microbial activity in sediments (Sundby et al., 1992; Reddy et al., 1996). Porewater SRP concentration was relatively low in the top oxidized layers and then increased with sediment depth (Table 2). The increase in SRP with depth is attributable to the release of phosphate from iron oxides as these are reduced and to the mineralization of organic P (Sundby et al., 1992). The SRP gradients at the sediment-water interface at all the sites indicate that P could diffuse from sediments into the overlying water under certain conditions. The diffusive fluxes of phosphorus can increase water P levels in Dianchi Lake, which is shallow and with a long hydraulic residence time. Sediment P would be the potentially dominant factor influencing water quality if the external load is reduced. Porewater SRP concentrations had no apparent difference between the sites with numerous anthropogenic sources (S6 and S1) and the ones in mid-lake (S4 and S5). The possible reasons was that: SRP in porewater was mobile and diffusion would transfer phosphorus from contaminant-rich to contaminant-poor layers in the lake. Moreover, porewater SRP concentrations were strongly dependent on redox conditions in sediment.

The actual release of phosphorus from lake sediments is governed not only by the reservoir of exchangeable phosphorus, but by other factors such as pH, redox-potential (Eh) and bioturbation (Zhou et al., 2001; Gao et al., 2004). In the summer, dissolved oxygen depleted and Eh decreased at the sediment-water interface, with the appearance of algal blooms in surface water and the degradation of fresh organic matter, upon which redox-sensitive P can be dissolved and released to the lake water. As a result of frequent algal blooms and increasing water pollution, the pH value in Dianchi Lake has increased with time (Zhe, 2002), which could accelerate the release of NaOH-P to the overlying water. As a whole, P released from the sediment was the P supply for the development of algae in water. On the other hand, algal blooms increased the pH value of lake water, and degradation of dead algae decreased the Eh value at the sediment-water interface and increased P concentrations in surface sediments, which could accelerate the P release. Release of phosphorus and appearance of the algal bloom were stimulated reciprocally, which may result in self-fueling. This was possibly one of the reasons that the water quality in Dianchi Lake has not improved for a long time. Therefore, sediment internal load should be taken into account in future restoration projects in Dianchi Lake.

References

Andrieux-loyer F, Aminot A. 2001. Phosphorus forms related to sediment grain size and geochemical characteristics in French coastal areas. Estuar. Coast. Shelf Sci., 52: 617-627

Boström B, Ahlgren I, Bell R. 1985. Internal nutrient loading in a eutrophic lake, reflected in seasonal variations of some sediment parameters. Vreh. Int. Ver. Limnol., 22: 3335-3339

Chen J N, Zhang T Z, Du R F.2002. Assessment of water pollution control strategies: a case study for the Dianchi Lake. J. Environ. Sci., 14: 76-78

Edlund G, Carman R. 2001. Distribution and diagenesis of organic and inorganic phosphorus in sediments of the Baltic proper. Chemosphere, 45: 1053-1061

Frankowski L, Bolalek J, Szostek A. 2002. Phosphorus in bottom sediments of Pomeranian Bay (southern Baltic-Poland). Estuar. Coast. Shelf Sci., 54: 1027-1038

Fu Y Q, Zhou Y Y, Li J Q. 2000. Sequential fractionation of reactive phosphorus in the sediment of a shallow eutrophic lake-Donghu lake, China. J. Environ. Sci., 12: 57-62

Gao L, Yang H, Zhou J M, et al. 2004. Lake sediments from Dianchi Lake: a phosphorus sink or source? Pedosphere, 14: 483-490

Gonsiorczyk T, Gasper P, Koschel R. 1998. Phosphorus binding forms in the sediments of an oligotrophic and an eutrophic hardwater lake of the Baltic district (Germany). Water Sci. Technol., 37: 51-58

Hieltjes A H M, Lijklema L. 1980. Fractionation of inorganic phosphates in calcareous sediments. J. Environ. Qual., 9: 405-407

Hisashi J .1983. Fractionation of phosphorus and releasable fraction in sediment mud of Osaka Bay. Bull. Jap. Soc. Sci. Fish., 49: 447-454

Hupfer H, Gächter R, Giovanoli R. 1995. Transformation of phosphorus species in settling seston and during early sediment diagenesis. Aquatic Sciences, 57: 305-324

Jin X C, Liu H L, Tu Q Y, et al. 1990. Lake Eutrophication in China. Beijing: China Environmental Science Press: 343-346

Kaiserli A, Voutas D, Samara C. 2002. Phosphorus fractionation in lake sediment-Lakes Volvi and Koronia, N. Greece. Chemosphere, 46: 1147-1155

Li X Y, Liu Y D, Song L R, Liu T T. 2003. Response of antioxidant systems in the hepatocytes of common carp (*Cyprinus carpio* L.) to the toxicity of Microcystin-cr. Toxicon, 42: 85-89

Lopez P, Morgui J A. 1993. Factors influencing fractional phosphorus composition in sediments of Spanish reservoirs. Hydrobiologia, 253: 73-82

Marsden W M. 1989. Lake restoration by reducing external phosphorus load: the influence of sediment phosphorus release. Freshwat Biol., 21: 139-162

NEPB (National Environmental Protection Bureau). 1998. Monitoring and Analysis of Water and Wastewater. Beijing: China Environmental Science Press: 246-286

Parson T R, Maita Y, Lalli C M. 1984. A Manual of Chemical and Biological Methods for Seawater Analysis. Sydney: Pergamon Press: 153-194

Pettersson K. 1998. Mechanisms for internal loading of phosphorus in lakes. Hydrobiologia, 371-374: 21-25

Psenner R, Boström B, Dinka M, et al. 1988. Fractionation of phosphorus in suspended matter and sediment. Arch. Hydrobiol. Beih., 30: 98-110

Ruttenberg K C. 1992. Development of a sequential extraction method for different forms of phosphorus in marine sediments. Limnol. Oceanogr., 37: 1460-1482

Rydin E. 2000. Potentially mobile phosphorus in lake Erken sediment. Water Res., 34: 2037-2042

Reddy K R, Fisher M M, Ivanoff D. 1996. Resuspension and diffusive flux of nitrogen and phosphorus in a hypereutrophic lake. J. Environ. Qual., 25: 363-371

Shen M H. 2003. The environmental variance of lake Dianchi drainage area and the social mechanism of environmental renovation. China Population, Resources and Environment, 13: 76-80

Søndergaard M, Windolf J, Jeppsen E. 1996. Phosphorus fractions and profiles in the sediment of shallow Danish lakes as related to phosphorus load, sediment composition and lake chemistry. Water Res., 30: 992-1002

Sundby B, Cobeil C, Silverberg N, et al. 1992. The phosphorus cycle in coastal marine sediments. Limnol. Oceanogr., 37: 1129-1145

Ulrich K U. 1997. Effects of land use in the drainage area on phosphorus binding and mobility in the sediment of four drinking-water reservoirs. Hydrobiologia, 345: 21-38

Whitmore T J, Brenner M, Jiang Z, et al. 1997. Water quality and sediment geochemistry in lakes of Yunnan Province, southern China. Environmental Geology, 32: 45-55

Wu D L, Qian B, He L H. 1992. Contributing factor analysis of eutrophication of Dianchi Lake. Research of Environmental Science, 5: 26-28

Yu Y T, Li P Q, Wu R, et al. 1996. Determinations of sedimentation rates in Kunming's Dianchi Lake. Oceanologia et Limnologia Sinica, 27: 41-45

Zhang X M, Zhang S, Ying W M, et al. 1996. Heavy metals pollution on the sediments in lakes Dianchi, Erhai and Poyanghu and historical records. Geo. Journal, 40: 201-208

Zhang Y, Peng B Z, Chen J et al. 2005. Evaluation of sediment accumulation in Dianchi Lake using ^{137}Cs dating. Acta Geographica Sinica, 60: 71-78

Zhe Y M. 2002. Eutrophication of Dianchi and its trend and treatment. Yunnan Environ. Sci., 21: 35-38

Zhou Q X, Gibson C E, Zhu Y M. 2001. Evaluation of phosphorus bioavailability in sediment of three contrasting lakes in China and the UK. Chemosphere, 42: 21-225

3.2 Removal of phosphate from aqueous solution by thermally activated palygorskite

3.2.1 Introduction

Phosphorous has long been recognized as an important element in plant nutrition. Recently, attention has been directed toward the role of phosphorous in the eutrophication of surface waters and the resultant degradation of water quality (Griffin and Jurinak, 1973). Significant sources of phosphorous entering freshwaters include drainage from agricultural land, excreta from livestock, municipal and industrial effluents, atmospheric deposition and diffuse urban drainage (Yeoman et al.,1988). Main lakes in China, Dianchi, Taihu, Chaohu and others, have been in an eutrophication condition to diverse extent (Feng and Wu, 2006). Many techniques have been applied to the removal of phosphate from wastewater (Karaca et al., 2006). Coagulation-precipitation and biological methods are widely accepted methods of phosphate removal at industrial level (Karaca et al., 2006; Kostantions et al., 2007). Extensive research has also been carried out to produce simplification of maintenance, stable running and removal efficiency. Among these researches, development of adsorbents with high selectivity and removal capacity for phosphate has been promoted (Ookubo et al.,1993; Shin et al., 1996; Haron et al., 1997; Zhang et al., 1997). The adsorbents tested include the following: fly ash (Grubb et al.,2000; Chen et al.,2006), red mud (Akay et al.,1998), iron and iron oxides (Seida and Nakano, 2002; Chitrakar et al.,2006), aluminum and aluminum oxides (Tanada et al., 2003), and silicates (Ye et al., 2006).

Palygorskite is a hydrated aluminum-magnesium silicate mineral with a fibrous morphology, and with a structure consisting of parallel ribbons of 2∶1 layers. It is characterized by a high viscosity, a high surface area, a moderate layer charge and a large number of silanol groups on its surface (Álvarez-Ayuso and Garcíía-Sánchez, 2007). And these physico-chemical characters make it a potentially attractive adsorbent. Palygorskite resource is abundant in China and the most abundant palygorskite deposits are located in the east of Anhui Province and the west of Jiangsu Province, China. About 60% palygorskite deposits of China are located at Xuyi County, Jiangsu Province, China (Ye et al., 2006). Palygorskite is excellent absorbents and the industrial applications of this mineral are numerous (drilling muds, agricultural carriers, industrial floor adsorbents, cat box absorbents, catalyst supports, gelling agents in paints, adhesives and cosmetics, reinforcing filler in rubbers and plastics and so on) (Álvarez-Ayuso and Garcíía-Sánchez, 2007). Recently, the application as environmental adsorbents has been paid more and more attention. It has been successfully used to remove colored component from vegetable oils (Murray, 2000), metal cations from water (Álvarez-Ayuso and Garcíía-Sánchez, 2007) and cationic dyes (Al-Futaisi et al., 2007). Its adsorption capacity can be increased by modifying its texture by means of chemical and/or thermal treatments (Ye et al., 2006). However, there is less systematic research in the dependence of thermal treatments on the phosphate adsorption capacity using palygorskite heated at different temperature. The objective of this work was to study the feasibility of using thermally activated palygorskite as adsorbents for phosphate removal from wastewater. The natural and heating palygorskite were characterized mineralogically by X-ray diffraction (XRD) and scanning electron microscopy (SEM), physically by specific surface area and micro-mesopore volume, and chemically by cation exchange capacity (CEC) and zeta potential (ζ-potential). The adsorption characteristics of these materials for phosphate removal from aqueous solutions were evaluated in batch experiments. Finally, an attempt is made to elucidate the influence mechanism of thermal activation on the adsorption behavior of phosphate.

3.2.2 Experimental Section

1. Adsorbents

The natural palygorskite (NPAL) used in the present experiments was mined from the Longwang Mountain at Xuyi County (China). The sample used was manually ground and selected for particles < 100 mesh. The bulk chemical analysis of the palygorskite (wt%): SiO_2, 27.09; CaO, 18.82; MgO, 13.36; Al_2O_3, 4.61; Fe_2O_3, 2.29; K_2O, 0.36; Na_2O, 0.06 and loss on ignition (at 950℃), 33.10. The modified palygorskite samples were prepared by thermal treatment at a period of 2 h over a range of heating temperatures (100, 200, 300, 350, 400, 450, 500, 600, 700, 800 and 1000℃). These were labeled H100, H200, etc.

2. Adsorbates

Artificial orthophosphate solutions were used throughout the adsorption tests. Initially, a stock solution of 1000 mg/L in orthophosphates was prepared by dissolving a certain amount of chemically pure $K_2HPO_4 \cdot 3H_2O$ in distilled water. An aliquot of the stock solution was mixed with a certain volume of water so that a phosphate solution was prepared at the desired experimental concentration.

3. Adsorption experiments

To evaluate the impact of activated palygorskite on the phosphate adsorption capacity, natural and heated palygorskite with different temperature were used. A defined volume of the phosphate stock solution, with concentrations of 200 mg P/L and 1000 mg P/L, was added to a weighted amount of adsorbents. The mixture was stirred at 200 rpm in thermostatic shaker for a defined period, then the orthophosphate removal was calculated and the optimum material was selected.

Phosphate adsorption isotherms were carried out with different initial concentrations of phosphate and a fixed concentration of the adsorbents at room temperature (25℃). Five hundred milligrams of the sample was loaded in 100 mL conical flasks, and 25 mL of PO_4^{3-} solution was then added. Thirteen levels of initial phosphate concentrations (5, 10, 20, 50, 100, 150, 200, 300, 400, 500, 600, 800, 1000 mg P/L) were used for natural palygorskite and the optimum heated sample. The flask was capped and stirred magnetically at 200 r/min for 24 h to ensure approximate equilibrium. After phosphate adsorption, the solution was filtered through a 0.22 μm membrane filter and then analyzed for PO_4^{3-}. The quantity of adsorbed phosphate (adsorption capacity) was calculated from the decrease of the phosphate concentration in solution.

Phosphate adsorption kinetics was evaluated at room temperature (25℃) and at original pH value. The initial phosphate concentration was 600 mg P/L while the adsorbent quantity was 0.5 g. Reaction solutions were sampled at intervals between 0 and 24 h of adsorption. The effect of pH on the removal of orthophosphate species, through adsorption onto natural and thermal activated palygorskite, was examined in the pH region between 3 and 9. The tests were carried out at constant experimental conditions with an initial solution concentration 600 mg P/L in orthophosphate and 0.5 g adsorbent. All the tests were carried out at constant retention time, fixed at 24 h.

4. Sample analysis

XRD patterns of the prepared samples were acquired with a D/MAX2200 X-ray diffractometer using CuKα radiation (40 kV, 40 mA) and a Ni filter. All XRD patterns were obtained from 3.0° to 60.0° with a scan speed of 4.0 °/min. A scanning electron microscope Hitachi X-650 was used for SEM analysis. The specific surface area was measured by the N_2 adsorption-desorption technique on an ASAP 2010 (Micrometrics) at liquid N_2 temperature (at −197℃). The cation exchange capacity (CEC) was determined by the ammonium acetate method (Tan, 1996). The Zeta-potential was measured by using a JS94H micro-electrophoresis meter.

The analysis of phosphate (as phosphorous) was done by the molybdenum-blue ascorbic acid method with a UV-Vis spectrophotometer (UV/VIS 721 model) (Soil Science Society of China, 2002). Each analysis point was an average of three independent parallel sample solutions. Triplicate tests showed that the standard deviation of the results was ±5%.

5. Phosphate fraction of the adsorbed phosphorus

A mass of 10.0 g of natural palygorskite and the optimum material was placed in a 1000 mL Erlenmeyer flask, and 500 mL of PO_4^{3-} solution with concentrations of 200 mg P/L was then added. The solutions were kept at room temperature (25℃), after 72 h of shaking, the shaker was stopped to reach 24 h equilibrium, and then the solutions were stirred for another 72 h. The exchanged samples were washed with distilled water until no phosphate can be checked in the filtrate and dried at 40℃ for 8 h. And phosphate-adsorption-saturated samples were acquired.

Phosphorous was defined into discrete chemical forms using a sequence of increasingly strong reagents to successively remove more recalcitrant forms according to the modified method of Chang and Jackson (1957) with the modifications proposed by Gu and Jiang (1990). Artificial orthophosphate solutions we used were inorganic phosphate, which was classified into four main groups: calcium phosphate (Ca-P), aluminum phosphate (Al-P), iron phosphate (Fe-P), and the occluded phosphate (O-P), then Ca-P was further classified into dicalcium phosphate (Ca_2-P), octacalcium phosphate (Ca_8-P) and ten-calcium phosphate (Ca_{10}-P). Briefly, the fractionation procedure was performed as follows: 1.0000 g of air-dried sample was transferred to 100 mL flasks and sequentially extracted with ① 0.25 mol/L $NaHCO_3$ (pH=7.5), 20-25℃, 1 h (Ca_2-P); ② 0.5 mol/L NH_4Ac (pH=4.2), 20-25℃, 1 h (Ca_8-P); ③ 0.5 mol/L NH_4F (pH=8.2), 20-25℃, 1 h (Al-P); ④ 0.1 mol/L NaOH-0.05 mol/L Na_2CO_3, 20-25℃, 4 h (Fe-P); ⑤ 0.3 mol/L $Na_3C_6H_5O_7·2H_2O$-0.5 mol/L NaOH, 80-90℃, 25 min (O-P); and ⑥ 0.25 mol/L H_2SO_4, 20-25℃, 1 h (Ca_{10}-P).

3.2.3 Results and Discussion

1. Mineralogy of the palygorskite

The XRD patterns of the natural and some heated samples are given in Fig. 1. According to the XRD peak intensities, the natural sample contains 40%-50% palygorskite with a d_{110} value of 1.04 nm and moderate reflections at 0.64, 0.42, 0.36 and 0.32 nm, a trace amount of quartz (q) (±5%) and feldspar (f) (±5%), a large amount of dolomite (d) (35%-45%), and the amount of non-clay minerals seems too small supported by the chemical analysis.

As seen in Fig. 1, heating to 300℃ did not result in any substantial collapse of d_{110} value. The 110 peak of palygorskite decreased from 1.04 to 0.91 nm while maintaining its position after heating between 300-400℃ with approximately half of its coordinated water removed, while the 0.36 nm reflection gained in intensity (Dixon and Weed, 1989). The position and intensity of the 110 peak disappeared after heating at 600℃. The position for the characteristic spacing for palygorskite was observed to change as the content of water was varied. Similar results have been previously reported (Hirsiger et al., 1975; Van-Scoyoc et al., 1979). It was assumed to be due to the effect of the water molecules on the channel dimensions (Van-Scoyoc et al., 1979). The intensity of dolomite was decreasing after heating from 700 to 1000℃. The peak for crystalline quartz remained virtually the same after each thermal treatment, indicating that the crystallinity of quartz is relatively little influenced by heating. Some new calcium-aluminum silicate with a *d* value of 0.21 nm appeared after heating at 700℃, and its peak intensifies greatly between 800 and 1000℃ with it becoming more crystalline, which may also be due to the decomposition of the palygorskite and dolomite.

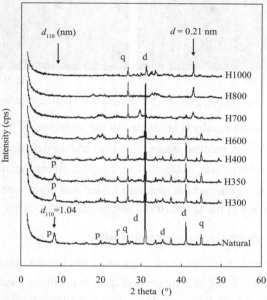

Fig. 1 XRD patterns of the natural palygorskite (NPAL) and some heated (H300, H350, etc.) samples
p: palygorskite, f: feldspar, q: quartz, d: dolomite

The effects of heating on SEM photos of natural palygorskite and some heated samples (H400, H700 and H1000) are shown in Fig. 2 as examples. Palygorskite displayed lath-like or fibrous forms (Fig. 2(a)), representing crystals elongated along the *a*-axis[2] (Dixon and Weed, 1989). The length of palygorskite fibers varies greatly, from < 1 μm up to nearly 20 μm, average lengths vary between 1 to 2 μm (Dixon and Weed, 1989). The individual fibers have diameters varying from 0.1 to 0.5 μm and often are arranged in bundles or sheaths. Transverse cross-sections of palygorskite fibers were reported to appear cylindrical, polygonal or six-sided (Dixon and Weed, 1989). As the temperature increases to 400℃, no apparent changes in crystal structure of the palygorskite occurred because of reversible dehydration which

(a) NPAL　　　　　　　　　　　　　(b) H400

(c) H700　　　　　　　　　　　(d) H1000

Fig. 2　SEM photos of natural palygorskite (NPAL) and some heated samples (H400, H700, H1000)

was in accordance with XRD results (Fig. 2(b)). As the temperature increases above 700℃, the space between the layers was reduced because of irreversible dehydration and dehydroxylation (Fig. 2(c)). At higher temperatures, from 800 to 1000℃, fibers became shrinked and curled, intra- and interparticles sintered, and the structure folded and pores were blocked resulting from decomposing and collapsing of the layers of the palygorskite (Fig. 2(d)).

2. Physico-chemical properties of the palygorskite

The specific surface area, S (m^2/g), were obtained from the standard Brunauer, Emmett and Teller (BET) procedure by using the adsorption data from the interval $0.05 < x\ (p/p_0) < 0.4$ (Brunauer et al., 1938; McClellan and Hornsberger, 1967; Sarikaya et al., 2002a). The adsorption capacities as liquid nitrogen volumes which were estimated from desorption isotherms at $x = 0.95$ are taken as the specific micro-mesopore volumes, V (cm^3/g), of the samples. As the temperature increases, the S and V values show a "zig-zag" change at the beginning, and there was no apparent change in the S and V values when the temperature increased to 700℃, then the values decreased rapidly from 47.0 to 3.7 m^2/g, from 201 to 15.7 m^3/g, respectively (Fig. 3). The zig-zag change in S and V originates mostly from the deformation of the palygorskite during dehydration over the interval of 80-540℃ and dehydroxylation over the range of 540-770℃ (Dixon and Weed, 1989). The decomposition of the layers of the palygorskite and collapsing of micro-mesopores by intra- and interparticle sintering, when the structure folds (Serna et al., 1975) and pores are blocked, causes the rapid decreases in S and V values at higher temperatures.

The variation of CEC by thermal treatment is given in Fig. 4. Note that the CEC of original palygorskite is about 4 cmol/kg and has no apparent change with increasing temperature to 400℃, then decreases rapidly and approaches zero at 700℃. The CEC of relatively pure palygorskite samples range from 5 to 30 cmol/kg (Dixon and Weed, 1989). The decrease in the CEC is a reflection of the deformation processes. As the temperature increased above 600℃, the space between the layers is reduced. The decomposition of the layers and collapsing of the interlayer causes the rapid decrease in CEC because it becomes increasingly difficult for the exchangeable cations to enter the interlayer of palygorskite. Fig. 4 also shows the changes in the zeta potential (ζ-potential) related to the heating temperature. The ζ-potential is related to the electrophoretic mobility of the particles by the Smoluchowski equation (Pashley, 1985).

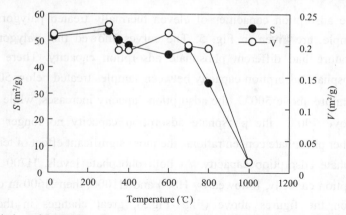

Fig. 3　The variation of the BET surface area and the pore volume as calculated from the adsorption and desorption data of N_2 at liquid N_2 temperature vs. the thermal-treatment temperature

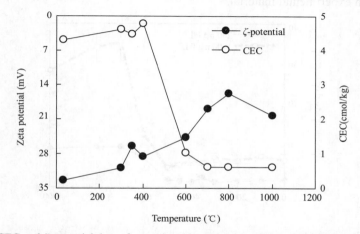

Fig. 4　CEC and ζ-potential data of natural and heated palygorskite with different temperature

There is a decrease in the particle mobility (ζ-potential becomes less negative) when the heating temperature increases. The decreases in ζ-potential suggest more aggregation of clay particles or the formation of large domains, and correspond to reduction in charge. In general, the ζ-potential did not change as much as expected on heating, possibly because charge reduction was due to a decrease in the CEC caused by irreversible collapse of the layers. The data on CEC and ζ-potential reflect the influence of thermal treatment on dispersion and particle charge. Heating results in severe charge reductions and the cations become non-exchangeable.

The surface area, the micro-mesopore volume, the CEC and the ζ-potential of the palygorskite change significantly under increased heating temperature. The quantitative XRD analysis indicates that the essential changes of the palygorskite occur at 700℃, and the SEM images clearly support this conclusion. As the temperature increases above 700℃, the layers of the palygorskite collapse, the structure was folded and pores were blocked, and all these cause the rapid decrease in surface area, pore volume values and CEC data.

Phosphate Adsorption Capacity of Heated Samples

The phosphate adsorption capacities of eleven thermally treated palygorskite, compared with natural sample, are given in Fig. 5. The curves showed that palygorskite treated at different temperature had different phosphate adsorption capacity. There is no apparent difference in phosphate adsorption capacity between samples treated below 500℃. when with the rising temperature above 500℃, the adsorption capacity increases, while the temperature increased to above 700℃, the phosphate adsorption capacity no longer increases, even reduces. The higher phosphate concentrations, the more significant effect of temperature above 500℃ on phosphate adsorption capacity. At both phosphate levels, H700 had the highest phosphate adsorption capacity, followed by H800 and H1000, then H600 in comparison with NPAL. Based on the figures above (Fig.1-Fig.5), great changes in the structure and physico-chemical properties of palygorskite occurred at 700℃, so we choose H700 as the next batch adsorption experimental material.

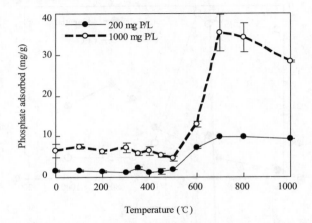

Fig. 5 Variation of the phosphate adsorption capacity of heated palygorskite vs. the heating temperature

Batch Sorption

The adsorption data of NPAL and H700 are summarized in Fig. 6, Table 1 and Table 2. Adsorption isotherms plotted as adsorbed amount (q, mg/g) vs. solution concentration (C_e, mg/L) at adsorption equilibrium are shown in Fig. 6(a) for selected samples. According to the results of the phosphate adsorption isotherm experiments, H700 had higher adsorption capacities than the natural sample, which were 42.0 mg/g and 12.9 mg/g, respectively, at the phosphate equilibrium concentration of 1000 mg/L. The adsorption data were fitted with the Freundlich isotherm:

$$q = KC^n$$

Where, q (mg/g) and C (mg/L) are the equilibrium concentrations of an adsorbate on the palygorskite and in the aqueous solution, respectively; K (mg^{1-n}L^n/g) is the Freundlich affinity coefficient; and n (unitless) is the Freundlich linearity index. The adsorption data agree with the Freundlich isotherm reasonably. It was reported that the isotherms of phosphate ion adsorption onto many adsorbents fit the Freundlich equation (Tanada et al.,2003). The two samples exhibited varied degrees of linearity with different n values. As indicated in Table 1, H700 has a much higher K value than NPAL. Thus, the adsorption of phosphate by H700 should have been much stronger than by NPAL.

Table 1 Estimated isotherm parameters for phosphate adsorption

Temperature (℃)	NPAL			H700		
	K	n	R^2	K	n	R^2
25	0.013	0.983	0.990	18.43	6.536	0.922

Table 2 Estimated kinetic model parameters for phosphate adsorption

	First-order equation			Second-order equation			The Elovich equation			Intraparticle diffusion equation	
	q_1 (mg/g)	k_1 (1/min)	R^2	q_2 (mg/g)	k_2 [g/(mg·min)]	R^2	α	β	R^2	k_{int} [mg/(g·min$^{1/2}$)]	R^2
NPAL	6.56	7.17	0.988	7.95	1.35	0.963	9.874	1.036	0.855	12.03	0.962
H700	6.79	6.94	0.689	26.88	9.66	0.999	162900	0.537	0.965	14.05	0.616

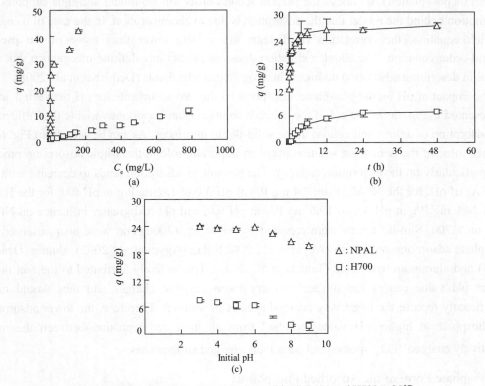

Fig. 6 Adsorption data of natural palygorskite (NPAL) and H700 at 25℃

(a) Isotherms plotted as equilibrium adsorbed amount (q) vs. aqueous phase concentration (C_e) for NPAL and H700. (b) Phosphate adsorption kinetic data for NPAL and H700. (c) Impact of pH on phosphate adsorption capacity for NPAL and H700

In Fig. 6(b), the results of phosphate adsorption kinetic data for NPAL and H700 at 25℃ are compared. The H700 had faster kinetics than the natural sample, the majority of phosphate adsorption on the H700 was completed in 4 h. Four typical kinetic equations, as described below in Eq. (1)-Eq.(4), were used for fitting the experimental data:

Pseudo first-order equation: $\ln(q_e - q_t) = \ln q_e - k_1 t$ (1)

Pseudo second-order equation: $t/q_t = 1/k_2 q_e^2 + t/q_e$ (2)

Simple Elovich equation: $q = \alpha + \beta \ln t$ (3)

Intraparticle diffusion equation: $q_t = k_i t^{1/2}$ (4)

Where, q_t and q_e are the amount adsorbed at time t and equilibrium (mg/g), respectively; and t is the adsorption time. The other parameters are different kinetic constants, which can be determined by regression of the experimental data. Four kinetic equations and estimated parameters with R^2 are shown in Table 2. Based on R^2, the kinetic of phosphate adsorption on NPAL can be satisfactorily described by the pseudo first-order and second-order function and Intraparticle diffusion equation. The pseudo first-order (and also the pseudo-second order) is based on the adsorption capacity: it only predicts the behavior over the "whole" range of studies supporting the validity, and is in agreement with chemisorption being the rate-control (Bulut et al., 2008). The results of intraparticle diffusion rate constants showed that the plots presented a multilinearity, which indicated that two or more steps occurred in the process. The adsorption system of phosphate/H700 obeys the pseudo second-order kinetic model and thus supports the assumption behind the model that the adsorption is due to chemisorption. In the case of using the Elovich equation, the correlation coefficients are a little lower than those of the pseudo second-order equation. The Elovich equation does not predict any definite mechanism, but it is useful in describing adsorption on highly heterogeneous adsorbents (Demirbas et al., 2008).

The impact of pH on the phosphate adsorption on the two adsorbents for pH between 3 and 9 is presented in Fig. 6(c). The pH of the aqueous solution is an important variable that influences the adsorption of anions and cations at the solid-liquid interfaces. As can be seen from Fig. 6(c), the pH value of the phosphate solution plays an important role in the whole adsorption process and particularly on the adsorption capacity. The phosphate adsorption tends to decrease with the increase of pH, for the NPAL, from 7.4 mg P/g at pH 3.0 to 1.6 mg P/g at pH 9.0; for the H700, from 24.1 mg P/g at pH 3.0 to 19.6 mg P/g at pH 9.0, and pH had greater influence on NPAL than on H700. Similar trends were reported by Ye et al., (2006) and were also observed for phosphate adsorption on fly ash (Agyei et al., 2002), slag (Agyei et al., 2002), alunite (Özacar, 2003) and aluminum hydroxide (Tanada et al., 2003). This is likely attributed to the fact that a higher pH value causes the surface to carry more negative charges and thus would more significantly repulse the negatively charged species in solution. Therefore, the lower adsorption of phosphate at higher pH values resulted from an increased repulsion between the more negatively charged PO_4^{3-} species and negatively charged surface sites.

Phosphate Form of the Adsorbed Phosphorus

As presented in Table 3, the highest amount of inorganic phosphorus observed was the $NaHCO_3$ extracted fraction Ca_2-P, followed by the NH_4Ac extracted Ca_8-P, and only a small percentage was measured in Al-P, Fe-P, Ca_{10}-P and O-P species in adsorbed phosphate on NPAL. The distribution of phosphorus in each fraction decreased in the order: Ca_2-P (87.7%) > Ca_8-P (10.1%) > Al-P (1.10%) > Fe-P (0.88%) > Ca_{10}-P (0.16%) > O-P (0.04%). But for H700, the distribution in each fraction was following the order: Ca_8-P (54.5%) > Al-P (29.8%) > Fe-P

(11.6%) > Ca_2-P (2.96%) > Ca_{10}-P (0.7%) > O-P (0.4%). In both samples, calcium bound phosphorus (Ca-P) is the main fraction of the adsorbed phosphorus, about 97.9% in NPAL and 58.1% in H700, and the Ca-P combination is considered to be the principle factor in the control of phosphate; while metal bound phosphorus (Al-P + Fe-P) differs totally, less than 2% in NPAL but more than 41.4% in H700. Another great difference lies in the percentage of Ca_2-P and Ca_8-P fraction in NPAL and H700, the binding force in Ca_8-P is stronger than that in Ca_2-P, it reflects that the phosphorus adsorbed on H700 is more difficult to transform than that adsorbed on NPAL.

Table 3 Phosphate fractionation in discrete chemical forms

Phosphorus fraction	Extractant	NPAL		H700	
		Amount extracted	Percent (%)	Amount extracted	Percent (%)
Ca_2-P	$NaHCO_3$	3013.6	87.7	167.4	2.96
Ca_8-P	NH_4Ac	346.5	10.1	3085.4	54.5
Al-P	NH_4F	37.8	1.1	1688.8	29.8
Fe-P	NaOH-$NaHCO_3$	30.2	0.88	656.8	11.6
O-P	$Na_3C_6H_5O_7 \cdot 2H_2O$-NaOH	1.5	0.04	22.4	0.4
Ca_{10}-P	H_2SO_4	5.5	0.16	39.4	0.7
Total amount (mg/kg)		3435.1		5660.2	

Proposed by several former papers, the Ca^{2+} ions are assumed to be absorbed in the tunnel or on the surface of palygorskite fibers to balance the charges in the octahedral and tetrahedral layers (Samtani et al., 2002), Mg^{2+} cations can be confirmed to principally occupy octahedral sites, Al^{3+} ions are generally assumed to occupy both octahedral and tetrahedral sites (García-Romero et al., 2004; Sánchez del Río et al., 2005), and Fe cations tend to occupy inner sites in octahedral layers (Samtani et al., 2002). Therefore, the phosphate is easy to be adsorbed by Ca in NPAL. Based on the XRD data, there is also a large amount of dolomite in the NPAL clay, and naturally occurring dolomite ($CaMg(CO_3)_2$) is a double salt of calcium and magnesium carbonate (Samtani et al., 2002). When the heating temperature increased to 700℃, the decomposition of dolomite occurred and it decomposed into MgO and CaO. Al-O-Si bonds broken at above 400℃, the lattice structure was strongly distorted when the —OHs were broken at temperatures higher than 600℃. When it was heated above 700℃, most of the —OHs was lost. The iron oxides and aluminum oxides also appeared, they have immobilized phosphate and have an important role in phosphate adsorption and they have been reacted with SiO_2 to produce some unknown calcium-aluminum silicate which was demonstrated by XRD figures.

The specific surface areas of NPAL and H700 were 51.3 and 47.0 m^2/g, respectively. Specific surface area maybe also a great contributor to phosphate adsorption capacity, but it is not the most important factor, for example, H300 didn't have the highest adsorption capacity, which had the highest surface area values and pore volume values. The adsorption findings presented suggest that the mechanisms of phosphate/NAPL and phosphate/H700 adsorption system are chemical sorption or chemisorption involving valence forces through sharing or exchange of electrons between adsorbent and adsorbate. It was believed that the chemical changes of the palygorskite play the most important role in the adsorption capacities of phosphate.

References

Agyei N M, Strydomb C A, Potgieter J H. 2002. The removal of phosphate ions from aqueous solution by fly ash, slag, ordinary Portland cement and related blends. Cement Concrete Res., 32: 1889-1897

Akay G, Keskinler B, Cakici A, et al. 1998. Phosphate removal from water by red mud using crossflow microfiltration. Water Res., 32: 17-726

Al-Futaisi A, Jamrah A, Al-Hanai R. 2007. Aspects of cationic dye molecule adsorption to palygorskite. Desalination, 214: 327-342

Álvarez-Ayuso E, Garcíía-Sánchez A. 2007. Removal of cadmium from aqueous solutions by palygorskite. J. Hazard. Mater., 147: 594-600

Brunauer S, Emmett P H, Teller E. 1938. Adsorption of gases in multimolecular layers. J. Am. Chem. Soc., 60: 308-319

Bulut E, Özacar M, Şengil I A. 2008. Equilibrium and kinetic data and process design for adsorption of Congo Red onto bentonite. J. Hazard. Mater., 154: 613-622

Chang S C, Jackson M L. 1957. Fractionation of soil phosphorus. Soil Sci., 84: 133-144

Chen J G, Kong H N, Wu D Y, et al. 2006. Removal of phosphate from aqueous solution by zeolite synthesized from fly ash. J. Colloid. Interface Sci., 300: 491-497

Chitrakar R, Tezuka S, Sonoda A, et al. 2006. Phosphate adsorption on synthetic goethite and akaganeite. J. Colloid Interface Sci., 298: 602-608

Demirbas E, Kobya M, Sulak M T. 2008. Adsorption kinetics of a basic dye from aqueous solutions onto apricot stone activated carbon. Bioresource Technol., 99: 5368-5373

Dixon J B, Weed S B. 1989. Minerals in Soil Environments. 2nd Ed. Wisconsin: Soil Science Society of America

Feng M W, Wu Y H. 2006. Measures for the control of lake eutrophication at home and abroad. Guangzhou Environ. Sci., 21: 8-11 (in Chinese)

García-Romero E, Suárez Barrios M, Bustillo Revuelta M A. 2004. Characteristics of a Mg-palygorskite in Miocene rocks, Madrid basin (Spain). Clay. Clay Miner., 52: 484-494

Griffin R A, Jurinak J J. 1973. The interaction of phosphate with calcite. Soil. Sci. Soc. Amer. Proc., 37: 847-850

Grubb D G, Guimaraes M S, Valencia R. 2000. Phosphate immobilization using an acidic type F fly ash. J. Hazard. Mater.,76: 217-236

Gu Y C, Jiang B F. 1990. An method for fractionation of inorganic phosphorus of calcarious soils. Soils, 22: 101-102 (in Chinese)

Haron M J, Wasay S A, Tokunaga S. 1997. Preparation of basic yttrium carbonate for phosphate removal. Water Environ. Res., 69: 1047-1051

Hirsiger W, Muller-Vonmoos M, Wiedemann H G. 1975. Thermal analysis of palygorskite. Thermochim. Acta, 13: 223-230

Karaca S, Gurses A, Ejder M, et al. 2006. Adsorptive removal of phosphate from aqueous solutions using raw and calcinated dolomite. J. Hazard. Mater., B128: 273-279

Kostantions K, Maximos P, Georgios N. A. 2007. Removal of phosphate species from solution by adsorption onto calcite used as natural adsorbent. J. Hazard. Mater., A139: 447-452

McClellan A L, Hornsberger H F. 1967. Cross-sectional areas of molecules adsorbed on solid surfaces. J. Colloid Interface Sci., 23: 577-599

Murray H H. 2000. Traditional and new applications for kaolin, smectite, and palygorskite: a general overview. Appl. Clay Sci., 17: 207-221

Ookubo A, Ooi K, Hayashi H. 1993. Preparation and phosphate ionexchange properties of a hydrotalcite-like compound. Langmuir., 9: 1418-1422

Özacar M. 2003. Adsorption of phosphate from aqueous solution onto alunite. Chemosphere, 51: 321-327

Pashley R M. 1985. Electromobility of mica particles dispersed in aqueous solutions. Clay. Clay Miner., 33: 193-199

Samtani M, Dollimore D, Alexander K S. 2002. Comparison of dolomite decomposition kinetics with related carbonates and the effect of procedural variables on its kinetic parameters. Thermochim. Acta, 392-393: 135-145

Sánchez del Río M, Suárez M, García Romero E, et al. 2005. Mg K-edge XANES of sepiolite and palygorskite. Nucl. Instr. and Meth. B., 238: 55-60

Sarikaya Y, Ada K, Alemdaroglu T, et al. 2002a. Bozdogan, I. The effect of Al^{3+} concentration on the properties of alumina powders obtained by reaction between aluminium sulphate and urea in boiling aqueous solution. J. Eur. Cream. Soc., 22: 1905-1910

Seida Y, Nakano Y. 2002. Removal of phosphate by layered double hydroxides containing iron. Water Res., 36: 1306-1312

Serna C, Ahlrichs J L, Serratosa J M. 1975. Folding in sepiolite crystals. Clay. Clay Miner., 23: 452-457.

Shin H S, Kim M J, Nam S Y, et al. 1996. Phosphorus removal by hydrotalcite-like compounds (HTLcs). Water Sci. Technol., 34: 161-168

Soil Science Society of China. 2002. Soil Agrochemistry Analysis Method. Beijing: China Agricultural Press (in Chinese)

Tan K H. 1996. Soil Sampling, Preparation and Analysis. New York: Marcel Dekker Inc.

Tanada S, Kabayama M, Kawasaki N, et al. 2003. Removal of phosphate by aluminum oxide hydroxide. J. Colloid Interface Sci., 257: 135-140

Van-Scoyoc G E, Serna C J, Ahlrichs J L. 1979. Structural changes in palygorskite during dehydration and dehydroxylation. Am. Mineral., 64: 215-223

Ye H P, Chen F Z, Sheng, Y. Q, et al. 2006. Adsorption of phosphate from aqueous solution onto modified palygorskites. Sep. Purif. Technol., 50: 283-290

Yeoman S, Stephenson T, Lester J N, et al. 1988. The removal of phosphorus during wastewater treatment: a review. Environ. Pollut., 49: 183-233

Zhang W X, Sakane H, Hatsushika T, et al. 1997. Phosphorus anion-exchange characteristics of a pyroaurite-like compound. Inorg. Mater., 4: 132-138

3.3 Short-term effects of copper, cadmium and cypermethrin on dehydrogenase activity and microbial functional diversity in soils after long-term fertilization

3.3.1 Introduction

Soil microbes respond rapidly to farm management and play important roles in many ecosystem processes such as biogeochemical cycling of nutrients, energy flow as well as transformation of contaminants (Doran and Zeiss, 2000; Sukul, 2006; Zhang et al., 2006). Thus maintenance of the biological activity in soil is generally regarded as a key feature of sustainable production to ensure ecosystem functions (Swift, 1994). Upon contaminants entering into soils, agricultural ecosystem was disturbed, and disturbance extent was affected by soil properties (Sannino and Gianfreda, 2001; Lu et al., 2004; Wang et al.,2004; Kong et al., 2006). Many studies have proved that increase of soil organic matter content by addition of organic manures can decrease the bioavailability of heavy metals and pesticides in soils (Pérez-de-Mora and Madrid, 2007; van Herwijnen et al., 2007; Sannino and Gianfreda, 2001). Long-term fertilization can result in microbial community shifts in soils (Marschner et al., 2003; Widmer et al., 2006; Chu et al., 2007), which would result in the varied sensitivity to soil pollution, because microbial community responded differently to soil contaminants (Gflier et al., 1998; Wang et al., 2004). Therefore, studies on the interaction between fertilization and contaminant bioavailability can develop approaches to eliminate or decrease the eco-environmental risks of those contaminants through disturbance of microbial communities.

Cu, Cd and pesticides, as the main soil contaminants, are entering agricultural field through waste irrigation (Hu et al., 2006), fertilization and agrochemicals application (Zheljazkov and Warman, 2003; Perkiömäki and Fritze, 2005; Wang et al., 2006). With the organophosphate products being phased out, the use of pyrethroid insecticides is increasing (Amweg et al., 2005). The synthetic pyrethroid cypermethrin is commonly found in rivers, sediments, soils and even foodstuffs (Allan et al., 2005; Amweg et al., 2005; Sannino et al., 2003), and thus concerns regarding its fate and effect on eco-environment are increasing. Soil dehydrogenase activity is an indicator of soil quality and microbial activity. It is the most frequently used to determining the influence of the various pollutants (heavy metals, pesticide) on the microbiological quality of soils (Brookes, 1995; Pascual et al., 2000; Sannino and Gianfreda, 2001). Determining community level substrate utilization (CLSU) patterns with the Biolog^{TM} system is one approach to characterize microbial communities (Garland and Mills, 1991). CLSU assays have been applied in many studies in order to gain information on responses of microbial communities to fertilization (Widmer et al., 2006), heavy metals (Akmal et al., 2005) and pesticide (de Lipthay et al., 2004). To our knowledge, the effects of cypermethrin on CLSU are still not well understood. In fact, microorganisms are usually exposed simultaneously or sequentially to a variety of pollutants via multiple exposure routes (Feron and Groten, 2002). Hence, it is of great significance for us to evaluate the combined effects of heavy metals and cypermethrin in soils to ensure agroecosystem sustainability.

Changes in soil quality may develop slowly and may adjust to a new long-term steady state after a change of management or conversion to a different farming system (Widmer et al., 2006). Therefore, long-term agricultural field experiments might reveal more changes that were undetectable in short-term fertilization studies. In our present study, a 17-year field experiment was conducted to elucidate ① the effect of long-term fertilization on soil microbial activity, and ② the responses of soil microbial community to Cu, Cd, cypermethrin and their combinations as well as their differences in three kinds of fertilized soils.

3.3.2 Materials and Methods

1. Soil description and preparation

This study was carried out in a long-term field experiment site, which was established in 1989 in the Fengqiu Ecological Experimental Station (35°04′N, 113°10′E) of the Chinese Academy of Sciences, Henan Province, China. The soil in this area was derived from alluvial sediments of the Yellow River, and classified as aquic inceptisol. It has a sandy loam texture (about 9% clay, 21.8% silt). Three fertilization modes, i.e., wheat straw compost (OM), fertilizer NPK (NPK) and no fertilization (CK) were employed in this study, which were under an annual crop rotation of winter wheat and summermaize. The annual application rates of N, P and K in both OM and NPK modes were 150 kg N/ha, 32.7kg P/ha and 124 kg K/ha for winter wheat, respectively, and 150 kg N/ha, 26.2 kg P/ha and 124 kg K/ha for maize, respectively. Compost was applied in the OM mode, which was made of wheat straw mixed with soybean cake and cotton seed cake to enrich N content. N, P and K contents were determined before application, and then the amount needed for application was calculated based on the N content. Inorganic fertilizers complement the shortage of P and K from the applied compost. Inorganic N, P and K fertilizers were applied as urea, superphosphate and $K_2(SO_4)$, respectively. For each fertilization mode, soil samples (0-20 cm) were collected from 16 points, then mixed and sieved through a 2 mm sieve to remove the roots in April 2006. The fundamental properties of soil samples were shown in Table 1. Soil pH was measured in 1 : 2.5 soil-water suspension. Soil organic matter content, total N and total P were determined by dichromate oxidation, Kjeldahl digestion, and sodium carbonate fusion, respectively (Lu, 2000). Available N was extracted with 2 mol/L KCl and analyzed with a segmented flow analyzer. Available P and available K were extracted by sodium bicarbonate and ammonium acetate, respectively (Lu, 2000).

Table 1 The physicochemical properties of soils with different fertilization mode

Fertilization mode	pH (H_2O)	SOM (g/kg)	Total N (g/kg)	Total P (g/kg)	Available N (mg/kg)	Available P (mg/kg)	Available K (mg/kg)
OM	8.03b	13.84a	0.94a	0.64a	30.57ab	45.34a	171.9a
NPK	8.26b	9.31b	0.58b	0.65a	37.20a	24.80b	147.0b
CK	8.51a	6.83c	0.44c	0.52b	12.75b	2.05c	57.1c

The background contents of Cu^{2+} and Cd^{2+} in three fertilized soils had no significant difference ($P > 0.05$) and were 14.2-14.7 mg/kg and 0.29-0.37 mg/kg, respectively. Pilot study showed that the critical doses of Cu and Cd which inhibit cypermethrin degradation in used soils were 100 mg Cu/kg and 5 mg Cd/kg, respectively.

2. Experimental design

Soil samples of each fertilization mode were amended with Cu^{2+} (100 mg/kg, $CuCl_2 \cdot 2H_2O$) and Cd^{2+} (5 mg/kg, $CdCl_2 \cdot 2.5H_2O$) dissolved in water, respectively. The contaminated soils were shaken and sieved through a 2 mm sieve again to distribute Cu^{2+} and Cd^{2+}

homogeneously. The water contents in both contaminated and uncontaminated soils were adjusted to 60% of soil water-holding capacity and stored for 30 days.

Cu^{2+}, Cd^{2+} contaminated soils and uncontaminated soils were amended with cypermethrin (purity >96%, Sigma) dissolved in acetone (0.5 mL) to obtain a pesticide concentration of 10 mg/kg, which is a field residue after pesticide application in China (Xie et al., 2008). The pesticide was mixed homogeneously when acetone evaporated off. In sum, there were six treatments for each fertilization soil, i.e., uncontaminated soil as control (B), Cu^{2+} contaminated soil (Cu), Cd^{2+} contaminated soil (Cd), cypermethrin contaminated soil (C), combination of Cu^{2+} and cypermethrin (Cu + C), and combination of Cd^{2+} and cypermethrin (Cd + C). Soil samples (dry weight equivalent) of 20 g in each treatment was adjusted to 60% of soil water-holding capacity and placed in beakers, covered with perforated aluminum foil. All treatments were made in triplicate and incubated at $(25 \pm 2)°C$ for 28 days.

3. *Soil dehydrogenase activity and Biolog Eco-plate analyses*

After incubation for 0, 7, 14, 21 and 28 days respectively, soil dehydrogenase activities (DHA) were determined by the reduction of triphenyltetrazolium chloride to triphenylformazan (Casida et al., 1964). At day 14, soil microbial functional diversity of each treatment was measured in Biolog Eco-plates (Biolog, Hayward, CA). The method was similar to the description by Waldrop et al. (2000) and Kong et al. (2006). Briefly, 10 g (dry weight equivalent) of incubated soils was suspended in 100 mL sterile saline solution (0.85%, m/v) with 5 g of 3 mm glass beads on a rotary shaker at 300 r/min for 10 min at 25°C. Suspensions (150 mL) from a 10^{-3} dilution were added to Biolog Eco-plates. Plates were incubated at 28°C and read on a BIOLOG^{TMM} Microplate Reader every 24 h over 7 days. The readings at 72 h incubation were used for subsequent analysis (see below). At this reading time, the microbial substrate utilization potential (OD) had stabilized and the greatest differences in utilization pattern were observed.

4. *Statistical analysis*

Well absorbance values were adjusted by subtracting the absorbance of the control well (water only) before data analysis, and substrates with an OD < 0 were excluded from further analysis. Average well color development (AWCD), calculated as the average optical density across all wells per plate, was used as an indicator of general microbial activity. Soil microbial community structure and functional diversity were determined using Shannon index and evenness (Zak et al., 1994; Staddon et al., 1997). Shannon index was measured by $H' = -\sum p_i \ln p_i$, and p_i was calculated by subtracting the control from each substrate absorbance and then dividing this value by the total color change recorded for all 31 substrates. Evenness was calculated as $E = H' / \ln(richness)$, where *richness* referred to the number of substrates utilized (Zak et al., 1994). According to their chemical nature, the substrates were divided into six substrate categories, i.e. carboxylic acids, amines and amides, amino acids, polymers, and miscellaneous (Preston-Mafham et al., 2002), and total absorbance of each category was calculated. DHA, AWCD, substrate utilization of each substrate category, Shannon diversity and evenness were compared by ANOVA with the SPSS 10.0 for Windows.

3.3.3 Results

1. *Soil dehydrogenase activities (DHA)*

The changes of DHA in all treatments are shown in Fig. 1. Due to long-term different fertilization, DHA differed significantly in three fertilized soils ($P < 0.01$). The highest value was observed in OM soil, followed by NPK, and the lowest in CK soil. Cd of 5 mg/kg had little effect on DHA, while Cu of 100 mg/kg suppressed DHA significantly ($P < 0.01$). During the entire period of incubation, Cu down-regulated DHA by 52.3%, 41.4% and 25.6% in NPK soil, CK soil and OM soil, respectively. The results indicated that DHA in NPK and CK soils were more sensitive to Cu than that in OM soil. Meantime, for each fertilization soil, DHA in the treatments of C, Cu + C and Cd + C were higher than those in the treatments of B, Cu and Cd, respectively ($P < 0.05$). Therefore, cypermethrin of 10 mg/kg can significantly improve DHA, which was in accord with the report by Rangaswamy et al. (1994).

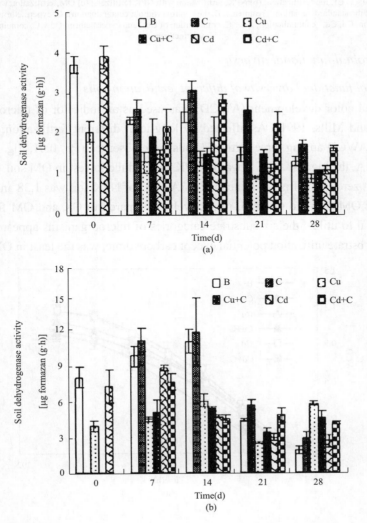

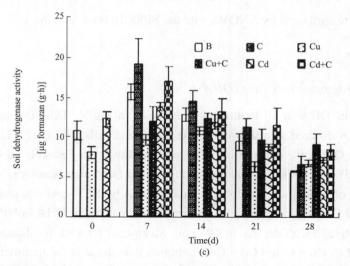

Fig. 1　Changes of soil dehydrogenase activities in three fertilized soils following addition of Cu, Cd and cypermethrin singly or in combination

Fertilization mode: (a) CK, no fertilization; (b) NPK, fertilization with NPK fertilizers; (c) OM, fertilization with organic manure. Error bars represent the standard deviation. Treatments: B, the control (without contamination); C, cypermethrin contamination; Cu, Cu^{2+} contamination; Cd, Cd^{2+} contamination; Cu + C, combination of Cu^{2+} and cypermethrin; Cd + C, combination of Cd^{2+} and cypermethrin

2. Soil microbial functional diversity

(1) Microbial functional diversity of different fertilization soils

Average well color development (AWCD) was used as an indicator of microbial activity in soil (Garland and Mills, 1991). As affected by long-term different fertilization, the significant differences of AWCD among three fertilized soils were observed ($P < 0.01$, Fig. 2). Contrary to changes of DHA, the highest AWCD was in NPK soil and the lowest in OM soil. AWCD of OM soil was almost zero in the first 24 h incubation. AWCD of NPK soil was 1.38 and 1.21 times as high as that of OM soil and CK soil, respectively. Compared to CK and OM fertilization, the highest potential to utilize the six substrate categories of microorganisms appeared in NPK soil (Fig. 3). The substrate utilization potential except carbohydrate, was the least in OM soil.

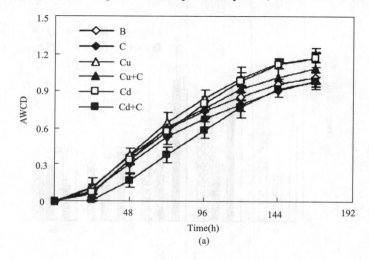

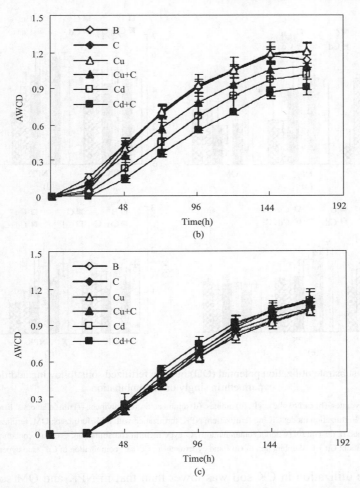

Fig. 2 AWCD of 31 carbon sources in different fertilized soils following addition of Cu, Cd and cypermethrin singly or in combination

Fertilization mode: (a) CK, no fertilization; (b) NPK, fertilization with NPK fertilizers; (c) OM, fertilization with organic manure. Error bars represent the standard deviation. Treatments: B, the control (without contamination); C, cypermethrin contamination; Cu, Cu^{2+} contamination; Cd, Cd^{2+} contamination; Cu + C, combination of Cu^{2+} and cypermethrin; Cd + C, combination of Cd^{2+} and cypermethrin

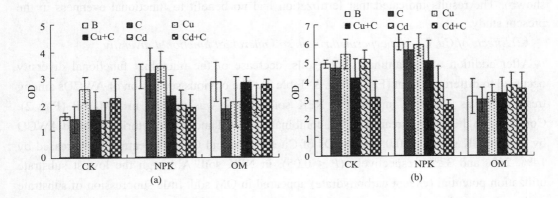

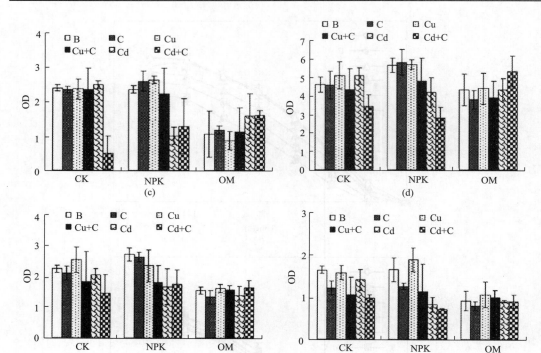

Fig. 3 Microbial substrate utilization potential (OD) in three fertilized soils following addition of Cu, Cd and cypermethrin singly or in combination

Substrates: (a) carbohydrates, (b) carboxylic acids, (c) amines, (d) amino acids, (e) polymers, (f) miscellaneous. Error bars represent the standard deviation. Fertilization mode: CK, no fertilization; NPK, fertilization with NPK fertilizers; OM, fertilization with organic manure. Treatments: B, the control (without contamination); C, cypermethrin contamination; Cu, Cu^{2+} contamination; Cd, Cd^{2+} contamination; Cu + C, combination of Cu^{2+} and cypermethrin; Cd + C, combination of Cd^{2+} and cypermethrin

Carbohydrate utilization in CK soil was lower than that in NPK and OM soils ($P < 0.05$). Compared to the NPK soil, polymer degradation in CK soil decreased significantly ($P < 0.05$), but the decrease of substrate utilization for other groups did not reach the significant level. Shannon indices of NPK and CK soils were higher than that of OM soil ($P < 0.05$), whereas Shannon evenness of CK soil was higher than that in NPK and OM soils ($P < 0.05$) (data not shown). The results indicated that fertilization had no benefit to functional evenness in the present study.

(2) *Effects of Cu, Cd and cypermethrin on soil microbial functional diversity*

After addition of contaminants into soils, decrease of the microbial functional diversity occurred to different extent (Fig. 2, Fig. 3, Table 2). The obvious separation of AWCDs among treatments was observed in CK and NPK soils, but it was unclear in OM soil (Fig. 2). Compared to the control treatment (B), combination of Cd and cypermethrin inhibited AWCD by 19% in CK soil ($P < 0.05$). AWCDs of Cu + C, Cd and Cd + C treatments decreased by 14%, 26% and 37%, respectively ($P < 0.05$), in NPK soil. Although the lowest substrate utilization potential (except carbohydrate) appeared in OM soil, little suppression of substrate utilization was found in contaminated treatments (Fig. 3, Table 2). In NPK soil, polymer degradation was significantly inhibited by combination of Cu and cypermethrin and substrate utilization for six groups significantly decreased in the treatments of Cd and Cd + C ($P <$

0.05). Except carbohydrate, the decrease of substrate utilization for other five groups in the treatment of Cd + C also reached the significant level in CK soil ($P < 0.05$), especially amines utilization (deceased by 78.9%). Cd and combination of Cd and cypermethrin in NPK soil decreased Shannon index by almost 9% ($P < 0.05$), whereas application of contaminants had slight influence in CK and OM soils (Table 2). Evenness of all contaminated treatments decreased significantly in CK soil ($P < 0.05$). Cd and combination of Cd and cypermethrin also reduced evenness to a significant level in OM soil ($P < 0.05$). However there was no significant change of evenness in NPK soil (Table 2). Therefore, single addition of Cu or cypermethrin had little influence on AWCD, substrate utilization, Shannon index (except CK soil) and evenness. But combinations of cypermethrin and heavy metals, especially Cd, could result in the great reduction of functional diversity (Table 2).

Table 2 Significance of effects of cypermethrin, Cu, and Cd singly or in combination on substrate utilization, Shannon index and evenness in each fertilization mode determined with two-way ANOVA

	Significance levels														
	CK					NPK					OM				
	C	Cu	Cd	C × Cu	C × Cd	C	Cu	Cd	C × Cu	C × Cd	C	Cu	Cd	C ×Cu	C ×Cd
Carbohydrates	-	-	-	-	-	-	-	*	-	-	-	-	-	-	-
Carboxylic acids	-	-	-	*	*	-	-	**	-	*	-	-	-	-	-
Amines	-	-	-	**	-	-	-	**	-	-	-	-	-	-	-
Amino acids	-	-	-	-	*	-	-	**	-	**	-	-	-	-	-
Polymers	-	-	-	-	*	**	-	*	-	-	-	-	-	-	-
Miscellaneous	**	-	*	*	*	-	-	-	*	-	-	-	-	-	-
Shannon index	-	-	-	-	-	-	-	*	-	-	-	-	-	-	-
Shannon evenness	*	*	*	-	-	-	-	-	-	-	-	-	*	-	-

Notes: Significance levels in same line in each fertilization mode: (-) $P > 0.05$; * $P < 0.05$; ** $P < 0.01$. Treatments: B, the control (without contamination); C, cypermethrin contamination; Cu, Cu^{2+} contamination; Cd, Cd^{2+} contamination; C×Cu, interaction between Cu^{2+} and cypermethrin; C×Cd, interaction between Cd^{2+} and cypermethrin. Fertilization mode: CK, no fertilization; NPK, fertilization with NPK fertilizers; OM, fertilization with organic manure.

3.3.4 Discussion

1. *Effect of fertilization on the soil microbial activity*

DHA is often used as the indicator of soil fertility and it also can denote the amount and activity of soil microbes (Casida et al., 1964; Gil-Sotres et al., 2005). Hence higher DHA in OM soi lindicated that long-term application of composed straw was more beneficial to microbial biomass and activity than the application of NPK and no fertilization. However, compared to the NPK and CK soils, the substrate utilization potential of OM soil decreased significantly except carbohydrate ($P < 0.05$) in our study (Fig. 3).These results suggested that soil microbial community shifts had occurred after long-term fertilization. Soil nutrients, especial organic matter, are important drivers of soil microbial community composition (Steenwerth et al., 2008). Fertilization-induced changes to the soil properties affected soil nutrient availability greatly, such as N, P and organic matter (Table 1). Previous studies have also reported that chemical nature and diversity of soil organic matter were influenced by fertilization (Ellerbrock et al., 1999; Vineela et al., 2008), which in turn induced changes of microbial substrate utilization (Cooksona et al., 2008; Steenwerth et al., 2008). In OM soil, the organic matter content was high because of the compost application, as well as its fraction of carbohydrate was likely to increase during wheat straw degradation. This soil environment had benefits to promote DHA (Gil- Sotres et al., 2005) and the growth of carbohydrate-utilized microbe, however the growth of other microorganisms would be suppressed due to nutrition limitation, therefore the microbial diversity reduced. In NPK and CK soils, the organic matter was mainly from root turnover and exudates and its contents were relatively low. These may create soil conditions to maintain or improve the ability of microbial substrate utilization (Malhi et al., 2006; Cooksona et al., 2008). The variation in the microbial functional diversity was from complex interactions between soil properties that mediated substrate availability and microbial nutrient demand (Cooksona et al., 2008).

There were some limitations of the Biolog method in application, mainly including its focusing on bacterial species that are able to respond rapidly to the substrates and carbon sources that are different from the contaminants found in soils (Preston-Mafham et al., 2002). But due to the high sensitivity and efficiency, the effects of contaminants on CLSU can be regarded as the early warning signals for soil evaluation (Kong et al., 2006). Therefore, the persistence of organic pollutants, in our point of view, might enhance with the lower carbon source utilizations in organic manured soil. Despite of no fertilization over a long period of time, the microbial capability of carbon source utilization is relative high in CK soil.

2. Effects of Cu, Cd and cypermethrin on soil microbial activity

Due to the widespread contamination by heavy metals and pesticides in cropland soils, it is essential to eliminate or decrease the potential risk of these compounds in the environment. In the present study, the responses of microorganisms in three fertilized soils to Cu, Cd and cypermethrin and their combinations were different. Cu of 100 mg/kg decreased DHA significantly ($P < 0.05$), and the decrease extents were in an order of NPK > CK > OM. In CLSU experiment, there was slight influence of contaminants on soil microbial diversity in OM soil with the exception of Shannon evenness. But in NPK and CK soils, combination of cypermethrin and Cd significantly suppressed AWCD, substrate utilization and Shannon index ($P < 0.05$) and the greater decrease was in NPK soil. Furthermore, in NPK soil, combination of Cu and cypermethrin and Cd reduced AWCD and Cd reduced substrate utilization significantly ($P < 0.05$). Therefore, the sensitivity of microorganisms in soils to pollutants followed the order as: NPK > CK > OM. The results indicated that fertilization mode had the great effect on contaminant bioavailability in soil. Little change of soil pH was observed among different treatments in each soil (data not shown). The previous studies have reported soil organic matter can decrease pollutant bioavailability via the formation of complexes between them, and organic manure application is an efficient way to reduce environmental risks of pollutants (Sannino and Gianfreda, 2001; Pérez-de-Mora and Madrid, 2007; van Herwijnen et al., 2007). Hence the low sensitivity in OM soil was a result of compost application. However, higher organic matter content in NPK soil than that in CK soil suggested the effect of contaminants on microbes was controlled by additional regulators. Microbial community responded differently to soil contaminants (Gflier et al., 1998; Wang et al., 2004), so microbial community shifts induced by varied fertilization mode (Marschner et al., 2003; Widmer et al., 2006; Chu et al., 2007) should be taken when assessing contaminant bioavailability. Additionally, in cypermethrin degradation study, the lowest dissipation rate was recorded in NPK soil (Xie and Zhou, 2008), which also contributed to the high sensitivity in NPK soil to some extent.

Although the influence of cypermethrin on the soil microbial diversity was slight, we found that a great synergistic effect could be provoked when applied together with heavy metals, especially with Cd (Table 2).Thus more attentions should be paid to the eco-environmental risk of cypermethrin in heavy metal polluted soils.

Heavy metal levels in the present study were the critical doses, which could significantly reduce cypermethrin degradation in soil. Cu of 100 mg/kg could significantly suppress DHA, while Cd of 5mg/kg had relative greater influence on soil microbial diversity. The results suggested that the effect of each pollutant on soil microbes was specific.

References

Akmal M, Xu J, Li Z, et al. 2005. Effects of lead and cadmium nitrate on biomass and substrate utilization pattern of soil microbial communities. Chemosphere, 60: 508-514

Allan I J, House W A, Parker A, et al. 2005. Diffusion of the synthetic pyrethroid permethrin into bed-sediments. Environ. Sci. Technol., 39: 523-530

Amweg E L, Weston D P, Ureda N M. 2005. Use and toxicity of pyrethroid pesticides in the central valley, California, USA. Environ. Toxicol. Chem., 24(4): 966-972

Brookes P C. 1995. The use of microbial parameters in monitoring soil pollution by heavy metals. Biol. Fert. Soils, 19: 269-279

Casida L E, Klein D A, Santoro T. 1964. Soil dehydrogenase activity. Soil Sci., 98: 371-376

Chu H Y, Lin X G, Takeshi F, et al. 2007. Soil microbial biomass, dehydrogenase activity, bacterial community structure in response to long-term fertilizer management. Soil Biol. Biochem., 39: 2971-2976

Cooksona W R, Murphya D V, Roperb M M. 2008. Characterizing the relationships between soil organic matter components and microbial function and composition along a tillage disturbance gradient. Soil Biol. Biochem., 40: 763-777

de Lipthay J R, Johnsen K, Albrechtsen H J, et al. 2004. Bacterial diversity and community structure of a sub-surface aquifer exposed to realistic low herbicide concentrations. FEMS Microbiol. Ecol., 49: 59-69

Doran J W, Zeiss M R. 2000. Soil health and sustainability: managing the biotic component of soil quality. Appl. Soil Ecol., 15: 3-11

Ellerbrock R, Höhn A, Gerke H. 1999. Characterization of soil organic matter from a sandy soil in relation to management practice using FT-IR spectroscopy. Plant Soil, 213: 55-61

Feron V J, Groten J P. 2002. Toxicological evaluation of chemical mixtures. Food Chem. Toxicol., 40: 825-839

Garland J L, Mills A L. 1991. Classification and characterization of heterotrophic microbial communities on the basis of patterns of community-level sole carbon source utilization. Appl. Environ. Microbiol., 57: 2351-2359

Gflier K E, Witter E, McGrath S P. 1998. Toxicity of heavy metals to microorganisms and microbial proceses in agricultural soils:a review. Soil Biol. Biochem., 30: 1389-1390

Gil-Sotres F, Trasar-Cepeda C, Leiró s M C, et al. 2005. Different approaches to evaluating soil quality using biochemical properties. Soil Biol. Biochem., 37: 877-887

Hu K L, Zhang F R, Li H, et al. 2006. Spatial patterns of soil heavy metals in urban-rural transition zone of Beijing. Pedosphere, 16: 690-698

Kong W D, Zhu Y G, Fu B J, et al. 2006. The veterinary antibiotic oxytetracycline and Cu influence functional diversity of the soil microbial community. Environ. Poll., 143: 129-137

Lu R K. 2000. Soil Agro-Chemical Analysis. Beijing: China Agricultural Scientech Press: 106-253

Lu Z M, Min H, Ye Y F. 2004. Short-term influence of herbicide quinclorac on enzyme activities in flooded paddy soils. Pedosphere, 14: 71-76

Malhi S S, Lemke R, Wang Z H, et al. 2006. Tillage, nitrogen and crop residue effects on crop yield, nutrient uptake, soil quality, and greenhouse gas emissions. Soil Till. Res., 90: 171-183

Marschner P, Kandeler E, Marschner B. 2003. Structure and function of the soil microbial community in a long-term fertilizer experiment. Soil Biol. Biochem., 35: 453-461

Pascual J A, Garcia C, Hernandez T, et al. 2000. Soil microbial activity as a biomarker of degradation and remediation processes. Soil Biol. Biochem., 32: 1877-1883

Pérez-de-Mora A, Madrid C F. 2007. Amendments and plant cover influence on trace element pools in a contaminated soil. Geoderma, 139: 1-10

Perkiömäki J, Fritze H. 2005. Cadmium in upland forests after vitality fertilization with wood ash—a summary of soil microbiological studies into the potential risk of cadmium release. Biol. Fert. Soils, 41: 75-84

Preston-Mafham J, Boddy L, Randerson P F. 2002. Analysis of microbial community functional diversity using sole-carbon-source utilization profiles-a critique. FEMS Microbiol. Ecol., 42: 1-14

Rangaswamy V, Reddy B R, Venkateswarlu K. 1994. Activities of dehydrogenase and protease in soil as influenced by monocrotophos, quinalphos, cypermethrin and fenvalerate. Agric. Ecosyst. Environ., 47: 319-326

Sannino F, Gianfreda L. 2001. Pesticides influence on enzymatic activities. Chemosphere, 45: 417-425

Sannino A, Bandini M, Bolzoni L. 2003. Determination of pyrethroid pesticide residues in processed fruits and vegetables by gas chromatography with electron capture and mass spectrometric detection. J. AOAC Int., 86: 101-108

Staddon W J, Duchesne L C, Trevors J T. 1997. Microbial diversity and community structure of post-disturbance forest soils as determined by sole-carbon-sourceutilization patterns. Microb. Ecol., 34: 125-130

Steenwerth K L, Drenovsky R E, Lambert J J, et al. 2008. Soil morphology, depth, and grapevine root frequency influence microbial communities in a Pinot noir Vineyard. Soil Biol. Biochem., 40: 1330-1340

Sukul P. 2006. Enzymatic activities and microbial biomass in soil as influenced by metalaxyl residues. Soil Biol. Biochem., 38: 320-326

Swift M J. 1994. Maintaining the biological status of soil: a key to sustainable land management. In: Greenland. D J, Szabolcs I. Soil Resilience and Sustainable Land Use. CAB International, Wallingford: 33-39

van Herwijnen R, Hutchings T R, Al-Tabbaa A, et al. 2007. Remediation of metal contaminated soil with mineral-amended composts. Environ. Poll., 150: 347-354

Vineela C, Wani S P, Srinivasarao C, et al. 2008. Microbial properties of soils as affected by cropping and nutrient management practices in several long-term manurial experiments in the semi-arid tropics of India. Appl. Soil Ecol., 40: 165-173

Waldrop M P, Balser T C, Firestone M K. 2000. Linking microbial community composition to functional in a tropical soil. Soil Biol. Biochem., 32: 1837-1846

Wang A, Chen J, Crowley D E. 2004. Changes in metabolic and structural diversity of soil bacteria community in response to cadmium toxicity. Biol. Fert. Soils, 39: 452-456

Wang F, Bian Y R, Jiang X, et al. 2006. Residual characteristics of organochlorine pesticides in Lou soils with different fertilization modes. Pedosphere, 16: 161-168

Widmer F, Rasche F, Hartmann M, et al. 2006. Community structures and substrate utilization of bacteria in soils from organic and conventional farming systems of the DOK long-term field experiment. Appl. Soil Ecol., 33: 294-307

Xie W J, Zhou J M. 2008. Cypermethrin persistence and soil properties as affected by long-term fertilizer management. Acta Agric. Scand. B., 58: 314-321

Xie W J, Zhou J M, Wang H Y, et al. 2008. Effect of nitrogen on the degradation of cypermethrin and its metabolite 3-phenoxybenzoic acid in soil. Pedosphere, 18: 638-644

Zak J C, Willig M R, Moorhead D L, et al. 1994. Functional diversity of microbial communities: a quantitative approach. Soil Biol. Biochem., 26: 1101-1108

Zhang X X, Cheng S P, Zhu C J, et al. 2006. Microbial PAH-degradation in soil: degradation pathways and contributing factors. Pedosphere, 16: 555-565

Zheljazkov V D, Warman P R. 2003. Application of high Cu compost to Swiss chard and basil. Sci. Total Environ., 302: 13-26

3.4 Index models to evaluate the potential metal pollution contribution from washoff of road-deposited sediment

3.4.1 Introduction

As point sources of pollution in China and many other countries decrease, urban runoff containing contaminated road-deposited sediment (RDS) is becoming an increasingly serious problem (Zhu et al., 2008; Zhao et al., 2010). RDS from impervious surfaces is an important carrier of contaminants, often containing metals at elevated concentrations (Aryal et al., 2009; Xiang et al., 2010). Large amounts of pollutants, such as nutrients, metals, and hydrocarbons, are usually transported in RDS washoff (Sartor and Boyd, 1972; Huber and Dickinson, 1988; Al-Khashman, 2007). Quantifying the relationship between RDS and washoff particles in urban runoff could provide a new method for estimating the pollution load that a waterway receives (Herngren et al., 2005; Zhao et al., 2011). A range of diffuse pollution models for urban areas have been proposed, including STORM, HSPF, DR3M-QUAL, SWMM, and SLAMM, and they have been used to estimate pollutant loads in runoff (Egodawatta et al., 2007; Wang et al., 2011). Despite the fact that the models mentioned above are widely used to estimate diffuse pollution sources in urban areas, they generally require substantial amounts of parameterization and calibration data (Shaw et al., 2006). Consequently, there is a need for improved runoff pollution estimates.

Identifying and prioritizing critical source areas could greatly improve the efficiency of RDS washoff pollution controls. Index models have been widely used for identifying diffuse critical source areas of phosphorus (P) on farms, and are less likely than other models to be constrained by a lack of input data (Heathwaite et al., 2005; Sharpley et al., 2008). Index models allow critical source areas to be identified by quantifying the relative pollution risk (e.g., as a probability) as opposed to the actual pollutant loading, because it is difficult to quantify transport factors (Buczko and Kuchenbuch, 2007; Buchanan et al., 2013). The index models mentioned above were initially devised for diffuse pollution in agricultural areas, and no index model has yet been specifically designed for diffuse pollution in urban areas that can be used to identify critical source areas of urban runoff. Characterizing RDS (e.g., the amount, grain sizes, associated metals, and mobility) can make it possible to identify critical source areas of urban runoff. A risk index (called RI_{RDS}) that combines RDS characteristics with a potential ecological risk index (RI) was developed by Zhao and Li (2013). The RI_{RDS} can be used to evaluate the amount of pollution present per unit area, but not the total pollution load. There is a clear need for the development of new simple but effective index models that use RDS characteristics to quantify or semi-quantify urban pollutant loadings.

The urban, suburban, and rural areas in and near large cities in China are generally divided into areas classed as urban districts, suburban county areas, rural townships, rural villages, and urban villages (Zhao et al., 2011). Previous studies have shown that RDS characteristics (including the amount of RDS present, its grain size distribution, metals associated with it, and its mobility) vary significantly along urban-rural gradients (Zhao et al., 2011). It is important to be able to evaluate the metal load associated with RDS correctly along urban-rural gradients, and developing such an ability will improve our understanding of the risks posed by heavy metal pollution in RDS being transported into urban water. The RDS characteristics mentioned above are critical to the assessment and evaluation of the role of RDS in metal pollution in urban runoff (Zhao and Li, 2013). However, there is still no index model available for identifying critical source areas of pollutants associated with RDS. Quantifying pollutant loadings will assist managers to focus remediation actions on decreasing RDS washoff into water bodies.

In this study, we aimed to ① determine the source and transport factors for heavy metals in RDS using observed and weighted RDS characteristics in a multiplicative index, and ②combine a number of functions for evaluating the pollution load and strength into one RDS index.

3.4.2 Materials and Methods

1. Study area and RDS sampling

In general, big cities in China are divided by the government into urban, suburban, and rural areas, and the classifications used are urban district, suburban county, rural township, rural village, and urban village. We collected RDS samples from areas that were typical of each of these administrative divisions within the Beijing metropolitan region. The Beijing metropolitan region includes 16 administrative sub-divisions, which are county-level units that are governed directly by the municipality. There are six such districts in the urban area, and ten (eight districts and two counties) in the suburban area. The whole region can be divided into urban, suburban, and rural areas. In suburban and rural areas, each county consists of a group of towns, and each town consists of a group of villages. We chose five sampling areas along an urban-suburban-rural gradient, including a central urban area (UCA), an urban village area (UVA), a central suburban county area (CSA), a rural town area (RTA), and a rural village area (RVA). In general, the population density, traffic density, energy consumption, and frequency of road sweeping all decrease moving along the urban-rural gradient. Roads in the UCA and CSA generally have relatively smooth and undamaged impervious surfaces, and are regularly (daily) mechanically swept, whereas roads in the RTA, RVA, and UVA generally have very damaged surfaces, and are rarely swept.

RDS samples were collected using a domestic vacuum cleaner (Philips FC8264; Philips, Amsterdam, Netherlands) between 2 and 10 September 2009, following a period of about two weeks of dry weather. Three sampling sites were selected in each of the study areas. An unspecified area at each site was vacuumed from the central road marking to the curb until a reasonable amount of RDS was collected, then the size of the area sampled was measured with a ruler. Each RDS sample was weighed with an electronic balance, and sample masses of 0.8-1.5 kg were found to have been collected at each site. The mass of RDS per unit area was calculated by dividing the RDS mass collected by the size of the sampling area, and it ranged from 2 to 570 g/m^2 for all of the samples that were collected. The amount of RDS collected at a site was generally larger in the RTA, RVA, and UVA than in the UCA and CSA. The samples were separated into the grain size fractions <44, 44-62, 62-105, 105-149, 149-250, 250-450, 450-1000, and >1000 μm using polyester sieves.

2. *Analytical methods and quality control*

Metals were measured after the RDS samples had been digested with a mixture of HF and $HClO_4$ on a hotplate (Tessier et al., 1979). All solutions were stored at 4℃ until they were analyzed. The Cr, Cu, Ni, Pb, and Zn concentrations were determined using an Elan 6000 inductively coupled plasma-optical emission spectroscopy instrument (Perkin-Elmer, Waltham, MA, USA). Certified geochemical soil reference materials (CRMs) GSS-1 and GSS-2 were also analyzed to provide quality assurance and quality control (QA/QC) information. No RDS CRMs are available, but using soil CRMs has been found to be an acceptable way of determining the quality of RDS analyses in previous studies (Sutherland, 2003; Zhao et al., 2010). The recoveries of the five metals that were analyzed were 75%-110% (75%-95% for Cr, 90%-108% for Cu, 90%-104% for Ni, 90%-110% for Pb, and 98%-107% for Zn). The detection limits for Cr, Cu, Ni, and Zn ranged from 0.1 to 1 μg/L, and the detection limit for Pb ranged from 1 to 10 μg/L. Duplicate aliquots of 2% of the RDS samples were analyzed, and the metal concentrations found in the duplicates were always within ±10% of the mean concentration. Reagent blanks were included in each batch of samples that was

3. Source and transport factors in the RDS indices

(1) Source factors in the RDS indices

The amount of pollutants that is present in RDS is important because it determines the amount of pollutants in surface runoff. The source factors used in the RDS index models included the amount of RDS, the grain sizes, and the amounts and types of pollutants associated with the RDS. To obtain the best possible RDS parameter estimates, we studied a large number of sites (167 in total, 97 of which were in areas with busy main roads and 70 of which were in residential areas) along an urban-rural gradient. The mass of RDS per unit area varied widely, with a range of 2-570 g/m^2 and a mean of 70 g/m^2. The metal concentrations were measured in each RDS grain size fraction.

(2) Transport factors in RDS indices

The particle mobility for each RDS grain size fraction was measured using simulated rainfall. A specially designed rainfall simulator and small impervious surface plots were used to estimate the RDS removal rates in surface runoff. The amount of each RDS size fraction in the runoff from the impervious surface was expressed as a percentage of the total mass of RDS in the runoff with respect to the initial total mass of RDS on the surface, calculated using Eq. (1) (Zhao et al., 2011).

$$F_w (\%) = \frac{M_{F_w}}{M_{initial}} \times 100\% = \frac{\int_0^1 C(t) \times Q(t) dt}{M_{initial}} \times 100\% \tag{1}$$

Where, F_w is the percentage of an RDS size fraction washed off the surface (%); M_{F_w} is the mass of the size fraction washed off the surface over the entire rain event (mg); $M_{initial}$ is the initial mass of the size fraction on the surface (mg); $C(t)$ is the concentration of the size fraction in the surface runoff water (mg/L) at a particular sampling time t; and $Q(t)$ is the surface runoff flow rate at the sampling time (m^3/min).

To obtain F_w, 35 different simulated washoff scenarios using seven RDS size fractions (<44, 44-62, 62-105, 105-149, 149-250, 250-450, and 450-1000 μm) and five simulated rainfall intensities (12.87, 32.70, 46.7, 82.67, and 101.37 mm/h) were each conducted for 1 h. F_w was used to calculate the amounts of RDS washoff for different particle sizes and to calculate the amounts of the heavy metals that were associated with the washed off particles. The F_w for each grain size particle was independent of the amount and size of the particles, but the F_w for the bulk RDS particles was dependent on the grain size distribution (Shaw et al., 2006; Zhao et al., 2011). We used the washoff percentage for each RDS grain size to calculate the bulk RDS washoff percentage and amount for the corresponding sampling area, to properly represent the RDS washoff scenarios in our calculations. We used the F_{wi} values for each RDS size fraction to calculate the amount of RDS washoff that occurred for the RDS samples taken from each of the urban-rural gradient area types, i.e. UCA, UVA, CSA, RTA, and RVA. Further details of the experimental design are available elsewhere (Zhao et al., 2011).

(3) *Ranking the source and transport factor ratings*

No standard methods are available from the literature for assessing RDS source and transport factors and to allow rankings to be allocated. We assigned weightings to the source and transport factors according to our in situ investigations into the RDS characteristics. The RDS mass per unit area was ranked into six levels ($M_{weighted}$), in which 0-30 g/m^2 was given a value of 1.00, 31-60 g/m^2 a value of 1.75, 61-90 g/m^2 a value of 2.50, 91-140 g/m^2 a value of 3.00, 141-190 g/m^2 a value of 3.50, and > 190 g/m^2 a value of 3.75 (Zhao and Li, 2013). The RDS mass per unit area ranking levels were primarily based on the results of our in situ investigation into the RDS mass per unit area (using a total of 167 sites, 97 of which were from areas on main roads and 70 of which were in residential areas), because no standard values are available for dividing RDS masses per unit area into levels. We found that places that had higher RDS masses per unit area also had higher proportions of relatively coarse particles. We assigned comparatively low values to the higher RDS masses per unit area because coarser particles give lower levels of risk associated with their metal contents and the mobility of those metals (Zhao and Li, 2013). Metals associated with RDS were ranked by comparing the measured concentrations of the metal of interest with its background concentrations. The background heavy metal concentrations that were used were determined in 120 soil samples that did not suffer from heavy metal pollution, and these had been selected from a total of 803 soil samples (Chen et al., 2004).

Transport factors can be ranked according to the RDS washoff percentages for each grain size fraction (F_{wi}), as described by Zhao et al. (2011). The transport factors were incorporated into the RDS index by giving the lowest F_w value for a certain grain size fraction a value of 1. Other F_w values for other grain size fractions were assigned by dividing the F_w value by the lowest F_w value (Zhao and Li, 2013). The transport factor ratings for each grain size fraction are listed in Table 1.

Table 1 Transport factor ratings in RDS index for pollutant strength

Transport factor	Grain size fraction of RDS (μm)						
	< 44	44-62	62-105	105-149	149-250	250-450	450-1000
$F_{wi,weighted}$	17.0	10.0	4.5	4.3	2.9	1.5	1.0

4. *Calculating the RDS index*

In this section, we will describe the general principles involved in calculating the RDS index. The RDS index was based on the phosphorus index that is used for modeling diffuse pollution in agricultural areas (Sharpley et al., 2008; Buchanan et al., 2013), but the RDS index was designed specifically to deal with the RDS characteristics found in diffuse pollution in urban areas.

The RDS index is used in this study not only to identify critical source areas by quantifying relative pollution risks, but also to quantify actual pollutant loadings. Both semi-quantitative and quantitative applications use the RDS characteristics and take source and transport factors into account. Semi-quantitative and quantitative pollutant loadings were calculated using the weighted RDS characteristics and the measured RDS characteristics, respectively. The RDS index was calculated using Eq. (2)

$$RDS_{index} = F_{source} \times F_{transport} \quad (2)$$

Where, F_{source} and $F_{transport}$ are the RDS source and transport factors, respectively.

(1) *RDS index model for the load*

The RDS index model for the load was calculated using the observed source and transport factor values, using Eq. (5) The source and transport factors that were included for the load estimations ($F_{source, load}$; $F_{transport, load}$) were calculated separately using Eq. (3) and Eq. (4).

$$F_{source,load} = \sum_{j}^{m}\sum_{i}^{n}(M_i \times C_{ij} \times A) \quad (3)$$

$$F_{trasnport,load} = F_{w,observed} \quad (4)$$

$$RDS_{index,load} = F_{source,load} \times F_{transport,load} = \sum_{j}^{m}\sum_{i}^{n}(M_i \times C_{ij} \times A \times F_{wi,observed}) \quad (5)$$

Where, $RDS_{index,load}$ is the potential amount of pollution in the runoff (kg); M_i is the mass of a particular RDS size fraction per unit area (mg/m^2); C_{ij} is the measured concentration of the metal of interest j in the RDS with a grain size i (mg/kg); i and j are the numbers of grain size fractions and metal species in the study, respectively; and A is the street surface area in the administrative division unit along the urban-rural gradient (10^4m^2).

(2) *RDS index model for pollution strength*

The RDS index for pollution strength was calculated using the weighted values for the source and transport factors, using Eq. (8). The source and transport factors that were included for the strength estimations ($F_{source,strength}$, $F_{transport,strength}$) were calculated separately using Eq. (6) and Eq.(7).

$$F_{source,strength} = \sum_{j}^{m}\sum_{i}^{n}(T_r^j \times \frac{C_{ij}}{C_{rj}} \times P_i \times M_{weighted}) \quad (6)$$

$$F_{transport,strength} = F_{w,weighted} \quad (7)$$

$$RDS_{index,strength} = F_{source,strength} \times F_{transport,strength} = \sum_{j}^{m}\sum_{i}^{n}(T_r^j \times \frac{C_{ij}}{C_{rj}} \times P_i \times M_{weighted} \times F_{wi,weighted}) \quad (8)$$

Where, $RDS_{index,strength}$ is the potential pollution strength in the runoff, T_r^j is the toxic response factor for the metal of interest according to Hakanson (1980) (the values for each element of interest were, in order, Zn = 1, Cr = 2, Ni = 3, and Cu = Pb = 5); C_{rj} is the background concentration of the metal in soil (31.1, 19.7, 27.9, 25.1, and 59.6 mg/kg for Cr, Cu, Ni, Pb, and Zn, respectively (Chen et al., 2004)); P_i is the amount of RDS with grain size i expressed as a percentage of the total RDS mass, and $M_{weighted}$ is the level of the bulk RDS mass per unit area for each sampling site. C_{ij}, i, and j have the same meanings as described above for Eq. (5).

Values and categories were established for the $RDS_{index,strength}$ method according to the products of the source and transport factors. $RDS_{index,strength}$ was classified as indicating low levels of risk when $RDS_{index,strength} \leqslant 150$, moderate levels of risk when $150 < RDS_{index,strength} \leqslant 300$, considerable levels of risk when $300 < RDS_{index,strength} \leqslant 600$, and high levels of risk when $RDS_{index,strength} > 600$ (Zhao and Li, 2013).

3.4.3 Results

In this section, we will present the results of using the RDS index in a case study in the Beijing metropolitan region. Developing the RDS index for use at the scale of an urban region could allow it to be used to reflect the spatial distributions of the potential environmental risks posed by metals in RDS along the urban-rural gradient. We will first describe the RDS characteristics (the amounts present, grain sizes, associated metals, and its mobility) along the urban-rural gradient.

1. *Source factors in the RDS index*

The source factors in the RDS index in this study included the amount of RDS present, the proportions of RDS particles of each grain size, and the metals associated with the RDS. The source factors were investigated along an urban-rural gradient in the Beijing metropolitan region. The mass of RDS per unit area clearly varied along the urban-rural gradient, with low masses per unit area being found in the UCA ((23.9 ± 18.8)g/m^2) and CSA ((24.2 ± 17.9) g/m^2), medium masses per unit area being found in the RTA ((54.1 ± 56.4) g/m^2), and high masses per unit area being found in the UVA ((134.5 ± 153.8) g/m^2) and RVA ((188.2 ± 202.6) g/m^2) (Fig. 1). About 70% of the $M_{weighted}$ values for the UCA and CSA sampling sites fell into level 1, and about 35%-40% of the $M_{weighted}$ values for the RVA and UVA sampling sites fell into levels 5 and 6. The RDS particle size distribution also varied along the urban-suburban-rural gradient (Fig. 2). The median RDS diameter, d_{50}, was less than 149 μm in the UCA and CSA, but more than 149 μm in the RTA, RVA, and UVA. In general, the rural areas had higher proportions of coarser RDS particles and larger amounts of RDS. Our site investigations indicated that the RVA and UVA were seldom swept (0.2-0.3 times/d) and had high proportions of bare soil or broken and rough road surfaces. The streets in the RVA and UVA are often swept with brooms, whereas mechanical sweepers, which give higher RDS removal efficiencies, are used in urban areas.

The metal concentrations in the RDS along the urban-rural gradient are shown in Fig. 3. The mean Cr, Cu, Pb, and Zn (i.e. excluding Ni) concentrations in the RDS samples from across Beijing were approximately two to four times higher than the background concentrations in soils from Beijing (Chen et al., 2004). The mean total concentrations of the metals analyzed decreased along the urban-rural gradient in the order UCA > CSA > UVA > RTA > RVA. Zn was the most abundant element in the RDS from all of the areas.

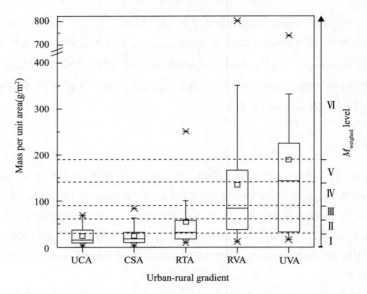

Fig. 1 Box plot of the amounts of road-deposited sediment (RDS) found along the urban-rural gradient in Beijing

Values are given as the mean ± the standard deviation. Dashed lines in the figure refer to the values for the $M_{weighted}$ levels. UCA = central urban area, UVA = urban village area, CSA = central suburban county area, RTA = rural town area, and RVA = rural village area

2. Transport factors in the RDS index

The transport factors that were assessed in this study were mainly chosen for their potential to affect surface runoff that would affect the RDS index. The percentage of each RDS size fraction washed off the surface (F_{wi}) was used to represent the transport factors in the RDS index. The $F_{wi,observed}$ values ranged from 1% to 66%, and are shown in Table 2 in a publication by Zhao et al. (2011). The washoff amount was calculated for each of the RDS size fractions from the $F_{wi,observed}$ values by adjusting the amounts and particle size compositions in the RDS to match those in the sampling areas (Shaw et al., 2006; Zhao et al., 2011). The RDS washoff amounts along the urban-rural gradient are shown in Fig. 4. The amount of particles washed off per unit area increased as the rainfall intensity increased, but the washoff as a percentage of the RDS decreased along the urban-rural gradient.

Table 2 Load of metals combined along the urban-rural gradient/ $RDS_{index,\ load}$ (Unit: kg)

Sampling site[①]	Rainfall intensity (mm/h)[③]				
	12.87[②]	32.7	46.7	82.67	101.37
UCA	47.9	197.3	192.6	373.0	444.0
CSA	39.8	136.9	141.9	258.1	318.0
RTA	33.7	111.1	116.3	210.0	258.1
RVA	104.8	344.7	363.3	661.9	805.3

Notes: ① Sampling site area: UCA=the central urban area; UVA=the urban village area; CSA=the central suburban county area; RTA=the rural town area; RVA=the rural village area. ② The area of main traffic road for each administrative division units along the urban-rural gradient (10^4 m^2), UCA: 9869, CSA: 12071, RTA: 5603, RVA: 11591. The areas above were calculated according to that the traffic road area accounting for 30% of the total traffic areas in Beijing city. The data of urban village area was not available. ③ The simulated rainfall duration was one hour, and surface runoff samples were collected manually at 2 min intervals during the first 10 min, at 5 min intervals during the second and third 10 min, and at 10 min intervals thereafter until there was no more surface runoff.

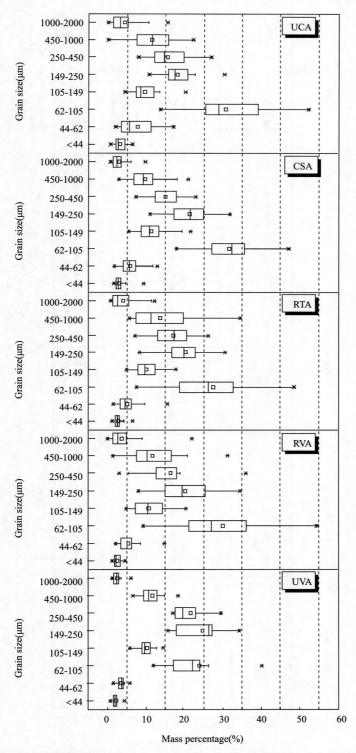

Fig. 2 Box plot of the mass percentages of road-deposited sediment (RDS) particles in each grain size fraction along the urban-rural gradient in Beijing

UCA = central urban area, UVA = urban village area, CSA = central suburban county area, RTA = rural town area, and RVA = rural village area

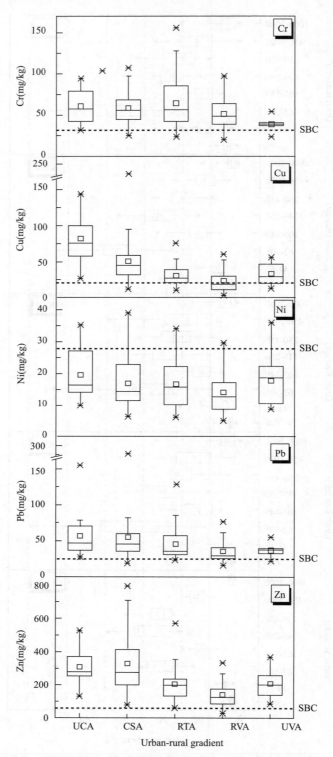

Fig. 3 Box plot of the metal concentrations in the road-deposited sediment (RDS) collected along the urban-rural gradient in Beijing

Dashed lines in each subfigure refer to the background concentrations in soils in Beijing area (SBC, C_{ri}) (Chen et al., 2004). UCA = central urban area, UVA = urban village area, CSA = central suburban county area, RTA = rural town area, and RVA = rural village area

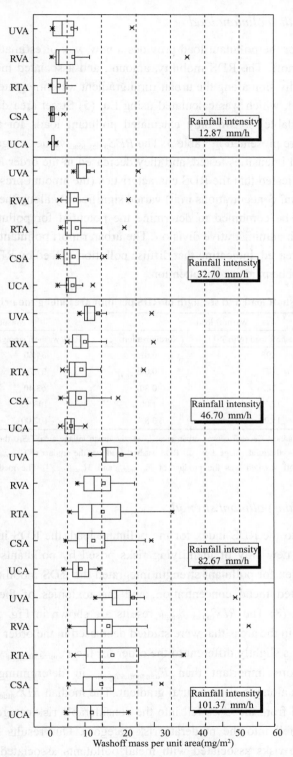

Fig. 4 Box plot of the amount of road-deposited sediment (RDS) washoff that will occur along the urban-rural gradient in Beijing

UCA = central urban area, UVA = urban village area, CSA = central suburban county area, RTA = rural town area, and RVA = rural village area. The simulated rainfall in the tests lasted 1 h

3. RDS index for the pollutant load

The RDS index for the pollutant load provides a new way of estimating pollutant loads in large-scale urban runoff. The RDS mobility, amount, and associated metals, and the area of each administrative division along the urban-rural gradient were integrated into the RDS index for the pollutant load, which was calculated using Eq. (5) Street area data for the residential areas were not available, so we only calculated pollutant loads for the main roads. The $RDS_{index,load}$ results are presented in Table 3. The $RDS_{index,load}$ values ranged from 33.7 to 805.3 kg at the five rainfall intensities tested, and they decreased in the order RVA > UCA > CSA > RTA. The results indicated that the RDS characteristics (the amount present, the grain size, its mobility, and the metal concentrations in it) varied significantly along the urban-rural gradient, and that they could be combined to determine the potential for pollution to be caused by surface runoff in each administrative division. The urban runoff pollutant load increased as the rainfall intensity increased, indicating that diffuse pollution caused by RDS washoff in heavier rainfall events should be paid special attention.

Table 3   The washoff load and strength of RDS per unit area along the urban-rural gradient

Sampling site[①]	Washoff load		Washoff strength	
	RDS washoff load (g/m²)[②]	Normalization[④]	RDS washoff strength[③]	Normalization
UCA	2.95	0.01	34.60	0.00
CAS	2.73	0.00	37.20	0.04
RTA	7.22	0.30	63.40	0.42
RVA	17.70	1.00	103.90	1.00
UVA	15.22	0.83	102.00	0.97

Notes: ① Sampling site area: UCA=the central urban area; UVA=the urban village area; CSA=the central suburban county area; RTA=the rural town area; RVA=the rural village area. ② RDS washoff load was the product of $F_{wi,\text{ observed}}$ and M_i at rainfall intensity = 46.7 mm/h. ③ RDS washoff strength was the product of $F_{w,\text{ weghted}}$ and M_{weighted}. ④ The method of normalization is (X−min)/(max−min).

4. RDS index for the pollutant strength

In a similar way to the RDS index for the pollutant load, the RDS index for the pollutant strength provided a new way of identifying risks posed by pollutants in large-scale urban runoff. The RDS index for pollutant strength integrated the RDS mobility, amount, grain size distribution, associated metal concentration, and the toxicities of the metals, and it was calculated using Eq. (8) The $RDS_{index,strength}$ results are shown in Fig. 5. The $RDS_{index,strength}$ values for the runoff in the areas that were studied decreased in the order UVA > RVA > RTA > CSA > UCA, which is slightly different to the order for the $F_{source,strength}$ values, indicating that $F_{source,strength}$ was more important than $F_{transport,strength}$ in determining the order for the $RDS_{index,strength}$ values along the urban-rural gradient. The median $RDS_{index,strength}$ for the metals in the RDS collected from the UVA fell into the considerable risk category, and the values in the other areas all fell into the moderate risk category. Our results help to improve our understanding of the risks associated with metal pollutants associated with RDS in urban water.

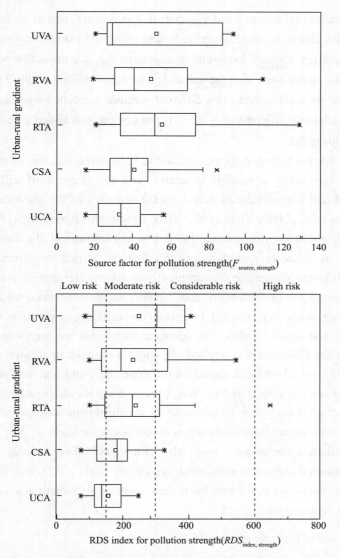

Fig. 5 Box plot of the source factors and the road-deposited sediment (RDS) index for pollution strength along the urban-rural gradient in Beijing

UCA = central urban area, UVA = urban village area, CSA = central suburban county area, RTA = rural town area, and RVA = rural village area

3.4.4 Discussion

1. *Further considerations for the RDS index*

There are no index models specifically designed for quantifying and identifying the potential for diffuse pollution caused by surface runoff in urban areas. In this study, we have summarized our research into the relationship between RDS and pollution in urban surface runoff, including in terms of risk assessment (Zhao and Li, 2013) and estimating pollution loads (Zhao et al., 2011), and we developed an RDS index model based on the phosphorus index that is used to rank diffuse pollution risks in agricultural areas. Although the RDS index model could be used for two purposes, quantifying pollution loadings and identifying risks from pollution, there are still many deficiencies in the source and transport factors that are used in the index.

In fact, the index cannot be directly verified using measured values because the index cannot be measured. The RDS washoff loads and strengths in the five areas that were studied were compared using the paired t-test after the values had been normalized (Table 3), and this showed that the values that were calculated using different methods correlated significantly ($P < 0.001$) but were not significantly different ($P = 0.132$). The comparison above indirectly analyzed the effectiveness of this index.

A key point is that the transport factor used in the RDS index was mainly derived from RDS washoff experiments using a rainfall simulator, and the impacts of differences between simulated rainfall and natural rainfall values on the accuracy of the transport factors are still unclear (Herngren et al., 2005; Zhao et al., 2010; Brodie and Egodawatta, 2011). The use of simulated rainfall could provide greater flexibility and control of the fundamental rainfall parameters such as intensity and duration and thereby helped to eliminate some of the variables which inherently increase the complexity of stormwater quality research. It can also overcome the constraints of variability and random nature associated with natural rainfall events. But further study was needed to reveal the differences of RDS washoff between simulated rainfall and actual rainfall. The transport factor that we used was focused only on the percentage of the RDS that was washed off, and did not take in account the distances that the particles could travel, being calculated for only one slope and one road surface roughness. The distance between the source and the target water body should be added to the model, so that an understanding of the hydraulic connectivity involved is included, to improve the ability of the RDS index to estimate large-scale urban runoff pollutant loads.

Using geo-statistical techniques may also allow pollution sources and the spatial distributions of released pollutants to be identified (Wang et al., 2012), which would be helpful in evaluating the release of pollutants from their sources, in targeting water bodies, and in defining hydraulic connectivity.

2. Important role of particle grain size

The particle grain size is a very important indicator because it links many aspects of the RDS, e.g., the mobilities and amounts of the particles present, the sources and sinks of the RDS, and the pollutant concentrations and species that are present, including the types of particles to which the pollutants are attached. It has been shown in many previous studies that the RDS particle grain size distribution has an important relationship with the RDS characteristics because it affects the amount of RDS present, the concentrations and speciation of the associated metals, and the mobility of the RDS (Sutherland, 2003; Deletic and Orr, 2005; Lau and Stenstrom, 2005; Zhao et al., 2010). The particle grain size distribution affects the source and transport factors in the RDS index model. We even think that a new RDS index, called the grain size threshold for RDS pollution (for $PM_{2.5}$, a particulate matter diameter $\leqslant$ 2.5 μm, for example) could be used in future research, and that such as index would provide a simple and effective way of assessing heavy metal pollution in RDS.

References

Al-Khashman O A. 2007. Determination of metal accumulation in deposited street dusts in Amman, Jordan. Environmental Geochemistry and Health, 29(1): 1-10

Aryal R, Kandasamy J, Vigneswaran S, et al. 2009. Review of stormwater quality, quantity and treatment methods, part 1: stormwater quantity modelling. Environmental Engineering Research, 14(2): 71-78

Brodie I M, Egodawatta P. 2011. Relationships between rainfall intensity, duration and suspended particle washoff from an urban road surface. Hydrology Research, 42(4): 239-249

Buchanan B P, Archibald J A, Easton Z M, et al. 2013. A phosphorus index that combines critical source areas and transport pathways using a travel time approach. Journal of Hydrology, 486: 123-135

Buczko U, Kuchenbuch R O. 2007. Phosphorus indices as risk-assessment tools in the USA and Europe—a review. Journal of Plant Nutrition and Soil Science, 170(4): 445-460

Chen T, Zheng Y, Chen H, et al. 2004. Background concentrations of soil heavy metals in Beijing. Chinese Journal of Environmental Science, 25(1): 117-122

Deletic A, Orr D W. 2005. Pollution buildup on road surfaces. Journal of Environmental Engineering, 131(1): 49-59

Egodawatta P, Thomas E, Goonetilleke A. 2007. Mathematical interpretation of pollutant wash-off from urban road surfaces using simulated rainfall. Water Research, 41(13): 3025-3031

Hakanson L. 1980. An ecological risk index for aquatic pollution control. A sedimentological approach. Water Research, 14(8): 975-1001

Heathwaite A L, Quinn P F, Hewett C J M. 2005. Modelling and managing critical source areas of diffuse pollution from agricultural land using flow connectivity simulation. Journal of Hydrology, 304(1): 446-461

Herngren L, Goonetilleke A, Ayoko G A. 2005. Understanding heavy metal and suspended solids relationships in urban stormwater using simulated rainfall. Journal of Environmental Management, 76(2): 149-58

Huber W, Dickinson R. 1988. Stormwater Management Model (SWMM) User's Manual, Version 4.0. United States Environmental Protection Agency, Athens, Georgia

Lau S L, Stenstrom M K. 2005. Metals and PAHs adsorbed to street particles. Water Research, 39(17): 4083-4092

Sartor J D, Boyd G B. 1972. Water pollution aspects of street surface contaminants. US EPA Office of Research and Monitoring. Repor EPA-R2-72-081

Sharpley A N, Kleinman P J, Heathwaite A L, et al. 2008. Integrating contributing areas and indexing phosphorus loss from agricultural watersheds. Journal of Environmental Quality, 37(4): 1488-1496

Shaw S B, Walter M T, Steenhuis T S. 2006. A physical model of particulate wash-off from rough impervious surfaces. Journal of Hydrology, 327(3): 618-626

Sutherland R A. 2003. Lead in grain size fractions of road-deposited sediment. Environmental Pollution, 121(2): 229-237

Tessier A, Campbell P G, Bisson M. 1979. Sequential extraction procedure for the speciation of particulate trace metals. Analytical Chemistry, 51(7): 844-851

Wang L, Wei J, Huang Y, et al. 2011. Urban nonpoint source pollution buildup and washoff models for simulating storm runoff quality in the Los Angeles County. Environmental Pollution, 159(7): 1932-1440

Wang M, Bai Y, Chen W, et al. 2012. A GIS technology based potential eco-risk assessment of metals in urban soils in Beijing, China. Environmental Pollution, 161: 235-242

Xiang L, Li Y, Yang Z, et al. 2010. Seasonal difference and availability of heavy metals in street dust in Beijing. Journal of Environmental Science and Health, Part A, 45(9): 1092-1100

Zhao H, Li X. 2013. Risk assessment of metals in road-deposited sediment along an urban-rural gradient. Environmental Pollution, 174: 297-304

Zhao H, Li X, Wang X. 2011. Heavy metal contents of road-deposited sediment along the urban-rural gradient around Beijing and its potential contribution to runoff pollution. Environmental Science and Technology, 45(17): 7120-7107

Zhao H, Li X, Wang X, et al. 2010. Grain size distribution of road-deposited sediment and its contribution to heavy metal pollution in urban runoff in Beijing, China. Journal of Hazardous Materials, 183(1): 203-210

Zhu W, Bian B, Li L. 2008. Heavy metal contamination of road-deposited sediments in a medium size city of China. Environmental Monitoring and Assessment, 147(1-3): 171-181

3.5 Phosphorus mobility in soil column experiment with manure application

3.5.1 Introduction

Land application of manure is a common waste management practice for animal production facilities and can improve soil fertility and physical properties (Ojekami et al., 2011). However, continuous application of manure to agricultural land to meet nitrogen requirements of crops can result in phosphorus (P) accumulation in soils. The accumulated P can be lost from soil to adjacent surface or ground water (Idowu et al., 2008), and this may have a detrimental impact on water quality.

A vast body of research has been focused on better understanding the movement of inorganic P which is the main available form for plant. Inorganic P is widely known to strongly interact with both organic and inorganic components of the soil leading to low P solution in soils. In calcareous soils, inorganic P is mainly bound to adsorption surfaces at low concentrations of orthophosphate in solution, whereas it is mainly precipitated as Ca phosphates at higher concentrations (Tunesi et al., 1999). In acid soils, Fe and Al phosphates are the typical precipitation products (Pierzynski et al., 1990). Therefore, surface runoff is generally regarded as the main pathway for the loss of inorganic P (Sims et al., 1998), and the loss by leaching through the soil profile is generally considered minimal (Ojekami et al., 2011).

Organic P can constitute a significant portion of total P in manure (Hansen et al., 2004; He et al., 2008; Idowu et al., 2008). However, few studies have focused on the fate and movement of organic P in the terrestrial system after manure application (Turner and Haygarth, 2000). As there are fewer retention sites for some forms of organic P compared to inorganic P in soils (Anderson and Magdoff, 2005), the loss of organic P from soils amended with manure to ground water is possible. This might have significant environmental implication, because some forms of organic P can be easily hydrolyzed by enzymes to become bioavailable (Shand and Smith, 1997; Anderson and Magdoff, 2005; Murphy, 2007). Therefore, it is important to have a clear understanding of the organic P forms supplied to soil from manure (Hansen et al., 2004). Fortunately, it becomes possible to study the P composition in manure by solution phosphorus-31 nuclear magnetic resonance (^{31}P-NMR) spectroscopy (Hansen et al., 2004; Cade-Menun, 2005; He et al., 2009), which was first introduced by Newman and Tate (1980) to characterize the structural composition of P in soils.

The hilly area south of the Yangtze River and Huang-huai-hai Plain are two main grain production areas in China. In some areas with intensive animal production, amounts of manure have been applied for optimum crop yield. Understanding the effects of manure amendment on organic P movement in soils is important for effective field management in these areas. Our objectives were to ① study the P forms in two different manures which are common in these areas by solution ^{31}P-NMR, and ② investigate and compare the effects of manure application on the vertical movement, with time, of organic P in two typical soils in these areas.

3.5.2 Materials and Methods

1. Soil and manure sampling

A calcareous silt loam soil (FAO: Cambisols) and an acid sandy loam soil (FAO: Ferralsol) were collected from the Hebei Province and Jiangxi Province, respectively. The two soils were separately selected to represent typical soil types in these areas. Soil samples were taken from the surface layer (0-20 cm). Cattle manure and poultry manure were conducted in this study. All the samples were air dried and ground to pass a 2 mm sieve.

2. Soil analysis

Soil texture was determined by the pipette method (Gee and Bauder, 1986). Soil pH was measured at 1 ∶ 2.5 soil to water ratio. Total P was determined by H_2SO_4-$HClO_4$ digestion (Lu, 1999). Organic C was analyzed using the Walkley-Black dichromate oxidation method (Lu, 1999). The initial properties of the soils are shown in Table 1.

Table 1 Initial properties of the two soils

Soil properties	Sand (g/kg)	Silt (g/kg)	Clay (g/kg)	pH	Organic C (g/kg)	Olsen P (mg/kg)	Total P (g/kg)
Calcareous soil	472	400	128	8.29	27.1	16.2	0.99
Acid soil	813	68	119	4.95	19.2	13.3	0.50

3. Manure analysis

Total P was determined by H_2SO_4-H_2O_2 digestion (Thomas et al., 1967). Water extractable inorganic and organic P was analyzed according to Agbenin and Igbokwe (2006). Manure pH was determined in a 1 ∶ 5 manure to water ratio. Organic carbon and ash were determined by the dry combustion method (Lu, 1999); total N was determined by the semi-micro Kjeldahl method (Bremner and Mulvaney, 1982). These values were expressed on an air-dry mass basis. The initial properties of the manures are presented in Table 2.

Table 2 Properties of the two animal manures

Material	pH	Organic C (%)	Total N (g/kg)	Ash (%)	WEP_I[①] (g/kg)	WEP_O[①] (g/kg)	Total P (g/kg)
Cattle manure	8.13	41.30	11.10	43.08	0.87	0.13	7.44
Poultry manure	8.49	53.60	35.89	29.58	3.36	0.52	17.19

Notes: ①WEP_I: water extractable inorganic P; WEP_O: water extractable organic P.

4. Solution ^{31}P-NMR spectroscopy

According to Turner (2004), 2 g of manure samples were shaken with 40 mL of a solution containing 0.25 mol/L NaOH and 0.05 mol/L Na_2EDTA for 4 h at 20℃. Extracts were centrifuged (10000×g for 30 min), rapidly frozen at −80℃, lyophilized and ground to a fine powder.

Immediately before NMR spectroscopy, each freeze-dried extract (approximately100mg) was redissolved in 0.9 mL of 1 mol/L NaOH and 0.1 mL of D_2O (for signal lock) and transferred to a 5 mm NMR tube. The pH of each dissolved sample was confirmed to be >13.

Solution ^{31}P-NMR spectra were obtained using a Bruker (Billerica, MA) Avance 400 MHz spectrometer operating at 161.955 MHz for ^{31}P. We used a 5 μs pulse (45°), a delay time of 5.0 s, an acquisition time of 0.5 s, 25℃, and broadband proton decoupling for all samples. The number of scans required to give an acceptable signal-to-noise ratio varied among the manures depending on total P concentration, being 3500 to 4500 for poultry manure extracts and 6000 to 8000 for cattle manure extracts. Chemical shifts (10^{-6}) of signals were determined relative to 85% H_3PO_4 and assigned to individual P compounds based on literature reports (Turner et al., 2003; Cade-Menun, 2005). Peak areas were calculated by integration on spectra processed with 5 Hz line-broadening using MestReNova software 8.0. The percentage of each P compound in solution can be calculated by the proportion of the total spectral area (Turner, 2004).

5. Soil column preparation

A modification of the miscible displacement technique described by Murphy (2007) was used for the leaching experiment. PVC columns with an internal diameter of 10 cm and length of 30 cm were conducted in this study. Each column was sealed at the base with a perforated circular plate of similar diameter. The bottom of the column was attached to a funnel to serve as a drainage port. Columns were initially packed with acid-washed quartzitic sand (5 cm) to ensure free drainage of the overlying soil, and then 15 cm of soil with no treatment (without being mixed with manure) was carefully packed above the sand to achieve a bulk density of 1.3g/cm^3. After that 5 cm of soil pre-mixed with manure was placed on top of the packed soil in each of the manured columns. For the control columns, 5 cm of soil without manure application was packed. A single P application rate, 350 kg total P/ha, was selected for the study. Based on the cross-sectional area of the columns (78.5 cm^2), the total amounts of P applied to individual columns were 275 mg. Having been packed in the way above reported, a depth of 20 cm of soil was completed and a 5 cm lip was left to prevent any loss of irrigation water from the soil surface. Care was taken to pack the columns as uniformly and as tight to the column wall as possible to reduce structural heterogeneities and the risk of side-flow. Rhizon soil solution samplers were inserted, as the soil was packed, through holes in the side of the columns at soil depth of 10 cm. Therefore, there were two leachate outlets (10 cm and 20 cm depth) of each column. Each treatment had three replicates. Columns were placed in a barrel and slowly filled with deionised water to saturate the columns. After that, columns were allowed to drain freely and placed in a cabinet with a constant temperature of 25℃. Soils were left to equilibrate for 7 weeks before leaching. A nylon net (300 mesh) was placed on the surface of the soil to reduce the impact of water drops.

6. Phosphorus leaching

A plastic inlet tube with a regulating valve was used to deliver the deionised water into each soil column. Columns were maintained under unsaturated flow conditions. The valve was adjusted to provide a drip of water with a flow rate of 0.2 mL/min. For each calcareous soil column, 150 mL deionised water (equivalent to 19 mm precipitation) was applied every two weeks, while 250 mL deionised water (equivalent to 32 mm precipitation) was added once a week to each acid soil column. The leaching period was 12 weeks (6 times of leaching) for the calcareous soil and 7 weeks (7 times of leaching) for the acid soil. This was designed to approximate precipitation rates of the two areas in summer. After each leaching, 30 mL leachate was collected at the depth of 10 cm and the remaining was collected at 20 cm.

7. *Leachate analysis*

Leachate was filtered through 0.45 μm membrane filters (<0.45 μm has been widely classified as dissolved species) immediately after sampling. Dissolved reactive P (DRP), which is considered to be largely inorganic orthophosphate, was analyzed using the malachite-green method (Ohno and Zibilske, 1991). Total dissolved P (TDP, total P of dissolved fraction in leachate) was determined by $K_2S_2O_8$ digestion in an autoclave at 121℃ for 30minutes followed by colorimetric determination (Ameel et al., 1993). Dissolved organic P (DOP) was calculated as the difference between TDP and DRP.

8. *Post leaching soil measurement*

In order to better understand how P moves into subsurface horizons, further studies have been conducted to examine the fractions of P that move within the soil. At the end of leaching, soils in the columns were sampled in depth sections of 0-2.5, 2.5-5, 5-7, 7-9, 9-11, 11-13, 13-15, 15-17 and 17-20 cm. Soil samples were air-dried and passed through a 2 mm sieve. Water extractable P was extracted by a soil: water ratio of 1∶10 (w/v) for 30 min (Kuo, 1996). Water extractable inorganic P (WEP_I) and water extractable total P (WEP_T) was determined before and after $K_2S_2O_8$ digestion followed by malachite-green colorimetric determination. Water extractable organic P (WEP_O) was calculated as the difference between WEP_T and WEP_I. Labile P, which was considered to be weakly-adsorbed and readily plant-available, was extracted with 0.5 mol/L $NaHCO_3$ (pH 8.5) solution (1∶20 soil to solution ratio) (Bowman and Cole, 1978). Labile inorganic P and labile total P were determined before and after persulfate and H_2SO_4 digestion (Kuo, 1996) followed by a colorimetric method (Murphy and Riley, 1962). Labile organic P was determined as the difference between labile total P and labile inorganic P.

9. *Statistical analysis*

All results in Figures and Tables were given as means of three replicates. Analysis of variance (ANOVA) was performed to evaluate differences among means with SPSS16.0 software. Results were considered significantly different at P <0.05. The error bars on the graphs represent one standard deviation above and below the mean.

3.5.3 Results and Discussion

1. *Phosphorus forms identified by ^{31}P-NMR spectroscopy*

The ^{31}P-NMR spectra for manures are shown in Fig. 1. Described by previous works (Turner et al., 2003; Turner, 2004; Cade-Menun, 2005), the strong signal appearing at 6.57×10^{-6} and 6.64×10^{-6} was assigned to inorganic orthophosphate, while signals between 4.0×10^{-6} and 6.0×10^{-6} were assigned to orthophosphate monoesters. Several signals were detected at 5.95×10^{-6}, 5.77×10^{-6}, 5.44×10^{-6}, 5.07×10^{-6}, 4.67×10^{-6} and 4.53×10^{-6}. The four signals at 5.95×10^{-6}, 5.07×10^{-6}, 4.67×10^{-6} and 4.53×10^{-6} in the ratio 1 : 2 : 2 : 1 were assigned to phytate (myo-inositol hexakisphosphate), although these were only clearly resolved in the more concentrated NaOH extracts (Turner, 2004) such as poultry manure. Some of the other signals in this region represent "other monoesters" (He et al., 2009), such as glucose-6-phosphate (5.44×10^{-6}). Signals between 2.5×10^{-6} and -2.0×10^{-6} were assigned to orthophosphate diesters. Moreover, signals between 0.5×10^{-6} and 1.9×10^{-6} were assigned to phospholipids, while signals between 0 and -2×10^{-6} were assigned to DNA. The signals of orthophosphate diesters were very weak, as these compounds degrade rapidly in alkaline solution (Turner et al., 2003).

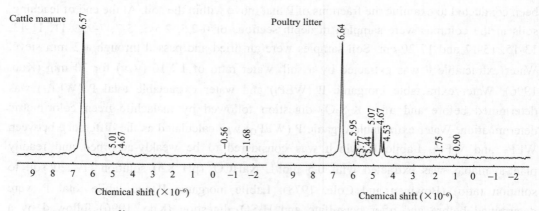

Fig. 1   Solution ^{31}P-NMR spectra of cattle manure (left) and poultry manure (right) extracted with NaOH-EDTA

Spectra are plotted with 5 Hz line broadening

The two organic materials differed in their P composition (Table 3, Fig. 1). The cattle manure contained mainly inorganic orthophosphate (93.9%), with a little of monoesters (5.5%) and only traces of DNA (<1%), and no detectable phospholipids. In contrast, the poultry manure contained mainly inorganic orthophosphate (68.8%) and monoesters (27.0%), with traces of phospholipids (2%) and no detectable DNA. The high P concentrations in these extracts meant that most of the low-concentration compounds were clearly detectable in the spectra (Turner, 2004).

Table 3 Phosphorus functional classes and proportions (%) of the total P in 0.25 mol/L NaOH-0.05 mol/L Na$_2$EDTA extracts of the two manures by solution ^{31}P nuclear magnetic resonance spectroscopy

Material	Inorganic orthophosphate	Phytate	Other monoesters	DNA	Phospholipids
Cattle manure	93.9	5.5	ND[①]	0.6	ND[①]
Poultry manure	68.8	27.0	2.2	ND[①]	2.0

Note: ① ND: not detected.

2. Phosphorus forms in leachate

For the calcareous soil, at 10 cm depth, manure treatments resulted in a significant increase of TDP in leachate (Fig. 2(a)). The TDP values ranged from 56.8 to 141.4 μg/L in control columns, from 101.9 to 951.3 μg/L in cattle manure columns, and from 115.1 to 573.5 μg/L in poultry manure columns. The large percentage (>90%) of the TDP eluted from calcareous soil columns was in the organic form. At 10 cm depth, there was a rapid increase of DOP after the first leaching in the soil columns treated with manure (Fig. 2(b)). The peak DOP values were 943 μg/L in cattle manure columns and 566 μg/L in poultry manure columns, which accounted for 99.2% and 98.8% of the effluent TDP at the second leaching, respectively. DOP declined in an exponential shape mode after the peak. Poultry manure columns showed much slower decline compared with cattle manure columns. No difference was observed in the total amounts of DOP leached (the sum of DOP in leachate of six times of leaching) from columns treated with the two organic materials. DRP concentrations in leachate collected at 10 cm depth of all columns remained constant and small (<12 μg/L) (Fig. 2(c)).

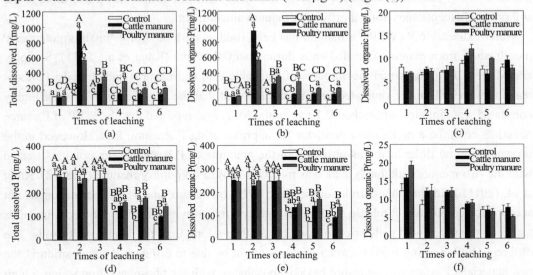

Fig. 2 Effects of manure application on total dissolved P, dissolved organic P and dissolved reactive P in leachate at depth of 10 cm ((a), (b), (c)) and 20 cm ((d), (e), (f)) for six leaching times in calcareous soil columns

Vertical bars represent standard deviations ($n=3$). Columns followed by the same lower case letter within a group were not significantly different at $P>0.05$. Columns followed by the same upper case letter across the groups for the same treatment were not significantly different at $P>0.05$. Analysis of variance (ANOVA) was not performed on DRP ((c), (f)) to evaluate differences among means as the concentration of P in leachate was very low

At 20 cm depth, dissolved P fractions showed a slow decline with leaching times in all treatments. TDP concentrations were 65.8-295.5 μg/L, 99.1-268.1 μg/L and 141.2-265.5 μg/L in control, cattle manure and poultry manure columns, respectively (Fig. 2(d)). High DOP values (>50 μg/L) were determined in leachate (Fig. 2(e)). In the first three times of leaching, DOP values of all columns were constant, and no significant difference was observed. DOP values showed an obvious decline in the last three leaching times, especially in control columns. The results showed that manure amendment could reverse the fall of DOP. This may be the result of desorption of organic P due to competitive sorption of organic acid for organic P binding sites (Kang et al., 2011). DRP concentrations were generally low, with means for all columns ranging from 0 to 20 μg/L (Fig. 2(f)).

Previous studies also reported that most of eluted P was inorganic forms. Idowu et al. (2008) found that there was a large percentage of organic P in leachate from columns amended with cattle manure in a short term experiment. Chardon et al. (1997) treated soil columns and field plots with animal manure and found that organic P represented >90% of the total P in leachate from soil columns, and >70% of the total P in leachate from lysimeters in field plots. Certainly, inorganic P represented large percentage of the total P in leachate was also reported. Ojekami et al. (2011) studied the movement of different forms of P in a clay loam soil with 33 years of manure application and found that inorganic P formed the bulk of total P in the leachate samples. Sharpley et al. (1984) reported a rapid decrease in soil organic P following years of cattle manure application. They held that organic P in manure had undergone chemical transformations upon addition to soils and was converted to inorganic P by mineralization. Therefore, organic P may be the main form of P which can move through the soil profile (except sandy soil) shortly after manure amendment.

Water extractable P (WEP) in manure has been used as an index for estimating potential P mobilization from manure-amended soils during rainfall events (Kang et al., 2011). Sharpley and Moyer (2000) measured the amounts of leachable P in six manures and compost (without reacting with soil) under simulated rainfall and reported that the amounts of WEP were highly correlated with amounts of leached P. Kang et al. (2011) also reported that the WEP of P source materials could be a preliminary predictor for approximating P leaching loss. However, in the present study no difference was observed in the amounts of P leached from columns treated with the two manures, although there was much more WEP in poultry manure (Table 2). Kang et al. (2011) held that this correlation between WEP of manure and P in leachate might not account for continuous P release from added manure mixed with soils. In our study, manures had been premixed with soils prior to leaching and had a long time for reacting with soils. The different DOP leaching trend between manures might be due to different P composition in the two materials. Phytate (myo-inositol hexakisphosphate), with six phosphate groups, has a high charge density and therefore has a strong soil fixation capacity (Celi et al., 1999).Thus, we could expect the phytate applied from manure to be retained in soil and have low mobility. Anderson and Magdoff (2005) studied different organic P compounds in soil columns and found that diester-P had the highest mobility. They concluded that diester-P might have a greater likelihood of leaching through a soil than monoester-P or soluble inorganic P. Therefore, the DOP in the leachate might be diester-P, to be specific, DOP eluted from cattle manure columns might be DNA, while DOP from poultry manure columns might be phospholipids. In this study, a much higher affinity of DRP over DOP to anion adsorption sites was observed in calcareous soil. The concentrations of DRP would not be of environmental concern as they were below the concentration (35-100 μg/L) required to trigger eutrophication in water (Haygarth and Jarvis, 1999). However, DOP concentrations were generally large enough to cause great environmental risk due to their demonstrated mobility. This is because diester-P compounds were considered to be fairly labile and susceptible to hydrolysis to inorganic P (Anderson and Magdoff, 2005; Murphy, 2007).

Compared with calcareous soil, the acid soil has a much larger ability to adsorb P and retard the vertical movement of dissolved P in soil profile (Fig. 3). Much smaller P concentrations were determined in this soil leachate compared with the calcareous soil. This was probably due to the greater P retention capacity of the acid soil (1516 mg P/Kg) compared with the calcareous soil (656 mg P/Kg) (sorption isotherm method, unpublished data). The TDP value in leachate ranged from 20 to 50 μg/L (Figs. 3(a), (d)). Similar to calcareous soil, organic P in leachate formed the bulk of TDP in acid soil. DOP accounted for more than 75% of total leaching P in all treatment (Figs. 3(b), (e)). DRP concentration was extremely low (<10 μg/L) (Figs. 3(c), (f)). No significant difference of TDP was observed among acid soil columns. It meant that the P of manure was not eluted from columns by leaching. Murphy (2007) believed that soil with large P sorption ability can act as a buffer to reduce the impact of applied P on ground water.

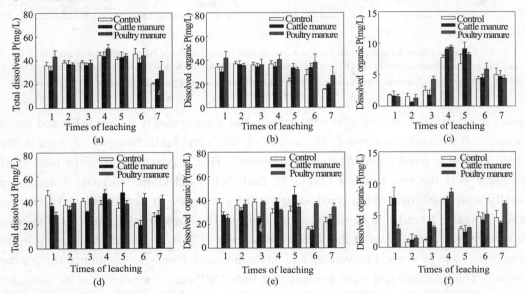

Fig. 3 Effects of manure application on total dissolved P, dissolved organic P and dissolved reactive P in leachate at depth of 10 cm ((a), (b), (c)) and 20 cm ((d), (e), (f)) for seven leaching times in acid soil columns

Vertical bars represent standard deviations ($n=3$). Analysis of variance (ANOVA) was not reported as the concentration of P in leachate was very low

3. *Phosphorus forms in soils*

The application of manure to soils had a direct effect on water extractable inorganic P (WEP$_I$) (Figs. 4(a), (c)), and labile inorganic P (Figs. 5(a), (c)). Soils sampled from 0-5 cm

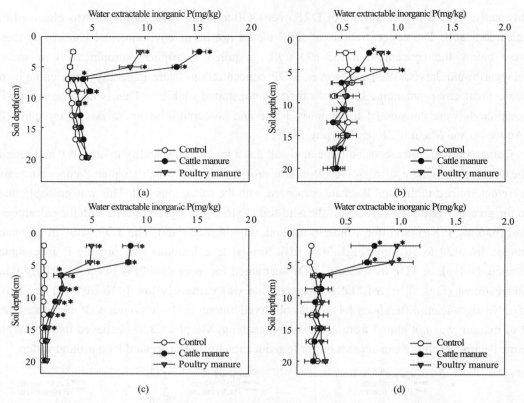

Fig. 4 Effects of manure application on water extractable inorganic P and water extractable organic P for each depth in calcareous soil ((a), (b)) and acid soil ((c), (d)) columns following the end of the experiment

Error bars represent standard deviations ($n=3$). Asterisk indicated treatment was significantly different at $P>0.05$ compared with control

layers in which soils were mixed with manure had a significantly higher WEP_I and labile inorganic P compared to control. The values of WEP_I and labile inorganic P were in the order: control < poultry manure< cattle 19manure. This is consistent with the content of inorganic P in the two manures (Table 3). Both WEP_I and labile inorganic P decreased rapidly with depth from the highest values at 0-5 cm depth. At depths below 10 cm soil labile inorganic P tended to be similar to the control. The large sorption capacity of the two soils appears to have delayed the vertical movement of P in the columns resulting in no more than 5 cm of vertical movement and labile inorganic P being concentrated toward the surface. This is consistent with the fact that little DRP was determined in the leachate. Similar results were reported in some previous studies (Malhi et al., 2003; Murphy, 2007). Murphy (2007) believed that the P could be lost from the soil, particularly in run-off, interacting with the P-enriched surface layer, and that little risk of P loss by leaching was expected because of large sorption capacity of the lower layer. Interestingly, WEP_I of soils in treatment columns was always higher than control at all depth. It was reported that organic acids might compete with P for sorption site and thereby affect the solubility of phosphate in soils (Iyamuremye et al., 1996). It was possible that the movement of organic acids enhanced the solubility of phosphate in soils at all depths.

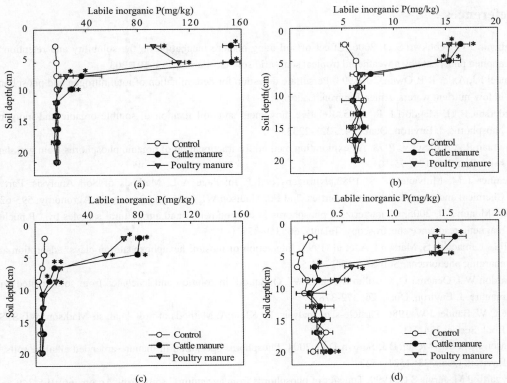

Fig. 5 Effects of manure application on labile inorganic P and labile organic P for each depth in calcareous soil ((a), (b)) and acid soil ((c), (d)) columns following the end of the experiment

Error bars represent standard deviations ($n=3$). Asterisk indicated treatment was significantly different at $P>0.05$ compared with control

Contrary to the fact that most of P in leachate was in the organic form, little organic P was determined in water extraction (<40%) (Figs. 4(b), (d)) and $NaHCO_3$ extraction (<30%) fractions (Figs. 5(b), (d)). Compared to the control columns, larger value of WEP_O and labile organic P were determined in surface soil (0-5 cm) of amendment columns. The values of WEP_O were in the order: control < cattle manure < poultry manure. It is because of much more organic P in poultry manure compared to cattle manure (Table 2 and Table 3). In calcareous soil, as much DOP has been observed in leachate at 10 cm depth, it was considered that WEP_O and labile organic P of soil layers (below 5 cm) would be larger than that of control. However, no difference was observed irrespective of manure types. WEP_O, along with labile organic P, is thought to be easily mineralizable organic P (Sharpley et al., 2004). It is possible that organic P must have been mineralized and/or converted to other forms of inorganic P in calcareous soil. Due to greater P retention capacity of the acid soil, much smaller WEP_O and labile organic P values were observed compared with calcareous soil. This result was consistent with the leachate experiment, in which little P was determined in leachate from acid soil columns. The soils such as this acid soil can act as a buffer to reduce the environmental risk of applied manure on soil surface. However, in this study soil was packed into columns and did not attempt to replicate the macropore structure of natural soils, which are important for preferential flow. Therefore, macropore structure should be taken into consideration when manure was applied on natural soils.

References

Agbenin J O, Igbokwe S O. 2006. Effect of soil-dung manure incubation on the solubility and retention of applied phosphate by a weathered tropical semi-arid soil. Geoderma, 133: 191-203

Ameel J J, Axler R P, Owen C J. 1993. Persulfate digestion for determination of total nitrogen and phosphorus in low-nutrient waters. Amer. Environ. Labor., 5: 1-11

Anderson B H, Magdoff F R. 2005. Relative movement and soil fixation of soluble organic and inorganic phosphorus. J. Environ. Qual., 34: 2228-2233

Bowman R A, Cole C V. 1978. An exploration method for fractionation of organic phosphorus from grassland soils. Soil Sci., 125: 95-101

Bremner J M, Mulvaney C S. 1982. Nitrogen-Total P. In: Page A L. Methods of Soil Analysis. Part 2. Chemical and Microbiological Properties. 2nd Ed. Madison WI: American Society for Agronomy: 595-624

Cade-Menun B J. 2005. Characterizing phosphorus in environmental and agricultural samples by ^{31}P nuclear magnetic resonance spectroscopy. Talanta, 66: 359-371

Celi L, Lamacchia S, Marsan F A, et al. 1999. Interaction of inositol hexaphosphate on clays: adsorption and charging phenomena. Soil Sci., 164: 574-585

Chardon W J, Oenema O, Castilho P, et al. 1997. Organic P in solution and leachates from soils treated with manure. J. Environ. Qual., 26: 372-378

Gee C W, Bauder J W. 1986. Particle-size analysis. In: Klute A. Methods of Soil Analysis Madison, WI: SSSA Book Series: 384-411

Hansen J C, Cade-Menun B J, Strawn D G. 2004. Phosphorus speciation in manure-amended alkaline soils. J. Environ. Qual., 33: 1521-1527

Haygarth P M, Jarvis S C. 1999. Transfer of phosphorus from agricultural soils. Adv. Agron., 66: 195-249

He Z Q, Honeycutt C W, Cade-Menun B J, et al. 2008. Phosphorus in poultry litter and soil: enzymatic and nuclear magnetic resonance characterization. Soil Sci. Soc. Am. J., 72: 1425-1433

He Z Q, Honeycutt C W, Griffin T S, et al. 2009. Phosphorus forms in conventional and organic dairy manure identified by solution and solid state P-31 NMR spectroscopy. J. Environ. Qual., 38: 1909-1918

Idowu M K, Ige D V, Akinremi O O. 2008. Elution of inorganic and organic phosphorus from surface-applied organic amendments. Can. J. Soil Sci., 88: 709-718

Iyamuremye F, Dick R P, Baham J. 1996. Organic amendments and phosphorus dynamics. I. Phosphorus chemistry and sorption. Soil Sci., 161: 426-435

Kang J, Amoozegar A, Hesterberg D, et al. 2011. Phosphorus leaching in a sandy soil as affected by organic and inorganic fertilizer sources. Geoderma, 161: 194-201

Kuo S. 1996. Phosphorus. In: Sparks D L. Methods of Soil Analysis. Madison, WI: SSSA Book Series: 869-919

Lu R K. 1999. Analytic Methods for Soil and Agro-Chemistry. Beijing: Chinese Agricultural Press (in Chinese)

Malhi S S, Harapiak J T, Karamanos R, et al. 2003. Distribution of acid extractable P and exchangeable K in a grassland soil as affected by long-term surface application of N, P and K fertilizers. Nutr. Cycl. Agroecosys., 67: 265-272

Murphy J, Riley J P. 1962. A modified single solution for the determination of phosphorus in natural waters. Anal. Chim. Acta, 27: 31-36

Murphy P N C. 2007. Lime and cow slurry application temporarily increases organic phosphorus mobility in an acid soil. Eur. J. Soil. Sci., 58: 794-801

Newman R H, Tate K R. 1980. Soil phosphorus characterization by ^{31}P nuclear magnetic resonance. Commun. Soil Sci. Plant Anal., 11: 835-842

Ohno T, Zibilske L M. 1991. Determination of low concentrations of phosphorus in soil extracts using malachite green. Soil Sci. Soc. Am. J., 55: 892-895

Ojekami A, Ige D, Hao X Y, et al. 2011. Phosphorus mobility in a soil with long term manure application. J. Agr. Sci., 3: 25-38

Pierzynski G M, Logan T J, Traina S J, et al. 1990. Phosphorus chemistry and mineralogy in excessively fertilized soils: descriptions of phosphorus-rich particles. Soil Sci. Soc. Am. J., 54: 1583-1589

Shand C A, Smith S. 1997. Enzymatic release of phosphate from model substrates and P compounds in soil solution from a peaty podzol. Biol. Fert. Soils, 24: 183-187

Sharpley A N, Moyer B. 2000. Phosphorus forms in manure and compost and their release during simulated rainfall. J. Environ. Qual., 29: 1462-1469

Sharpley A N, McDowell E W, Kleinman P A. 2004. Amounts, forms, and solubility of phosphorus in soils receiving manure. Soil Sci. Soc. Am. J., 68: 2048-2057

Sharpley A N, Smith S J, Stewart B A, et al. 1984. Forms of phosphorus in soil receiving cattle feedlot waste. J. Environ. Qual., 13: 211-215

Sims J T, Simard R R, Joern B, C. 1998. Phosphorus loss in agricultural drainage: historical perspective and current research. J. Environ. Qual., 27: 277-293

Thomas R L, Sheard R W, Moyer J R. 1967. Comparison of conventional and automated procedures for nitrogen, phosphorus and potassium analysis of plant material using a single digestion. Agron. J., 59: 243-340

Tunesi S, Poggi V, Gessa C. 1999. Phosphate adsorption and precipitation in calcareous soils: the role of calcium ions in solution and carbonate minerals. Nutr. Cycl. Agroecosys., 53: 219-227

Turner B L, Haygarth P M. 2000. Phosphorus forms and concentrations in leachate from four grassland soil types. Soil Sci. Soc. Am. J., 64: 1090-1097

Turner B L, Mahieu N, Condron L M. 2003. Phosphorus-31 nuclear magnetic resonance spectral assignments of phosphorus compounds in soil NaOH-EDTA extracts. Soil Sci. Soc. Am. J., 67: 497-510

Turner B L. 2004. Optimizing phosphorus characterization in animal manures by solution phosphorus-31 nuclear magnetic resonance spectroscopy. J. Environ. Qual., 33: 757-766

3.6 Rapid determination of isotope labeled nitrate using Fourier transform infrared attenuated total reflection spectroscopy

3.6.1 Introduction

Nitrogen is one of the most important elements in nature and the indispensable elements for plant. Nitrogen is absorbed mainly as ammonium nitrogen (NH_4^+-N) and nitrate (NO_3^--N) by plant (Killham, 1994). Generally, the determination methods of nitrate nitrogen content contain reduction distillation, electrode method, phenol disulfonic acid method, cadmium column method and so on (China Soil Society Agrochemistry Speciality Committee, 1989; Lu, 1999). However, there are several shortcomings for these methods, including low efficiency, complicated operation, application difficulties in the agricultural production (Hu, 2006; Chen et al.,1955). Although flow injection analysis can overcome these shortcomings to certain degree and are widely used in agricultural production, it still demands a lot of pretreatment, including extraction, filtration, and a series of reagents destroying samples. Despite advantages and disadvantages in the forementioned methods, one common problem is that all the methods can not distinguish the ^{15}N labeled nitrate nitrogen from the common nitrate nitrogen. The most common method for tracking of ^{14}N and ^{15}N is mass spectrometry, which demands a series of chemical treatments for conversion of different N species suitable for isotope ratio mass spectrometry measurements (Baggs, 2008). Gardner et al. (1991) found another method, cation-exchange and fluorometric-based detection method, for the determination of $^{14}N/^{15}N$ ratios of NH_4^+, unlike mass spectrometry, which used small volumes of samples, relatively low-cost equipment, and minimal instrumental manipulations. However, the method necessitated a relatively long time of analysis, and thus unsuitable for the dynamic determination of nitrate-N.

Given the problems of the above methods, Fourier transform infrared attenuated total reflection (FTIR-ATR) spectroscopy was applied to determine nitrate. Compared with conventional methods, this method requires minimal sample preparation, is non-destructive and regardless of sample size, shape, moisture content (Huang and Yin, 2011). In addition, the characteristic peak of nitrate is located around 1350 cm^{-1} in the infrared spectrum (Shaviv et al., 2003; Yang et al., 2013), which provides a theoretical basis for the direct determination of nitrate content in solution and soil.

This study used the Paddy soil as experimental materials and Fourier transform infrared attenuated total reflection (FTIR-ATR) spectroscopy to determine $^{14}NO_3^-$-N and $^{15}NO_3^-$-N in solution and soil. Principal component analysis (PCA) and partial least squares regression (PLSR) were used in data analysis. This work is aimed at providing a new method for rapid determining ^{14}N and ^{15}N labeled nitrate nitrogen.

3.6.2 Experiment

1. Experimental material

The normal KNO_3 and ^{15}N-labeled KNO_3 (produced in Sigma, USA) were used as experimental reagents. The soil samples studied were Paddy soils collected in Changshu Ecological Station Institute of Soil Science. The physical and chemical properties of soil were shown in Table 1.

Table 1 Physical and chemical properties of paddy soil

Soil Type	pH	NH_4^+-N (mg/kg)	NO_3^--N (mg/kg)	$CaCO_3$ (g/kg)	HNO_3^- (g/kg)	Organic matter (g/kg)
Paddy soil	7.09	10.10	4.74	12.30	0.031	41.8

2. Standard determination of nitrate in soil

Three milliliters (3 mL) KNO_3 solution with different NO_3^--N concentrations (0, 20, 40, 80, 120, 160, 200 mg/L) were added to 5.00 g of air-dried soil, in triplicate. After that, 2 mol/dm^3 KCl solution was added (with the ratio of water to soil 10 : 1). The shaking time was 1 hour, and suspensions were placed for 30 minutes and filtered. Then, the filtrate was measured by intermittent automatic chemistry analyzer (Smartchem 200, Italy).

3. ATR method

The spectra were recoded using a Fourier transform infrared (FT-IR) spectrometer (Nicolet 6700, Thermo Fisher Scientific, USA) equipped with a 45℃ ZnSe ATR accessory (Bruker, Germany) and a deuterated triglycine sulfate (DTGS) detector.

The KNO_3 solutions with different NO_3^--N concentrations (0, 20, 40, 60, 80, 100, 120, 140, 160, 180, 200 mg/L) were directly added onto the ATR crystal slot. Air-dried soil was in prior made into soil paste with proper distilled water (Linker et al., 2005, 2006), since air-dried soil can not form close contact with the crystal slot, leading to low spectrum signal and poor reproducibility. 3 mL of KNO_3 solution with different NO_3^--N concentrations (0, 20, 40, 80, 120, 160 and 200 mg/L) was added to 5 g air-dried soil to form soil paste which was spread onto the ATR crystal slot. The spectra of the ATR crystal and distilled water were recorded per hour, for spectral baseline correction and subtraction of water interference (Linker et al., 2004, 2005). The determining method of different $^{14}N/^{15}N$ ratio which were 1 : 0, 0 : 1, 1 : 1, 1 : 2, 2 : 1 labeled nitrate in KNO_3 solution (with the NO_3^--N concentrations 0, 40, 80, 120, 160, and 200 mg/L) and soil samples was the same as above. The scans were recorded in the wavenumber region of 800-4000 cm^{-1} with a resolution of 4 cm^{-1}, and the average of 32 successive scans per sample was used in data analysis.

4. Data analysis

Spectral data was processed using MATLAB R2009a based home-made procedures. The linear regression was implemented using Excel 2003. First, the absorption of water was removed, after that the random noise in spectra was filtered out, then principal component regression (PCR) and partial least squares regression (PLSR) were utilized for model calibration.

3.6.3 Results and Discussions

1. Fourier transform infrared attenuated total reflection (FTIR-ATR) spectra of KNO_3 solution

Fig. 1(a) showed the FTIR-ATR spectra of KNO_3 solutions. The overall trend of spectra of different concentrations KNO_3 solution was basically the same. The spectra were strongly interfered by water absorption, and the two strong bands around 3000-3800 cm^{-1} and 1500-1800 cm^{-1} were mainly caused by moisture. After water subtraction, the characteristic absorption band was situated from 1200 cm^{-1} to 1500 cm^{-1} (Du et al., 2009). The peak intensity around 1200-1500 cm^{-1} was well proportional to the nitrate concentration, and the linear correlation between PCA1 and nitrate was excellent, and the relation coefficient was 0.9912 (Yang et al., 2013). Thus, the result demonstrated the good capacity of FTIR-ATR spectra in quantifying nitrate in solution.

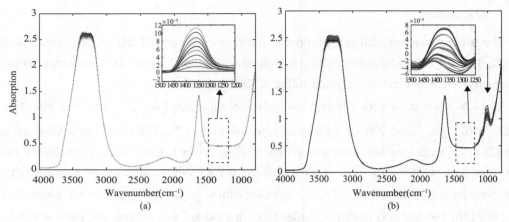

Fig. 1   FTIR-ATR spectra of KNO₃ solutions (a) and soil (b)

2. *Fourier transform infrared attenuated total reflection (FTIR-ATR) spectra of soil samples*

The FTIR-ATR spectra of soil samples were shown in Fig. 1(b). The spectra of soil paste with different nitrate content were basically in the same shape. Similarly, the water characteristic absorption range were around 3000-3800 cm^{-1} and 1500-1800 cm^{-1}. However, an absorption band around 1000 cm^{-1} was associated with minerals and organic matters, which was different from the KNO₃ solution, and the peak shape and absorption intensity varied with different samples. This area is mainly attributed to C—C, C—O stretching vibration and skeleton vibrations of some saturated hydrocarbons, as well as stretching vibration of some halide containing P, S and other chemical substances (Deng et al., 2009). After water subtraction, the characteristic absorption band was obtained. Since the spectra at 1200 cm^{-1} was interfered inferred by absorption band of mineral and organic matter, the characteristic absorption band of nitrate was defined from 1250 cm^{-1} to 1500 cm^{-1}. The band intensity around 1200-1500 cm^{-1} was proportional to the nitrate concentration. But the absorption band of soil paste with low nitrate-N content was much narrower and shifted slightly to the direction of high wavenumber.

In order to explain the peak shift, the second derivatives of characteristic absorption band of nitrate in KNO₃ solution and soil paste were obtained (shown in Fig. 2). The characteristic absorption peak was divided into three peaks. The left one, about 1460 cm^{-1}, was caused by N=O vibration; the right one was induced by N—O vibration, at around 1300 cm^{-1}; the middle one resulted from vibration of bond between N=O and N—O. Fig. 2(a) showed the second derivative spectra of nitrate characteristic absorption band in KNO₃ solution. The vibration intensity associated with N—O increased with the nitrate concentration, so did the vibration intensity by N=O, except 200 mg/kg treatment. The vibration intensity by N=O in low nitrate solution (0, 20, 40 mg/L) was greater than that by N—O, but for the high nitrate solution the opposite occurred. Similarly, the vibration intensity by N—O increased with the nitrate concentration, and N=O vibration intensity was greater than that by N—O. Therefore, absorption band caused by N=O was higher than that induced by N—O, which leaded to the shift of the characteristic absorption band to high wavenumber direction.

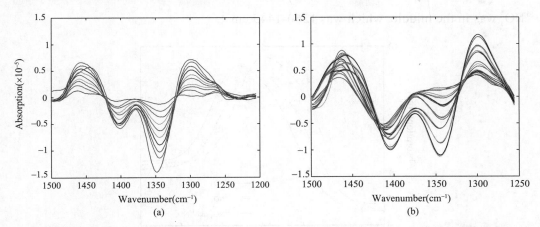

Fig. 2 Second derivatives of characteristic absorption band of KNO$_3$ solutions (a) and soil samples (b)

Principal component analysis (PCA) was used to extract principle components from raw spectra to quantify nitrate. For FTIR-ATR spectra of soil paste with different nitrate concentrations, the first principle component (PCA1) explained 98.55% of total variance, and the second principle component (PCA2) explained 0.93%. Obviously, PCA1 was sufficient enough to quantify nitrate. The principle component regression (PCR) was conducted by between PCA1 (independent variables) and nitrate concentration (dependent variables). The linear fittings were shown in Fig. 3, in which an excellent relation coefficient of 0.9840 was observed, indicating that FTIR-ATR spectroscopy could be well used as quantification of nitrate.

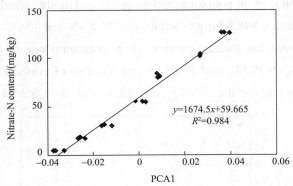

Fig. 3 The linear regression between nitrate nitrogen content and FTIR-ATR spectra of soil

3. Prediction of $^{14}NO_3^-$-N/$^{15}NO_3^-$-N in solution

Fig. 4 showed the FTIR-ATR spectra of different isotope labeled KNO$_3$ solution of 200 mg/L. It could be seen that the absorption bands of $^{14}NO_3^-$ and $^{15}NO_3^-$ were similar, but the absorption position was significantly different. The absorption region of $^{14}NO_3^-$ was 1275-1460 cm^{-1}, and the absorption band of $^{15}NO_3^-$ was 1240-1425 cm^{-1}.

Compared with $^{14}NO_3^-$, the absorption band of $^{15}NO_3^-$ obviously shifted to the direction of low wavenumber by 35 cm^{-1}. The absorption band of the 1 : 1 mixture of $^{14}NO_3^-$ and $^{15}NO_3^-$ was in the middle, which was 1260-1445 cm^{-1}.

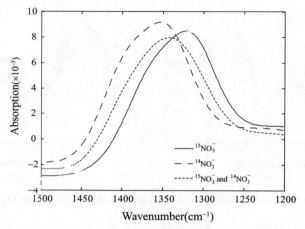

Fig. 4　FTIR-ATR spectra of $^{14}N/^{15}N$ labeled KNO$_3$ in solutions

Partial least squares (PLS) method was used to establish the model based on 1200-1500 cm^{-1}. Before modeling, cross-validation was applied to select the optimal number of principal components. The optimal number of principal components of $^{14}NO_3^-$ in solution was 10 and corresponding root-mean-square error of cross validation (RMSECV) was 8.36 mg/kg(shown in Fig. 5). Therefore, the first 10 principal components were used for model calibration. Fig. 6 showed the regression between predicted values obtained from PLSR and experimental values of $^{14}NO_3^-$-N and $^{15}NO_3^-$-N in potassium nitrate solution. Highly significant correlation with R^2 values of 0.9980 and 0.9982 for prediction of $^{14}NO_3^-$-N and $^{15}NO_3^-$-N, respectively, was achieved. Table 2 gived the statistic parameters in potassium nitrate solution achieved by PLSR. For the $^{15}NO_3^-$-N PLSR model, the optimal number of principal components was 12 and RPD was 4.76; while for the $^{14}NO_3^-$-N PLSR model, the optimal principal number of

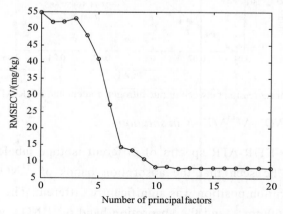

Fig. 5　RMSECV (root mean square error in cross validation) of nitrate–N with ^{14}N labeled $^{14}NO^-_3$-N in solution

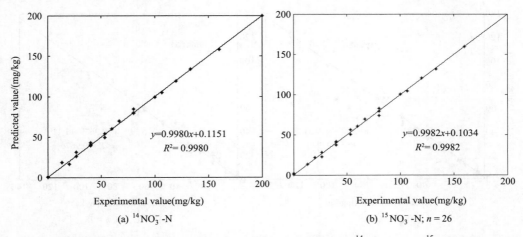

Fig. 6 Linear regression between measured values and predicted values of $^{14}NO_3^-$-N and $^{15}NO_3^-$-N in solution

components was 10 and RPD was 6.44. In infrared spectroscopy, RPD is an important criterion for model performance, which is defined as the ratio of standard deviation to root mean square error (RMSE). Generally, RPD > 3 is considered acceptable (Malley et al., 1999). The RPD values reported above are much higher than 3, indicating that prediction models are excellent and appropriate for nitrate quantification.

4. Prediction of $^{14}NO_3^-$-N/$^{15}NO_3^-$-N in soil

Similarly, cross-validation was applied to decide the optimal number of principal components (Table 2), and then PLSR model was established and the regressions between predicted values and experimental values of $^{14}NO_3^-$-N and $^{15}NO_3^-$-N in soil were obtained (Fig. 7). The results showed that the determination coefficients (R^2) of $^{14}NO_3^-$-N and $^{15}NO_3^-$-N in soil were 0.9794 and 0.9679, respectively. For the $^{15}NO_3^-$-N PLSR model, the optimal number of principal components was 5 and RPD was 4.78; while for the $^{14}NO_3^-$-N PLSR model, the optimal number of principal components was 6 and RPD was 5.75. In soil spectral analysis, prediction models could be graded into six categories based on the RPD value (Viscarra et al., 2006; Du and Zhou, 2007): the poor model with RPD lower than 1.0, the fair model with RPD between 1.0 and 1.4, an intermediate model with RPD in the range of 1.4

Table 2 Comparison of $^{14}NO_3^-$-N and $^{15}NO_3^-$-N determination based on PLSR in solution and soil

Isotope	Statistic parameters	Solution	Soil
$^{14}NO_3^-$-N	Relation coefficient (R^2)	0.9980	0.9794
	RPD Ratio of standard deviation to predicted error	6.44	5.75
	Optimized PLSR factor number	10	6
$^{15}NO_3^-$-N	Relation coefficient (R^2)	0.9982	0.9679
	RPD Ratio of standard deviation to predicted error	4.76	4.78
	Optimized PLSR factor number	12	5

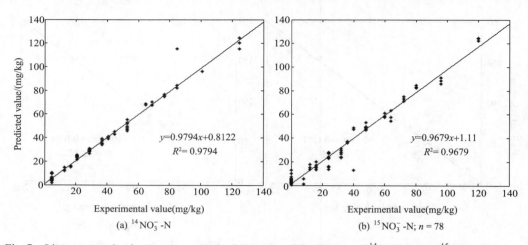

Fig. 7 Linear regression between measured values and predicted values of $^{14}NO_3^-$-N and $^{15}NO_3^-$-N in soils

and 1.8, a good model with RPD from 1.8 to 2.0, a very good model with RPD between 2.0 and 2.5, and an excellent model with RPD more than 2.5. RPD of the $^{14}NO_3^-$-N and $^{15}NO_3^-$-N model were both above 2.5 in our study, which indicated the resulting models were excellent and could be used for fast determination of $^{14}NO_3^-$-N and $^{15}NO_3^-$-N in soil.

References

Baggs E M. 2008. A review of stable isotope techniques for N₂O source partitioning in soils: recent progress, remaining challenges and future considerations. Rapid Communications in Mass Spectrometry, 22(11): 1664-1672

Chen M C, Zhang Q, Yang J L. 1995. Selection and verification of methods for determination of nitrate in soil extracts. Journal of Shanxi Agricultural Sciences, 23(1): 31-36

China Soil Society Agrochemistry Speciality Committee. 1989. Agrochemistry Conventional Analysis. Beijing: Science Press: 91-93

Deng J, Du C W, Zhou J M. 2009. Characterization of ions in greenhouse soils using infrared attenuated reflectance spectroscopy. Acta Pedologica Sinica, 46(4): 704-709

Du C W, Zhou J M, Linker R, et al. 2009. In situ evaluation of net nitrification rate in Terra Rossa Soil using a Fourier transform infrared attenuated total reflection N-15 tracing technique. Applied Spectroscopy, 63: 1168-1173

Du C W, Zhou J M. 2007. Prediction of soil available phosphorus using Fourier transform infrared-photoacoustic spectroscopy. Chinese J. Anal. Chem., 35: 119-122

Gardner W S, Herche L R, St. John P A, et al. 1991. Analytical Chemistry, 63(17): 1838-1843

Hu Y C. 2006. Development of quick on farm test to determine nitrate level and its potential use in China. Phosphate and Compound Fertilizer, 3: 7-10

Huang H Y, Yin Q H. 2011. Fundamentals and application advances in attenuated total internal reflectance Fourier transform infrared spectroscopy (ATR-FTIR). Journal of the Graduates Sun YAT-SEN University, 32(1): 20-31

Killham K. 1994. Soil Ecology. Cambridge: Cambridge University Press: 108-141

Linker R, Kenny A, Shaviv A, et al. 2004. Fourier transform infrared-attenuated total reflection nitrate determination of soil pastes using principal component regression, partial least squares, and cross-correlation. Applied Spectroscopy, 58: 516-520

Linker R, Shmulevich I, Kenny A, et al. 2005. Soil identification and chemometrics for direct determination of nitrate in soils using FTIR-ATR mid-infrared spectroscopy. Chemosphere, 61 (5): 652-658

Linker R, Weiner M, Shmulevich I, et al. 2006. Nitrate determination in soil pastes using attenuated total reflectance mid-infrared spectroscopy: improved accuracy via soil identification. Biosystems Engineering, 94(1): 111-118

Lu R K. 1999. Soil Agricultural Chemistry Analytical Method. Beijing: China Agricultural Science and Technology Press: 156-162

Malley D F, Yesmin L, Wray D, et al. 1999. Application of near-infrared spectroscopy in analysis of soil mineral nutrients. Communications in Soil Science and Plant Analysis, 30: 999-1012

Shaviv A, Kenny A, Shmulevitch I, et al. 2003. Direct monitoring of soil and water nitrate by FTIR based FEWS or membrane systems. Environmental Science and Technology, 37: 2807-2812

Viscarra R R A, Walvoort T D, McBratney A B, et al. 2006. Visible, near infrared, mid infrared or combined diffuse reflectance spectroscopy for simultaneous assessment of various soil properties. Geoderma, 131: 59-75

Yang J B, Du C W, Zhou J M, et al. 2013. Rapid determination of nitrate in chinese cabbage using fourier transforms mid-infrared spectroscopy. Chinese J. Anal. Chem., 41: 1264-1268

4 Nutrition diagnosis and quality control of crops

4.1 Effects of different nitrogen forms on the growth and cytokinin content in xylem sap of tomato (*Lycopersicon esculentum* Mill.) seedlings

4.1.1 Introduction

Mixed nitrate (NO_3^-) and ammonium (NH_4^+) nutrition are documented to be superior for most plants to either sole NO_3^--N or NH_4^+-N source (Marschner, 1995). The optimal proportions of NO_3^- to NH_4^+ for plant growth depend on plant species, environmental conditions, developmental stages and the total concentration of supplied nitrogen (N) (Chaillou et al., 1991; Xu et al., 1992; Stratton et al., 2001; Xu et al., 2001, 2002; Claussen, 2002; Kotsiras et al., 2002; Zou et al., 2005). Usually, maximum growth and yield of tomato or pepper was obtained by an optimal NH_4^+ concentration not more than 30% of total N in hydroponics in greenhouse (Takács and Técsi, 1992; Jung et al., 1994; Jung and Tadashi, 1994; Sandoval-Villa et al., 2001; Xu et al., 2001, 2002; Claussen, 2002). The superiorities of mixed N forms to either form alone were related to nearly neutral pH maintenance (Marschner, 1995), higher uptake of phosphorous, potassium, calcium and iron, etc., in mixed N forms either than single NO_3^- or NH_4^+ (Stratton et al., 2001; Kotsiras et al., 2002; Zou et al., 2005), avoiding excessive consumption of carbohydrate on NH_4^+ assimilation at the expense of root growth (Salsac et al., 1987; Schortemeyer et al., 1993; Schortemeyer and Feil, 1996) and cytokinin enhancement (Smiciklas and Below, 1992; Fetene and Beck, 1993; Wang and Below, 1996). Cytokinin has been widely accepted as a long-range signal in response to NO_3^- for regulation of leaf growth (Takei et al., 2002; Rahayu et al., 2005; Sakakibara et al., 2006) in a whole plant level. Cytokinins are considered to be synthesized mainly in the root tips and translocated to the shoot meristematic cells via xylem vessels (Bernier et al., 1977). Stimulatory effect on shoot growth by NO_3^- was found to be associated with a corresponding increase in zeatin (Z) and zeatin riboside (ZR) levels in leaves and xylem exudates, while inhibitory effect by exclusive NH_4^+ supply was just in a contrary way (Walch-Liu et al., 2000; Rahayu et al., 2005). However, the effect of mixed N form on the cytokinin concentrations either in leaves, roots or in xylem sap of tomato seedlings was little reported.

Root development is known to be remarkably sensitive to variations of the supply and distribution of inorganic nutrients in the soil (Forde and Lorenzo, 2001) and root morphology is closely involved in nutrient acquisition (Sattelmacher et al., 1993; Sorgonà and Cacco, 2002). Therefore, homogeneous or localized supply of NO_3^- and NH_4^+ could have a great effect on the regulation of plant growth. When the roots were completely supplied to the sole NH_4^+

nutrition under moderate N concentration, the growth of roots, shoots and even fruits was usually significantly inhibited in many studies (Chaillou et al., 1994; Feil, 1994; Schotemeyer et al., 1997; Claussen, 2002; Wang et al., 2003; Zou et al., 2005). However, if the roots were spatially half supplied to sole NH_4^+ and half to sole NO_3^-, where the roots in the sole NH_4^+ suffered from a depressed growth (Chaillou et al., 1994; Guo et al., 2002, 2007), the shoots remained to accumulate high dry matter (DM). It is indicated that plants have the ability to control the development in a whole-plant level (Gansel et al., 2001) to adapt themselves to uneven N-distribution environments (Zhang and Forde, 1998). To our knowledge, there are some comparisons of shoot and root growth responding to both homogeneous and localized supply of different N forms on tobacco and soybean plantlet (Chaillou et al., 1994; Walch-Liu et al., 2000), but few reports have described the difference in root morphology in response to different NO_3^- : NH_4^+ ratios both under homogeneous and split-root systems.

The changes in plant growth and development induced by NO_3^- and NH_4^+ were considered to be linked to alterations in hormonal balance, but there is much contradictory evidence in literatures (Britto and Kronzucker, 2002). Among the hormonal, cytokinins and auxin were especially concerned. Z, ZR, dihydrozeatin riboside [(diH)ZR], N^6-(Δ^2-isopentenyl) adenine (iP), isopentenyl adenosine (iPA) and kinetin (KT) are considered as major endogenous cytokinins in higher plants. However, these substances were not systemically examined under different ratios of NO_3^- to NH_4^+. Auxin (IAA) was also involved in regulating plant growth and root development but it received relatively little attention in response to N forms (Guo et al., 2005; Tamaki and Mercier, 2007). The fact that plants are always exposed to a heterogeneous NO_3^- and NH_4^+ soil environment (Schortemeyer and Feil, 1996) suggests that their roots could have the ability to acclimatize to mixed and heterogeneous distribution of N forms. The split-root system was well adopted to study the response of plants to a simultaneous but spatially separated supply of NO_3^- and NH_4^+, in which half of the root system is exposed to NO_3^- only and the other half to NH_4^+ only (Schortemeyer et al., 1993; Cramer and Lewis, 1993; Chaillou et al., 1994; Feil, 1994; Schortemeyer and Feil, 1996). This system has been approved to be superior to the homogeneous system to study the effects on physiological and molecular responses to different N forms (Chaillou et al., 1994; Walch-Liu et al., 2000, 2001; Gansel et al., 2001; Guo et al., 2002, 2007; Schmidt et al., 2003; Rahayu et al., 2005; Boukcima et al., 2006). Using the split-root system on Chaillou et al. (1994), NO_3^- or NH_4^+ was independently taken up, when supplied separately since it was difficult to assess the quantitative effects on assimilate partitioning when the two ions were supplied in combination.

The aims of this study were: ① to make a detailed description and comparison of the root morphology; ② to examine the concentrations of 6 cytokinins species in xylem sap; ③ to compare the NO_3^- and NH_4^+ levels in xylem sap of the hydroponically-grown tomato plants under both homogeneous and localized supply of different NO_3^- : NH_4^+ ratios.

4.1.2 Materials and Methods

1. *Plant treatment and growth condition*

Tomato (*Lycopersicon esculentum* Mill. Variety 'Xinpinbaoguan') seeds were sterilized using 8.82 mM H_2O_2 for 30 min and then germinated on soaked filter paper in a dish at 25°C. After germination the seeds were sown in clean and moist silica and 15 days afterwards uniform seedlings were selected. Tap roots were excised from 1 cm below the hypocotyls and transplanted into a 1/4 strength nutrient solution according to Hoagland and Arnon (1950) for induction of lateral roots. Modified, half-strength Hoagland and Arnon nutrient solution (0.5 mmol/dm^3 KH_2PO_4, 3 mmol/dm^3 K provided as KNO_3, KH_2PO_4 or KCl, 2.5 mmol/dm^3 Ca as $Ca(NO_3)_2$ or $CaCl_2$, 1.0 mmol/dm^3 $MgSO_4$, and the micronutrients 24.1 μmol/dm^3 B, 20.0 μmol/dm^3 Fe as Fe-EDTA, 9.1 μmol/dm^3 Mn, 2.0 μmol/dm^3 Zn, 0.5 μmol/dm^3 Cu, and 0.1 μmol/dm^3 Mo) was used in all treatments. NO_3^- was provided as KNO_3 and $Ca(NO_3)_2$, and NH_4^+ as $(NH_4)_2SO_4$. The total N concentration was equal to 5 mmol/dm^3. A nitrification inhibitor, $C_2H_4N_4$ (7 μmol/dm^3), was added to all the nutrient solutions to prevent nitrification. Throughout the experiment, the pH of each nutrient solution was adjusted to 6.2 ± 0.05 using a pH meter (Orion 86801) with 0.1 mol/dm^3 NaOH or 0.1 M HCl, with the solution being completely renewed every 2 days. Two weeks afterwards, 4-foliate seedlings with uniform roots were selected for the designated treatments.

Whole-root system was conducted together with split-root system. For whole-root system, the whole root was placed in a 3-L pot and supplied with combined NO_3^- and NH_4^+ in 3 ratios: NO_3^- : NH_4^+ =100 : 0 (100-0NA, control), 75 : 25 (75-25NA) and 50 : 50 (50-50NA). For the split-root system, lateral roots were evenly separated and spatially placed into double vessels with 2 L nutrient solution involved, with one plant in each double vessel and 4 lateral roots in each vessel. 3 treatments were included: sole NO_3^- supply (N|N), separated supply of NO_3^- and NH_4^+ (N|A), and separated supply of NO_3^- and the mixture of 75% NO_3^- plus 25% NH_4^+ (N|AN). Total N concentration in both systems was 5 mmol/L. All treatments were conducted with four replications.

Plants were grown in greenhouse at 25/15°C±3°C day/night temperature, 30%-35% relative humidity, and 14/10 h day/night photoperiod.

2. *Collection of xylem sap*

The xylem sap was collected as described by Rahayu et al. (2005) and Dodd et al. (2004) and slightly revised as follows. For collection of xylem sap, plant shoots were cut into 2 cm above the root-shoot interface below the cotyledons. After 3 min, the cut stem was cleaned with distilled water and then blotted with filter paper to avoid contamination of wounded cells and phloem sap. Then a silicon tube was fixed over the stem according to Rahayu et al. (2005). The xylem sap driven by root pressure in the silicon tube was removed with the help of an injector to the pipette and immediately stored on ice for short intervals to avoid tube overflow. The collection was lasted for 3h and subsequently frozen at −20°C until analyses.

3. Root parameters analysis

Root total length (RL), average diameter (AD), root surface area (RS) and root volume (RV) were determined by a root automatic scanning apparatus (EPSON® COLOR IMAGE SCANNER LA1600+, Canada), equipped with WinRHIZO 2003B software (Regent Instruments).

4. Plant harvest

Fifty days after transplanting, the 9-foliate plants were harvested. At harvest, plant shoots were cut into 2 cm above the root-shoot interface below the cotyledons for the collection of xylem sap. The shoots were then dried at 70℃ until constant weight was achieved and DM was determined. After the collection of xylem sap, the root parameters were immediately determined and then dried as shoots for further DM determination.

5. Phytohormone analysis by HPLC

Cytokinin and IAA concentrations in the xylem sap were determined essentially as described by Walch-Liu et al. (2000) and Takei et al. (2001) with some modifications given by Lu et al. (2007). Xylem sap was purified by adjusting the pH to 8.5, adding 0.2g polyvinylpyrrolidone (PVP) and vibrating for 30min, and then filtered through a 0.45 μm filter over an OASIS®HLB cartridge (Waters, USA). The cartridge was initially washed with 0.1 mol/dm^3 acetic acid, eluted with 4 mL of a mixture of 25% (v/v) methanol and 0.1 mol/dm^3 acetic acid, and eventually with 70% (v/v) methanol only. After vacuum evaporation, the purified samples were loaded on a reverse-phase column.

Standard Cytokinin-fractions and IAA samples were from Sigma-Aldrich, and chromategraphic conditions were described by Lu et al. (2007): Waters 600-2487; Hibar® column RT 250×4.6 mm; Purospher® STAR RP-18 (5 μm); column temperature 45℃; fluid phase: methanol-1% acetic acid (40 : 60, v : v), isocratic elution; fluid rate: 0.6 mL/min; UV detector, λ = 269 nm; and the injection volume 20 μL. A 0.22 μm filter was used for filtration of both the buffer and the samples before HPLC analysis. The Chromatogram of 6 Cytokinin and IAA fractions of both standards and samples analyzed by HPLC were depicted in Fig. 1.

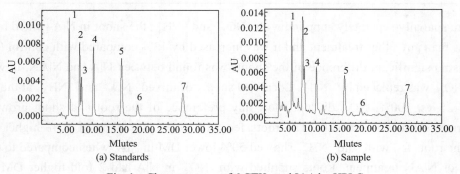

Fig. 1 Chromatogram of 6 CTKs and IAA by HPLC

Notes: 1, Zeatin. 2, Zeatin riboside. 3, Dihydrozeatin riboside. 4, Kinetin. 5, N^6-(Δ^2-isopentenyl)adenine. 6, Indole-3-acetic acid. 7, Iso-pentenyl adenosine

6. NO_3^- and NH_4^+ analysis

The concentrations of NO_3^- and NH_4^+ in xylem sap were measured by using continuous-flow autoanalyzer (Autoanalyzer 3, Bran+Luebbe GmbH, Germany) (Schortemeyer et al., 1993).

7. Statistical analysis

All data were tested according to one-way ANOVA followed by Fisher's test from SAS8.2 software.

4.1.3 Results

1. Effects of N nutrition modes on plant growth

When different ratios of NO_3^- to NH_4^+ nutrition were homogeneously supplied, both tomato roots and shoots gained maximum biomass in 75-25NA treatment (Table 1), and they were increased by 10.3% and 15.6%, respectively, compared to the control (100-0NA). Minimum root biomass was obtained at 50-50NA treatment, and it was decreased by 17.9% compared with the control. The difference between the shoot DM and root DM caused by different ratios of NO_3^- to NH_4^+ led to 16% increase in shoot DM/root DM (S/R) under 50-50NA in comparison to that under the other two treatments (6.22).

Table 1 The effects of different N forms on the shoot dry matter (DM) (g/plant), root DM (g/plant) and shoot DM/root DM (S/R) of tomato plants in both whole- and split-root system. Different small letters indicate significant differences (*P*<0.05) between treatments in one row, and capital letters indicate significant differences (*P*<0.05) among root DM in split-root system in the two rows by Fisher's-test (LSD)

Treatment		Shoot DM	Root left DM	Root right DM	Root total DM	Shoot/Root (S/R)
Whole-root system	100-0NA	10.51 d	-	-	1.638 c	6.41 ab
	75-25NA	11.59 c	-	-	1.893 b	6.10 ab
	50-50NA	9.78 d	-	-	1.345 d	7.25 c
Split-root system	N|N	11.78 bc	0.975 B	0.993 B	1.968 ab	5.95 a
	N|A	13.32 a	1.523 A	0.518 C	2.040 a	6.54 b
	N|AN	12.62 ab	1.008 B	0.988 B	1.995 a	6.33 ab

When spatially separately supplied with NO_3^- and NH_4^+, the shoot in N|A gained higher biomass than any other treatment and it was increased by 13% compared with control (N|N) whereas no significant difference of the shoots was found between N|A and N|AN, in which 25% NO_3^- was replaced by NH_4^+. Localized supply of mixed NO_3^- and NH_4^+ at the ratio of 75 : 25 vs. 100% NO_3^- did not show any preference of the root and shoot growth in comparison with those in N|N. The roots fed with sole NO_3^- showed 50% higher DM, while the roots fed with sole NH_4^+ showed 50% lower DM in N|A, when compared to those in N|N or N|AN treatment. Roots supplied with NO_3^- in N|A had 2 fold higher DM than that with sole NH_4^+. However, expressed on the total root DM, there was no significant difference among the three treatments. A higher S/R was obtained also in N|A, with the ratio of NO_3^- : NH_4^+ as 50 : 50.

2. Effects of N nutrition modes on root morphology

From Table 2, when NO_3^- plus NH_4^+ was provided in combination, 25% replacement of NO_3^- by NH_4^+ resulted in a significant increase in RS and RV compared to 100% NO_3^-, whereas 50% repplacement performed just in the contrary way. RS and RV in 75-25NA were increased by 29% and 19% compared to 100-0NA, respectively, and they were both increased by 40% when compared to 50-50NA. Both RL and AD in 75-25NA and 50-50NA treatments showed no significant difference comparing with 100-0NA, however, 30% decline was found in RL and AD when 25% NH_4^+ was increased to 50% in the mixed N forms.

Table 2 The effects of different N forms on total root length (RL, m), average root diameter (AD, mm), root surface area (RS, cm^2) and root volume (RV, mm^3) supplied with different ratios of NO_3^- and NH_4^+ in solution in whole root system. Different letters indicate significant differences ($p<0.05$) between treatments by Fisher's-test (LSD)

	RL(m/plant)	AD(mm/plant)	RS (cm^2/plant)	RV(cm^3/plant)
100-0NA	115.4 ab	1.13 ab	1279 b	14.19 b
75-25NA	129.7 a	1.30 a	1655 a	16.84 a
50-50NA	97.98 b	0.97 b	1179 c	11.96 c

Spatially separated supply of NO_3^- and NH_4^+ led to a dramatic change in root morphology (Table 3). Roots fed with sole NH_4^+ were significantly inhibited whereas those supplied with NO_3^- were significantly promoted. RL, AD, RS and RV of roots supplied with NH_4^+ in N|A treatment were found to be decreased by 79.9%, 44.7%, 74.7% and 66%, respectively, compared to that in control (N|N). In contrast, root grown under NO_3^- condition showed a significant stimulus in RL (49.5), AD (85.6), RS (62.7) and RV (72.9) to controls (N|N). In addition, when the comparison occurred between the two separate roots supplied either by NO_3^- or NH_4^+ in N|A treatment, we could see NH_4^+-fed root received no more than 30% of the RL (13.4), RS (15.5), RV (19.7) and AD (29.8) fed with NO_3^-.

Table 3 The effects of different N forms on total root length (RL, m), average root diameter (AD, mm), root surface area (RS, cm^2) and root volume (RV, mm^3) supplied with different ratios of NO_3^- and NH_4^+ in solution in split root system. Different letters indicate significant differences ($p<0.05$) between treatments by Fisher's-test (LSD).

	N|N		N|A		N|AN	
	N	N	N	A	N	AN
RL (m/plant)	79.60 c	80.13 c	119.4 a (49.5)	16.03 d (−79.9)	86.64 bc (8.5)	93.90 b (17.6)
AD (mm/plant)	0.81 c	0.78 c	1.47 a (85.6)	0.44 d (−44.7)	0.90 bc (13.4)	1.04 b (30.4)
RS (cm^2/plant)	907.0 b	915.6 b	1483 a (62.7)	230.6 c (−74.7)	936.1 b (2.7)	1016 b (11.5)
RV (cm^3/plant)	8.270 b	8.390 b	14.40 a (72.9)	2.833 c (−66.0)	8.395 b (0.8)	8.889 b (6.7)

Note: The figures in parentheses indicated the percentage (%) of the root parameters (RL, AD, RS and RV) in NO_3^-, NH_4^+ and mixed NO_3^- and NH_4^+ containing pot in N|A and N|AN, respectively, higher (+) or lower (−) than those in control (N|N).

Under N|AN treatment, however, only RL and AD of 75-25NA-fed root showed 17.6% and 30.4%, respectively, longer and higher than those in control (N|N). No significant difference was detected in RS and RV. Unlike roots supplied by NO_3^- in N|A, roots receiving NO_3^- in N|AN were not significantly stimulated. No significant difference was found in RL, AD, RS and RV between the two separate roots grown in N|AN treatment.

3. Effects of N nutrition modes on cytokinins and IAA in xylem exudates

Application of different N forms in both whole root and split-root experiments led to a significant difference in the total concentration of cytokinin fractions while in terms of IAA concentration no significant differences were found in hydroponically-grown tomato xylem exudates (Table 4). ZR was the dominant fraction of cytokinins in xylem exudates and accounted for up to 60%-80% of total cytokinins, which was in accordance with Takei et al. (2001). In a homogeneous supply, when roots were uniformly exposed to mixed NO_3^- and NH_4^+ nutrition solutions at a ratio of 75 : 25 (75-25NA), the increase of NH_4^+ resulted in a significant decrease of the level of Z (31.8%), iP (53.4%) and iPA (58.6%) in xylem sap. No significant decrease was observed in the ZR+(diH)ZR level. Under this condition, the KT level was found to be highest, i.e. 42.9% and 1.5-fold higher than that of controls or 50-50NA treatment. A sharp decrease in the levels of Z, ZR+(diH)ZR, iP and iPA in xylem sap was found when 50% NO_3^- was replaced by NH_4^+ compared to 25% replacement. In comparison to the control (100-0NA), a 50% replacement of NO_3^- resulted in a reduction to 81.8%, 31.5%, 42.9%, 83.5% and 74.1% in Z, ZR+(diH)ZR, KT, iP and iPA, respectively. Considering the total concentration of cytokinins, it decreased significantly as the ratio of external NH_4^+ increased, i.e. by 18.7% and 47.5%, respectively, under 25% or 50% replacement of NO_3^- by NH_4^+.

Table 4 The effects of different N forms on the concentrations (mg/L) of zeatin (Z), zeatin riboside and dihydrozeatin riboside (ZR+(diH)ZR), kinetin (KT), N^6-(Δ^2-isopentenyl)adenine (iP), iso-pentenyl adenosine (iPA), the total cytokinins (CTKs) and indole-3-acetic acid (IAA) in xylem sap in both whole- and split-root system. Different letters indicate significant differences ($P<0.05$) between treatments by Fisher's-test (LSD)

	Treatment	Z	ZR+(diH)ZR	KT	iP	iPA	CTKs	IAA
Whole-root system	100-0NA	0.044a	0.397ab	0.056b	0.103a	0.058a	0.657a	0.099a
	75-25NA	0.030b	0.353b	0.080a	0.048b	0.024b	0.534b	0.095a
	50-50NA	0.008e	0.272c	0.032c	0.017c	0.015d	0.345c	0.091a
Split-root system	N|N	0.025c	0.423a	0.031c	0.123a	0.014cd	0.616a	0.096a
	N|A	0.017d	0.388ab	0.038c	0.098a	0.013d	0.554b	0.103a
	N|AN	0.022c	0.349b	0.034c	0.111a	0.016c	0.532b	0.081a

In the localized supply treatment, the total concentration of cytokinin fractions was also significantly decreased in the NH_4^+-containing solution (N|A and N|AN). No significant difference was documented in the concentration of cytokinin fractions except Z and iPA. Localized single NH_4^+ supply led to a significantly reduced Z concentration, which was 32% lower than that in control (N|N). Except a reduced ZR concentration, no significance in the concentration of the other cytokinin species was found when 25% NO_3^- was replaced by NH_4^+ (N|AN). Plant with the roots exposed in half to the mixture of 75% NO_3^- plus 25% NH_4^+ (N|AN) revealed 30% and 20% higher Z and iPA level in xylem sap, respectively, than those with the roots exposed in half to sole NH_4^+ (N|A) and this treatment also led to a significantly lower ZR concentration by 17% than that in N|N.

Although both 50-50NA in the homogeneous supply and N|A in the localized supply had the ratio of NO_3^- to NH_4^+ of 50 : 50, a striking difference was observed in the total concentration of cytokinins and some fractions, where Z, ZR, and iP were included. The relative decline of Z, KT, iP, iPA and total cytokinins of 50-50NA to 100-0NA were all significantly higher than those of N|A to N|N.

4. *Effects of N nutrition modes on NO_3^- and NH_4^+ concentrations in xylem exudates*

Irrespective of the culture mode and the ratio of NO_3^- to NH_4^+, NO_3^- concentration was all higher than NH_4^+ in xylem sap. For the whole-root system exposed to a combination of NH_4^+ and NO_3^-, the NH_4^+ level in xylem exudates was found to increase as the external NH_4^+ increased (Fig. 2). 25% replacement of NO_3^- by NH_4^+ resulted in no significant effect on NO_3^- concentration. However, 50% NH_4^+ in solution led to a sharp decrease of 64% in NO_3^- level, compared to control (100-0NA). Localized supply of sole NH_4^+ led to a significant decrease of NO_3^- concentration compared to the other two treatments.

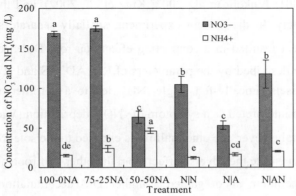

Fig. 2 Effects of different N forms on NO_3^- and NH_4^+ concentrations in xylem sap in whole-root system and split-root system

Different letters indicate significant differences ($P<0.05$) between treatments by Fisher's-test (LSD)

No matter what culture mode was, a closely negative correlation was found between the concentration of NH_4^+ and total cytokinins in xylem sap ($r= -0.951*$, $n=6$, $r_{0.01,4}= 0.917$), and ZR ($r= -0.963*$, $n=6$, $r_{0.05,4}= 0.811$). However, a significantly positive correlation was found between the xylem-NO_3^- and Z, in which the coefficient was 0.879* ($n=6$, $r_{0.05,4}= 0.811$) irrespective of culture mode. Under homogeneous application, NO_3^- was positively correlated with the concentrations of Z, ZR+(diH)ZR and total cytokinins. Localized supply of NO_3^- and NH_4^+ resulted in a significantly negative correlation ($r= -0.997*$, $n=3$, $r_{0.05,1}= 0.997$) between NH_4^+ concentration and ZR+(diH) ZR level. In the whole-root application, KT concentration in xylem sap was found to be positively correlated with the root DM ($r= 0.999$, $n=3$, $r_{0.05,1}= 0.997$).

4.1.4 Discussion

Nitrogen distribution in soils is not homogenous since two major forms, NO_3^- and NH_4^+, can appear anywhere and anytime, and their ratios depend on the soil environmental conditions. In this experiment, the split-root system was conducted synchronously comparing with the whole-root system to investigate the effects of different NO_3^- and NH_4^+ regimes on root growth and cytokinins concentrations in hydroponically-grown tomato seedlings. Plant roots responded differently to separately supplied NO_3^- and NH_4^+, with description of root and shoot growth (Table 1), root morphology (Table 3) and cytokinins species concentrations in xylem exudates (Table 4). Therefore, the split-root system seems to be more suitable for imitating the condition of uneven nutrients distribution traits in soil.

1. Effects of N forms on root morphology and plant growth

Plant root systems perform many essential adaptive functions including water and nutrient uptake (Sattelmacher et al., 1993). Roots can perceive efficiently the concentration of nutrients in soil, for example, nitrate, phosphate, sulfate and iron (Johnson et al., 1996; Zhang and Forde, 1998; Zhang et al., 1999; Ford and Lorenzo, 2001; Schmidt and Schikora, 2001; Williamson et al., 2001; Linkohr et al., 2002; Kutz et al., 2002), and then alternate the root architecture accordingly. In the present experiment, spatially separate supply of NH_4^+ and NO_3^- in N|A treatment resulted in a contrasting effects on root DM accumulation (Table 1) and root morphology described by the parameters of RL, AD, RS and RV (Table 2 and Table 3) in the double vessels. Locally-fed single NH_4^+ led to a detrimental effect on the root growth, which can be interpreted as a symptom of NO_3^- deprivation (Walch-Liu et al., 2000). In the same treatment, however, the other half roots exposed to the sole-NO_3^- were stimulated and hence, could compensate the adverse effect from the other half roots fed by NH_4^+. Also in this condition, the common shoot gained a higher DM accumulation (Table 1) and the maximal soluble carbohydrate as well as protein content (data not shown). However, if the roots were exposed to a combined NO_3^- and NH_4^+ nutrition, it was documented that the uptake of NO_3^- by the roots would be inhibited by NH_4^+ whereas a contrast effect, that is, a stimulation of translocation and assimilation of NH_4^+ was found in the presence of NO_3^- (Kronzucker et al., 1999a, 1999b). When the roots partially received the favorable nutrient condition as the combined 75% NO_3^- and 25% NH_4^+ versus 100% NO_3^- simultaneously (N|AN), almost no significant promotion was found, though it had been approved that this ratio was optimal for tomato growth in the homogeneous supply (Table 2; Sandoval-Villa et al., 2001; Claussen, 2002; Dong et al., 2004). Thus it was indicated that plant roots were able to adapt to the unfavorable environment, such as sole NH_4^+ for aerobic crops, efficiently by alternating the root architecture in NO_3^--rich vessel accordingly, and this was also approved by Forde and Lorenzo (Forde and Lorenzo, 2001).

On the other hand, despite that the ratio of NO_3^- to NH_4^+ in 50-50NA in homogeneous supply and N|A in split-root supply was both 50 : 50, 50-50NA treatment led to a significant decline in root growth and root parameters, whereas a significant higher shoot DM and a constant total root DM were obtained in N|A treatment. These responses agreed with López-Bucio et al. (2003) and Lynch (1995) that changes in the root architecture for essential uptake of nutrients are of fundamental importance for their adaptation to the environment. Plants both received a higher S/R at the ratio of NO_3^- to NH_4^+ at 50 : 50 irrespective of culture mode. In homogeneous supply, both plant roots and shoots decreased significantly and the higher S/R was at the expense of even lower root DM than shoot DM. However, in the localized supply, the higher S/R was due to the higher shoot DM but a non-changeable root DM. It was suggested that plant roots were able to develop some efficient mechanisms for the regulation of carbon and N metabolism at the whole plant level. NO_3^- has been identified as a signal to initiate coordinated changes in carbon and N metabolism (Scheible et al., 1997). *trans*Z-type cytokinins was also presumed to be important for the integration of nutrient signal level from roots to shoots at the whole plant (Hwang and Sakakibara, 2006). No matter what signal was involved in the plant growth regulation, the pathway of signal transduction in the whole root system might be significantly different from that in the split-root system. It was reasonable to presume that the signal involved in the whole root system might be different from that in the split root system, mainly depending on whether the shoot was involved in the signal's transduction or not (in whole root system).

2. *Long-distance regulation mechanism mediated by phytohormones*

Among the plant hormones, cytokinin was the most highlighted one involved in N nutrition (Wang and Below, 1996; Zhang et al., 1999; Sakakibara, 2003; Sakakibara et al., 2006). Not only essential for regulation of root development, cytokinin was also regarded as the root-to-shoot signal communicating N availability to mediate the leaf expansion (Walch-Liu et al., 2000; Takei et al., 2001, 2002; Sakakibara, 2003; Rahayu et al., 2005; Sakakibara et al., 2006). Cytokinins were mainly synthesized in the root and translocated via xylem vessels (Marscher, 1995) although some studies indicated an additional phloem transport (Takei et al., 2004) and was found to take effect on the leaf expansion through shoot-to-root signal (Collier et al., 2003). Root-borne cytokinin has been well accepted as a long-distance signal for leaf growth as a response to NO_3^- (Rahayu et al., 2005; Sakakibara et al., 2006), though Dodd et al., (2004), Dodd and Beveridge (2006) proposed that leaf growth responding to N deprivation was independent of xylem-borne cytokinins. In the present study, when the whole roots were supplied with combined NO_3^- and NH_4^+ homogeneously (Table 4), the total concentration of cytokinins in xylem exudates decreased as external NH_4^+ ratio increased, indicating the negative effect of NH_4^+ on the production and transportation of root-borne cytokinins. It was reasonable to predict that this hormone might be decreased to even lower levels if NO_3^- were completely replaced by NH_4^+. But previous observations pointed out that cytokinin levels were increased by the mixed N forms (Smiciklas and Below, 1992; Fetene and Beck, 1993) and even by NH_4^+ as the sole N source (Wang and Below, 1996). It was reported that not only N forms, but also cultivars had significant effects on the exudation rate, cytokinin concentration and cytokinin mass transfer (Wang and Below, 1996). No significant difference in the xylem-cytokinin concentration but a significant increase in cytokinin mass transfer between the ratios of NO_3^- : NH_4^+ at 50 : 50 and 100 : 0 indicated that some replacement of NO_3^- by NH_4^+ led to a significant increase of the exudates rate. But inconsistent opinion was estimated that NH_4^+ supply might play a negative effect on the root exudates rate due to the reduced water uptake rate (Wang et al., 2003). A relatively low concentration of cytokinin fractions obtained both in 75-25NA treatment in homogenous supply and N|A in localized supply of NO_3^- and NH_4^+ might be related to the exudation rate. Further study on the exudation rate and cytokinin mass transfer should be investigated.

NH_4^+ concentration in the xylem sap was found to be significantly negatively associated with total cytokinins concentration and ZR fraction, while NO_3^- positively correlated to Z level irrespective of culture mode. This was in accordance with Walch-Liu et al. (2000) and Rahayu et al. (2005) who proposed that the inhibition induced by exclusive NH_4^+ supply was associated with decreased zeatin+zeation riboside (Z+ZR) concentrations in roots, xylem sap and leaves. Stimulatory effects on shoot growth by NO_3^- was also found to be related to the corresponding increase in Z+ZR levels in leaves and xylem exudates (Rahayu et al., 2005). Notwithstanding, 25% NO_3^- replacement by NH_4^+ in this experiment led to an undiminished ZR level in xylem sap in significance compared to sole NO_3^- nutrition. Some mechanisms might be involved in maintaining the ZR level and thus supported a higher DM accumulation both in shoots and in roots. KT, one of the cytokinin fractions, was known from animal cells and just recently was identified in coconut water (Ge et al., 2005; Barciszewski et al., 2007). Unlike the other cytokinin species, 75-25NA treatment led to the highest KT concentration in the xylem sap whereas 50-50NA resulted in the lowest level. It seemed that the KT concentration was very closely correlated to shoot and root DM accumulation in homogeneous supply, which indicated that KT might play a key role in plant growth regulation.

The results in the present study demonstrated that root morphology was modulated in response to not only N forms and ratios but also N distribution whether even or not in nutrient solution. The facts that the examination of cytokinin fractions concentration in xylem exudates was not always consistent with the plant growth might come from the interacted influence of N forms and culture modes on the exudation rate. Therefore, it is still an open question as to whether there are differences in cytokinin mass transfer and static concentration in the root and shoot parts in response to homogeneous and localized supply of different NO_3^- to NH_4^+ ratios in hydroponics.

References

Barciszewski J, Massino F, Clark B F C. 2007. Kinetin — a multiactive molecule. Int. J. Biol. Macromol., 40:182-192

Bernier G, Kinet J-M, Jacqmard A, et al. 1977. Cytokinin as a possible component of the floral stimulus in *Sinapis alba*. Plant Physiol., 60:282-285

Boukcima H, Pagèsb L, Mousaina D. 2006. Local NO_3^- or NH_4^+ supply modifies the root system architecture of *Cedrus atlantica* seedlings grown in a split-root device. J. Plant Physiol., 163:1293-1304

Britto D T, Kronzucker H J. 2002. NH_4^+ toxicity in higher plants: a critical review. J. Plant Physiol., 159:567-584

Chaillou S, Vessey J K, Morot-Gaudry J F, et al. 1991. Expression of characteristics of ammonium nutrition as affected by pH and of the root medium. J. Exp. Bot., 235:189-196

Chaillou S, Rideout J W, Raper C D, et al. 1994. Responses of soybean to ammonium and nitrate supplied in combination to the whole root system or separately in a split-root system. Physiol. Plant, 90:259-268

Claussen W. 2002. Growth, water use efficiency, and proline content of hydroponically grown tomato plants as affected by nitrogen source and nutrient concentration. Plant Soil, 247:199-209

Collier M D, Fotelli M N, Nahm M, et al. 2003. Regulation of nitrogen uptake by *Fagus sylvatica* on a whole plant level-interactions between cytokinins and soluble N compounds. Plant Cell Environ., 26:1549-1560

Cramer M D, Lewis O A M. 1993. The influence of nitrate and ammonium nutrition on the growth of wheat (*Triticum aestivum*) and maize (*Zea mays*) plants. Ann. Bot., 72:359-365

Dodd I C, Ngo C, Turnbull C G N, et al. 2004. Effects of nitrogen supply on xylem cytokinin delivery, transpiration and leaf expansion of pea genotypes differing in xylem cytokinin concentration. Funct. Plant Biol., 31(9):903-911

Dodd I C, Beveridge C A. 2006. Xylem-borne cytokinins: still in search of a role? J. Exp. Bot., 57(1):1-4

Dong C X, Shen Q R, Wang G. 2004. Tomato growth and organic acid changes in response to partial replacement of NO_3^--N by NH_4^+-N. Pedosphere, 14(2):159-164

Feil B. 1994. Growth and ammonium: nitrate uptake ratio of spring wheat cultivars under a homogeneous and a spatially separated supply of ammonium and nitrate. J. Plant Nutr., 17:717-728

Fetene M, Beck E. 1993. Reversal of the direction of photosynthate allocation in *Urtica dioica* L. Plants by increasing cytokinin import into the shoot. Bot. Acta, 106:235-240

Forde B, Lorenzo H. 2001. The nutritional control of root development. Plant Soil, 232:51-68

Gansel X, Muños S, Tillard P, et al. 2001. Differential regulation of the NO_3^- and NH_4^+ transporter genes AtNrt2.1 and AtAmt1.1 in *Arabidopsis*: relation with long-distance and local controls by N status of the plant. Plant J., 26(2):143-155

Ge L, Yong J W H, Goh N K, et al. 2005. Identification of kinetin and kinetin riboside in coconut water using a combined approach of liquid chromatography-tandem mass spectrometry, high performance liquid chromatography and capillary electrophoresis. J. Chromatogr., 829:26-34

Guo S, Brueck H, Sattelmacher B. 2002. Effects of supplied nitrogen form on growth and water uptake of French bean (*Phaseolus vulgaris* L.) plants. Plant Soil, 239:267-275

Guo S, Kaldenhoff R, Uehlein N, et al. 2007. Relationship between water and nitrogen uptake in nitrate and ammonium supplied *Phaseolus vulgaris* L. plants. J. Plant Nutr. Soil Sci., 170:73-80

Guo Y F, Chen F J, Zhang F S, et al. 2005. Auxin transport from shoot to root is involved in the response of lateral root growth to localized supply of nitrate in maize. Plant Sci., 169:894-900

Hoagland D R, Arnon D I. 1950. The water-culture method for growing plants without soil. Univ. Calif. Agric. Exp. Stn. Circ., 347:1-32

Hwang I, Sakakibara H. 2006. Cytokinin biosynthesis and perception. Physiol. Plant, 126:528-538

Johnson J F, Vance C P, Allan D L. 1996. Phosphorus deficiency in *Lupinus albus*. Altered lateral root development and enhanced expression of phosphoenolpyruvate carboxylase. Plant Physiol., 112:657-665

Jung H B, Tadashi I, Toru M. 1994. Effects of shading and NO_3 : NH_4 ratios in the nutrient solution on the growth and yield of pepper plants in nutrient film technique culture. J. Jpn. Soc. Hortic. Sci., 63(2):371-377

Jung H B, Tadashi I. 1994. Effects of day temperature and ammonium-N addition to the nutrient solution on the growth and yield of pepper plants grown in hydroponics. Environ. Control. Biol., 32(1):41-46

Kotsiras A, Olympios C M, Drosopoulos J, et al. 2002. Effects of nitrogen form and concentration on the distribution of ions within cucumber fruits. Sci. Hortic., 95:175-183

Kutz A, Müller A, Hennig P, et al. 2002. A role for nitrilase 3 in the regulation of root morphology in sulphur-starving *Arabidopsis thaliana*. Plant J., 30:95-106

Kronzucker H J, Siddiqi M Y, Glass A D M, et al. 1999a. Nitrate-ammonium synergism in rice. A subcellular flux analysis. Plant Physiol., 119:1041-1045

Kronzucker H J, Glass A D M, Siddiqi M Y. 1999b. Inhibition of nitrate uptake by ammonium in barley analysis of component fluxes. Plant Physiol., 120:283-291

Linkohr B I, Williamson L C, Fitter A H, et al. 2002. Nitrate and phosphate availability and distribution have different effects on root system architecture of *Arabidopsis*. Plant J., 29:751-760

López-Bucio J, Cruz-Ramírez A, Herrera-Estrella L. 2003. The role of nutrient availability in regulation root architecture. Curr. Opin. Plant Biol., 6:280-287

Lu Y L, Dong C X, Dong Y Y, et al. 2007. Simultaneous determination of six endogenous cytokinins components and auxin in plant tissue by high performance liquid chromatography. Plant Nutr. Fert. Sci., 13(1):129-135 (in Chinese)

Lynch J P. 1995. Root architecture and plant productivity. Plant Physiol., 109:7-13

Marschner H. 1995. Mineral Nutrition of Higher Plants. 2nd Ed. New York : Academic Press

Rahayu Y S, Walch-Liu P, Neumann G, et al. 2005. Root-derived cytokinins as long-distance signals for NO_3^--induced stimulation of leaf growth. J. Exp. Bot., 56(414):1143-1152

Sakakibara H. 2003. Nitrate-specific and cytokinin-mediated nitrogen signaling pathways in plants. J. Plant Res., 116(3):253-257

Sakakibara H, Takei K, Hirose N. 2006. Interactions between nitrogen and cytokinin in the regulation of metabolism and development. Trends Plant Sci., 11(9):440-448

Salsac L, Chaillou S, Morot-Gaudry J F, et al. 1987. Nitrate and ammonium nutrition in plants. Plant Physiol. Biochem., 25:805-812

Sandoval-Villa M, Guertal E A, Wood C W. 2001. Greenhouse tomato response to low ammonium-nitrogen concentrations and duration of ammonium-nitrogen supply. J. Plant Nutr., 24(11):1787-1798

Sattelmacher B, Gerendas J, Thoms K, et al. 1993. Interaction between root growth and mineral nutrition. Environ. Exp. Bot., 33:63-73

Scheible W R, Lauerer M, Schulze E D, et al. 1997. Accumulation of nitrate in the shoot acts as a signal to regulate shoot-root allocation in tobacco. Plant J., 11: 671-691

Schmidt W, Michalke W, Schikora A. 2003. Proton pumping by tomato roots. Effect of Fe deficiency and hormones on the activity and distribution of plasma membrane H^+-ATPase in rhizodermal cells. Plant Cell Environ., 26:361-370

Schmidt W, Schikora A. 2001. Different pathways are involved in phosphate and iron stress-induced alterations of root epidermal cell development. Plant Physiol., 125:2078-2084

Schortemeyer M, Feil B, Stamp P. 1993. Root morphology and nitrogen uptake of maize simultaneously supplied with ammonium and nitrate in a split-root system. Ann. Bot., 72:107-115

Schortemeyer M, Feil B. 1996. Root morphology of maize under homogeneous or spatially separated supply of ammonium and nitrate at three concentration ratios. J. Plant Nutr., 19:1089-1097

Schortemeyer M, Stamp P, Feil B. 1997. Ammonium tolerance and carbohydrate status in maize cultivars. Ann. Bot., 79:25-30

Smiciklas K D, Below F E. 1992. Role of cytokinins in enhanced productivity of maize supplied with NH_4^+ and NO_3^-. Plant Soil, 142:307-313

Sorgonà A, Cacco G. 2002. Linking the physiological parameters of nitrate uptake with root morphology and topology in wheat (*Triticum durum*) and citrus (*Citrus volkameriana*) rootstock. Can. J. Bot., 80:494-503

Stratton M L, Good G L, Barker A V. 2001. The effects of nitrogen source and concentration on the growth and mineral composition of privet. J. Plant Nutr., 24(11):1745-1772

Takács E, Técsi L. 1992. Effects of NO_3^-/NH_4^+ ratio on photosynthetic rate, nitrate reductase activity and chloroplast ultrastructure in three cultivars of red pepper (*Capsicum annuum* L.). J. Plant Physiol., 140:298-305

Takei K, Takahashi T, Sugiyama T, et al. 2002. Multiple routes communicating nitrogen availability from roots to shoots: a signal transduction pathway mediated by cytokinin. J. Exp. Bot., 53:971-977

Takei K, Sakakibara H, Taniguchi M, et al. 2001. Nitrogen-dependent accumulation of cytokinins in root and the translocation to leaf implication of cytokinin species that induces gene expression of maize response regulator. Plant Cell. Physiol., 42(1):85-93

Takei K, Ueda N, Aoki K, et al. 2004. AtIPT3 is a key determinant of acronutrientresponsive cytokinin biosynthesis. Plant Cell. Physiol., 45:1053-1062

Tamaki V, Mercier H. 2007. Cytokinins and auxin communicate nitrogen availability as long-distance signal molecules in pineapple (*Ananas comosus*). J. Plant Physiol., 164:1543-1547

Walch-Liu P, Neumann G, Bangerth F, et al. 2000. Rapid effects of nitrogen form on leaf morphogenesis in tobacco. J. Exp. Bot., 51(343):227-237

Walch-Liu P, Neumann G, Engels C. 2001. Response of shoot and root growth to supply of different nitrogen forms is not related to carbohydrate and nitrogen status of tobacco plants. J. Plant Nutr. Soil Sci., 164(1):97-103

Wang X, Below F E. 1996. Cytokinins in enhanced growth and tillering of wheat induced by mixed nitrogen source. Crop. Sci., 36:121-126

Wang G Y, Li C J, Zhang F S. 2003. Effects of different nitrogen forms and combination with foliar spraying with 6-benzylaminopurine on growth, transpiration, and water and potassium uptake and flow in tobacco. Plant Soil, 256:169-178

Williamson L, Ribrioux S, Fitter A, et al. 2001. Phosphate availability regulates root system architecture in *Arabidopsis*. Plant Physiol., 126:1-8

Xu G H, Wolf S, Kafkafi U. 2001. Effect of varying nitrogen form and concentration during growing season on sweet pepper flowering and fruit yield. J. Plant Nutr., 24(7):1099-1116

Xu G H, Wolf S, Kafkafi U. 2002. Mother plant nutrition and growing contion affect amino and fatty acid compositions of hybrid sweet pepper seeds. J. Plant Nutr., 25(4):719-734

Xu Q F, Tsai C L, Tsai C Y. 1992. Interaction of potassium with the form and amount of nitrogen nutrition on growth and nitrogen uptake of maize. J. Plant Nutr., 15:23-33

Zhang H, Jennings A, Barlow P W, et al. 1999. Dual pathways for regulation of root branching by nitrate. Proc. Natl. Acad. Sci. USA, 96:6529-6534

Zhang H, Forde B. 1998. An *Arabidopsis* MADS box gene that controls nutrient-induced changes in root architecture. Science, 279:407-409

Zou C Q, Wang X F, Wang Z Y, et al. 2005. Potassium and nitrogen distribution pattern and growth of flue-cured tobacco seedlings influenced by nitrogen form and calcium carbonate in hydroponic culture. J. Plant Nutr., 28:2145-2157

4.2 Effects of N fertilizer application time on dry matter accumulation and yield of Chinese potato

4.2.1 Introduction

Potato is one of the most important crops in China, occupying an area of about 5.2 million ha and producing 8.15 million tonnes of tubers per year. The total planting area and production of China hold the 1st position in the world but yield in China is lagging far behind many other countries.

Nitrogen (N) is recognized as the most limiting nutrient to potato crops (Li et al., 1999). Potato yield is greatly affected by nutrient availability, which is why research on appropriate fertilizer regimes has received much attention worldwide (Hamouz et al., 2005; Westermann 2005; El- Sirafy et al., 2008; Kulhnek et al., 2008; Poljak et al., 2008; Vos, 2009). Inadequate N fertilization leads to poorer potato growth and yield while excessive N application leads to delayed maturity, poor tuber quality and occasionally a reduction in tuber yield (Jacobs and Clarke, 1993; Alva, 2004; Shari et al., 2005; Zebarth et al., 2006; Haase et al., 2007; Sincik et al., 2008; Arriaga et al., 2009; Cerny et al., 2010).

With rising environmental concerns for current practices of N fertilizer management in the world, efficient N use is important for the economic sustainability of cropping systems (Goffart et al., 2008; Shrestha et al., 2010). The implementation of the best management practices (BMPs) can bring about increases in potato yields and tuber quality sometimes even after reducing N input (Delgado et al., 2007). Development of BMPs for N management must consider timing of N supply and crop demand (Zebarth and Rosen, 2007).

In recent years, improved understanding of yield responses to alterations in availability of assimilates during different phenological phases has been a major advance in crop physiology. There are many factors that can affect the source-sink relations during the different growth phases; particularly N is one of the most important nutrients which affect production and distribution of assimilates, thus affecting directly or indirectly the source-sink relation (Muchow, 1988; Arduini et al., 2006; Christos 2009). N can influence the leaf area development and maintenance as well as photosynthetic efficiency and dry matter partitioning to reproductive organs (Prystupa et al., 2004). The presence of a strong tuber sink led to a lower level of leaf carbohydrates, thus avoiding feedback inhibition of photosynthesis (Basu et al., 1999), while in contrast the photo assimilates will be deposited in stems and leaves instead of tuber if the sink was not powerful enough (Borkhard et al., 2005). Ierna et al. (2011) demonstrated that the source/sink ratio dropped as the N fertilization level decreased from 300 kg N/ha to 50 kg N/ha. Higher N pre-treatment of the plantlets resulted in significantly lower tuber fresh yield and the fraction of tubers in the total plant dry matter (Tadesse et al., 2001). High level N tends to reduce tuber development by promoting shoot growth while treatments which inhibit or reduce shoot growth, such as a growth suppressor, promote tuber formation (Peres et al., 2005). High levels of N also delay tuber initiation. The more N available to the plant, the lower percentage of plant dry matter will be partitioned to tubers early in the season. This will favor haulm growth but postpone tuber bulking. However, N may increase the number of tubers by enhancing individual stem vigour, although effects are not always consistent (Tekalign and Hammes, 2005).

Because the requirement of N for plant growth and development is persistent, applying all the N fertilizer at one time is not reasonable, even for loamy soils. Scientific work should be done to elucidate the best N management for potato cultivars under specific conditions（Li et al., 2003; Srek et al., 2010). Vos (1999) stated there was little or no uptake of N in potatoes 28-42 days after planting and after about 60 days after emergence. Iritani (1978) had advised that 1/3 or 1/2 of all N fertilizer should be applied at planting and the remainder should be dressed in the season. Zelalem et al. (2009)got a yield of 45 t/ha by applying one half N at planting and the other half at 45 day after planting. Both Ferreira and Goncalves (2007) and Darwish et al. (2006) demonstrated that the tuber yield was depressed when the N fertilization increased from 160 kg

N/ha to 240 kg N/ha and from 125 kg N/ha to 500 kg N/ha, respectively.

Although extensive work has been done on N fertilization timing in potato, very little has been done to determine the specific requirements of Chinese potato cultivars grown under Chinese conditions. To address this problem, the experiment reported here was conducted at one N application level with 4 application timing patterns on the widely grown variety KX 13. Dry matter accumulation and distribution at different growth stages, as well as the yield and quality of tubers influenced by N application time were measured and discussed.

4.2.2 Materials and Methods

This study was conducted in 2010 at the experimental station of Northeast Agriculture University (126°44′58.35″E、45°43′14.65″N) in Harbin, Heilongjiang Province, PRC, on a black soil with loam-clay texture. The soil was sampled pre-planting from surface to a depth of 30 cm before the application of the fertilizers. The soil characteristics were determined according to Bao (2005). Briefly, the soil contained 24.8 g/kg organic matter, 81 g/kg P (Olsen), 174 g/kg exchangeable K and pH 7.22 (1 ∶ 2.5 in H_2O).

The potato genotype used in the study, KX 13 (*Solanum tuberosum* L.), is one of the most widely cultivated mid-later-maturing varieties in China, and has a growth period of around 100 days after emergence.

The experimental design consisted of randomized complete block with four treatments and four replications. Individual plots contained 7 rows, each 12 m long and 0.7 m between rows. Potato seed pieces were planted manually 30 cm apart at a depth of 10 cm on the 17th of May. Plants emerged on June 4th. Tubers were harvested on September 16th.

The four treatments were as follows: all the 150 kg N/ha applied at planting (T1), 100 kg N/ha applied at planting and 50 kg N/ha applied one week before tuber initiation (20 days after emergence, DAE) (T2), 100 kg N/ha applied at planting and 50 kg N/ha applied one week before tuber bulking stage (35 DAE) (T3), and 100kg N/ha applied at emergence and 50 kg N/ha applied one week before tuber bulking stage (35 DAE) (T4). For all treatments, 90 kg P_2O_5/ha (($NH_4)_2HPO_4$)) and 150 kg K_2O/ha (K_2SO_4) were applied at planting. Plants were grown without supplemental irrigation. The field was kept free of weeds by hand-hoeing when necessary.

The six-plant sample was taken from each plot at 30, 42, 54, 70, 90 and 101 DAE. All the plant samples were separated into leaves, stems and tubers, washed in distilled water, and then tubers were sliced into 10 mm wide strips of differing lengths. All the plant organs were deactivated at 105℃ for 30 minutes and dried at 80℃ to constant weight for dry matter (DM) determination.

$10\ m^2$ of each plot was left for yield measurement at harvest. Tubers diameter more than 30 mm were used for determining total yield and tubers classified as marketable by weight (>75 g) were counted and weighed. Specific gravity of marketable tubers was determined by the weight in water -weight in air method on a composite sample subsampled from marketable tuber. Data analyses were performed using the DSP (DPS v 7.05 version). The Duncan test ($P \leqslant 0.05$) was used to find significant differences among means. Some calculations were based on Tadesse et al. (2001) and Dordas (2009): harvest index (HI) = tuber DM / total DM at harvest; source/sink = vine DM at harvest / tuber DM at harvest; transport of vine DM accumulated before tuber bulking (t/ha) = tuber DM at harvest − (whole plant DM at harvest − whole plant DM at early tuber bulking stage); transport percentage of vine DM accumulated before tuber bulking (%) = transport of vine DM accumulated before tuber bulking / whole plant DM at early tuber bulking stage × 100%; transport of vine DM accumulated during tuber bulking (t/ha) = vine DM at early tuber bulking stage − vine DM at harvest; transport percentage of vine DM accumulated during tuber bulking (%) = transport of vine DM accumulated during tuber bulking / vine DM accumulated at early tuber bulking stage × 100%.

4.2.3 Results

Dry matter accumulation in potato plants followed a classical sigmoid growth pattern (Fig. 1). T3, applying N at planting and dressing 35 DAE, accumulated more dry matter during mid-tuber bulking stage than did the other treatments, although the differences were significant only at 70 DAE at the level of $P \leqslant 0.05$. Dry matter accumulation in potato vine followed a parabolic curve (Fig. 2). T1, applying all N at planting, had the highest vine dry matter accumulation of all treatments but differences were not significant. Dry matter accumulation in potato tubers increased linearly with time (Fig. 3). T3 accumulated more dry matter than the other treatments from tuber bulking stage to harvest and the differences were significant at each time at the level of $P \leqslant 0.05$. Once the tuber appeared, the sink should be transferred from vine to tubers and tubers biomass gradually came to dominate the biomass of the whole plant. The faster dry matter accumulation in tubers of T3 led to higher total biomass than other treatments after tuber initiation. This phenomenon can be seen in Fig. 4, where T3 had a significantly higher HI than T1 and T4 since tuber formed. At harvest, T2 also had a significantly higher HI than T1 and T4.

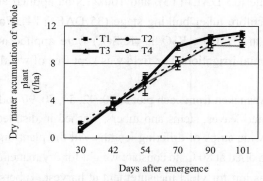

Fig. 1 Dry matter accumulation of whole plants (vine + tuber)

T1 = all the 150 kg N/ha applied at planting; T2 = 100 kg N/ha applied at planting and 50 kg N/ha applied at one week before tuber initiation (20 days after emergence, DAE); T3 = 100 kg N/ha applied at planting and 50 kg N/ha applied at one week before tuber bulking stage (35 DAE); and T4 = 100 kg N/ha applied at emergence and 50 kg N/ha applied at one week before tuber bulking stage (35 DAE). Each data point is the mean value of four replications of six-plant dry matter samples. Vertical bars represent the standard error

4 Nutrition diagnosis and quality control of crops

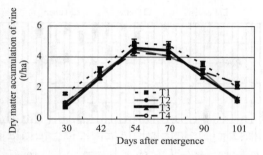

Fig. 2 Dry matter accumulation in vine followed a parabolic curve

T1 = all the 150 kg N/ha applied at planting; T2 = 100 kg N/ha applied at planting and 50 kg N/ha applied at one week before tuber initiation (20 DAE); T3 = 100 kg N/ha applied at planting and 50 kg N/ha applied at one week before tuber bulking stage (35 DAE); and T4 = 100 kg N/ha applied at emergence and 50 kg N/ha applied at one week before tuber bulking stage (35 DAE). Each data point is the mean value of four replications of six-plant dry matter samples. Vertical bars represent the standard error

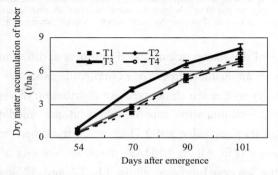

Fig. 3 Dry matter accumulation in potato tubers increased linearly with time

T1 = all the 150 kg N/ha applied at planting; T2 = 100 kg N/ha applied at planting and 50 kg N/ha applied at one week before tuber initiation (20, DAE); T3 = 100 kg N/ha applied at planting and 50 kg N/ha applied at one week before tuber bulking stage (35 DAE); and T4 = 100 kg N/ha applied at emergence and 50 kg N/ha applied at one week before tuber bulking stage (35 DAE). Each data point is the mean value of four replications of six-plant dry matter samples. Vertical bars represent the standard error

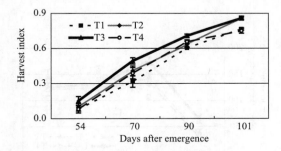

Fig. 4 T3 had a significant higher HI than other treatment all the time

Harvest index (HI) = tuber DM / total DM at harvest. T1 = all the 150 kg N/ha applied at planting; T2 = 100 kg N/ha applied at planting and 50 kg N/ha applied at one week before tuber initiation (20 DAE); T3 = 100 kg N/ha applied at planting and 50 kg N/ha applied at one week before tuber bulking stage (35 DAE); and T4 = 100 kg N/ha applied at emergence and 50 kg N/ha applied at one week before tuber bulking stage (35 DAE). Each data point is the mean value of four replications of six-plant dry matter samples. Vertical bars represent the standard error

The distribution of dry matter among different organs changed throughout the season (Fig 5). Once tuberization initiated, tubers gradually became the main sink for photosynthates. T3 had a significantly ($P \leqslant 0.05$) higher dry matter proportion in tubers than the other treatments.

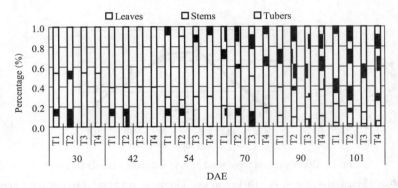

Fig. 5 Effects of N application time on the dry matter allocation proportion of different organs

Dry matter allocation proportion of leaves= leave22s dry matter / total dry matter×100%. The same calculated formulas to stems and tubers. T1 = all the 150 kg N/ha applied at planting; T2 = 100 kg N/ha applied at planting and 50 kg N/ha applied at one week before tuber initiation (20 DAE); T3 = 100 kg N/ha applied at planting and 50 kg N/ha applied at one week before tuber bulking stage (35 DAE); and T4 = 100 kg N/ha applied at emergence and 50 kg N/ha applied at one week before tuber bulking stage (35 DAE). Each data point is the mean value of four replications of six-plant dry matter samples. Vertical bars represent the standard error

The rate and quantity of dry matter accumulated in tubers at different growth stages can be affected significantly by N application time. For example, the excessive vine production of T1 consumed considerable dry matter which delayed the distribution of assimilates to tubers. T1 had a slower dry matter accumulation rate in tubers at the middle tuber bulking stage (70 DAE) and arrived at its peak value at 90 DAE while the other three treatments reached peak values 20 days earlier (Fig. 6). At 70 DAE, the tuber biomass of T3 was 92.3 g/plant, representing 54% of the harvest biomass, while T1, T2 and T4 had tuber biomasses of 46.9 g/plant, 60.8 g/plant and 56.1 g/plant, representing 31%, 42% and 40% of harvest biomass, respectively (Fig. 7). Tubers of T3 accumulated more dry matter than others, result from the high transportation efficiency of assimilates from vine to tubers after tuberization (Table 1).

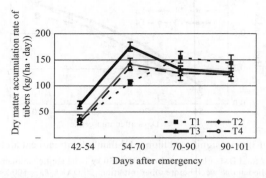

Fig. 6 Dry matter accumulation rate refer to the amount of dry matter accumulated in tubers per day per hectare

T1 = all the 150 kg N/ha applied at planting; T2 = 100 kg N/ha applied at planting and 50 kg/N ha applied at one week before tuber initiation (20 DAE); T3 = 100 kg N/ha applied at planting and 50 kg N/ha applied at one week before tuber bulking stage (35 DAE); and T4 = 100 kg N/ha applied at emergence and 50 kg N/ha applied at one week before tuber bulking stage (35 DAE). Each data point is the mean value of four replications of six-plant dry matter samples. Vertical bars represent the standard error

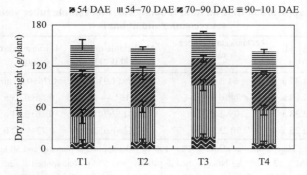

Fig. 7 Effects of N application time on dry matter accumulated in tubers at different stage

Each column was separated according to the amount of dry matter accumulated during different growth periods. T1 = all the 150 kg N/ha applied at planting; T2 = 100 kg N/ha applied at planting and 50 kg N/ha applied at one week before tuber initiation (20 DAE); T3 = 100 kg N/ha applied at planting and 50 kg N/ha applied at one week before tuber bulking stage (35 DAE); and T4 = 100 kg N/ha applied at emergence and 50 kg N/ha applied at one week before tuber bulking stage (35 DAE). Each data point is the mean value of four replications of six-plant dry matter samples. Vertical bars represent the standard error

Table 1 Effects of N application time on DM translocation from vine to tuber before tuber bulking stage and during tuber bulking stage

Treatments	Transportation of vine DM accumulated before tuber bulking (t/ha)	Transportation percentage of vine DM accumulated before tuber bulking (%)	Transportation of vine DM accumulated during tuber bulking (t/ha)	Transportation percentage of vine DM accumulated during tuber bulking (%)
T1	3.9±0.65 ab	72.2±4.28 a	2.9±0.06 bc	58.9±3.32 c
T2	3.9±0.44 a	78.0±1.98 a	3.0±0.32 b	70.1±3.61 b
T3	4.5±0.46 a	83.3±3.08 a	3.7±0.18 a	78.1±2.77 a
T4	2.2±0.43 b	46.8±8.46 b	2.3±0.46 c	51.2±9.67 c

Notes: T1 = all the 150 kg N/ha applied at planting; T2 = 100 kg N/ha applied at planting and 50 kg N/ha applied at one week before tuber initiation (20 DAE); T3 = 100 kg N/ha applied at planting and 50 kg N/ha applied at one week before tuber bulking stage (35 DAE); and T4 = 100 kg N/ha applied at emergence and 50 kg N/ha applied at one week before tuber bulking stage (35 DAE). Transport of vine DM accumulated before tuber bulking (t/ha) = tuber DM at harvest − (whole plant DM at harvest − whole plant DM at early tuber bulking stage); transport percentage of vine DM accumulated before tuber bulking (%) = transport of vine DM accumulated before tuber bulking / whole plant DM at early tuber bulking stage × 100%; transport of vine DM accumulated during tuber bulking (t/ha) = vine DM at early tuber bulking stage − vine DM at harvest; transport percentage of vine DM accumulated during tuber bulking (%) = transport of vine DM accumulated during tuber bulking / vine DM accumulated at early tuber bulking stage × 100%. Each data is the mean value of four replications of six-plant dry matter samples ± standard error. Means followed by the same letters are not significantly different according to Duncan's multiple range test at $P=0.05$.

Total tubers yields of T1 and T3 were higher than that of the other two treatments, but the differences were not significant, while the marketable tubers yield of T3 was significantly higher than that of the other three treatments (Table 2). T1 had a relatively higher tuber yield but the lowest marketable tuber ratio (64.8%). Compared with applying all N at planting, applying N at planting as well as in season can enhance the HI from 0.76 (T1) to 0.86 (T3) which indicated that assimilates of T3 were transported from vines to tubers efficiently at maturity.

In-season N dressing caused slight but insignificant increases in specific gravity and dry matter content (Table 3). Compared with applying all N at planting, applying N at planting in conjunction with dressing before tuber bulking significantly increased the average marketable tuber weight per plant from 503 g to 630 g.

Table 2 Effects of N application time on total tubers yield, marketable tuber yield, harvest index and source/sink ratio at harvest

Treatments	Tuber yield (t/ha)	Marketable tuber ratio %	Marketable tuber yield (t/ha)	Harvest index (g/g)	Source / sink (g/g)
T1	35.95±1.47 a	64.8±2.5 c	23.95±1.03 b	0.76±0.06 ab	0.33±0.05 b
T2	33.04±1.68 a	74.0±1.6 ab	24.21±1.44 ab	0.86±0.03 a	0.17±0.04 a
T3	38.04±2.38 a	79.2±2.1 a	30.01±2.74 a	0.86±0.03 a	0.17±0.04 a
T4	31.91±2.25 a	70.2±0.9 bc	22.51±1.65 b	0.73±0.02 b	0.37±0.02 b

Notes: T1 = all the 150 kg N/ha applied at planting; T2 = 100 kg N/ha applied at planting and 50 kg N/ha applied at one week before tuber initiation (20 DAE); T3 = 100 kg N/ha applied at planting and 50 kg N/ha applied at one week before tuber bulking stage (35 DAE); and T4 = 100 kg N/ha applied at emergence and 50 kg N/ha applied at one week before tuber bulking stage (35 DAE). Marketable tuber ratio % = weight of fresh tubers ⩾75g / weight of total fresh tubers×100%. Harvest index (HI) = tuber DM / total DM at harvest. Source/sink = vine DM at harvest / tuber DM at harvest. Each data is the mean value of four replications of six-plant dry matter samples± standard error. Means followed by the same letters are not significantly different according to Duncan's multiple range test at $P=0.05$.

Table 3 Effects of N application time on dry matter contents, specific gravity of marketable tubers and average weight of marketable tuber at harvest

Treatments	Dry matter content(%)	Specific gravity (g/cm^3)	Average weight of tubers >75g (g/plant)
T1	20.0±0.1 a	1.087±0.005 a	503±55 b
T2	21.2±0.4 a	1.091±0.003 a	508±33 b
T3	21.0±0.3 a	1.093±0.002 a	630±33 a
T4	21.2±0.4 a	1.092±0.001 a	472±68 b

Notes: T1 = all the 150 kg N/ha applied at planting; T2 = 100 kg N/ha applied at planting and 50 kg N/ha applied at one week before tuber initiation (20 DAE); T3 = 100 kg N/ha applied at planting and 50 kg N/ha applied at one week before tuber bulking stage (35 DAE); and T4 = 100 kg N/ha applied at emergence and 50 kg N/ha applied at one week before tuber bulking stage (35 DAE). Each data is the mean value of four replications of six-plant dry matter samples ± standard error. Means followed by the same letters are not significantly different according to Duncan's multiple range test at $P=0.05$.

4.2.4 Discussion

The accumulation and distribution of dry matter within plants are important processes determining crop productivity. Insufficient allocation of assimilates to the vegetative organs may give a poor crop, but an excess in N early in the season may result in high total biomass but with a relatively low proportion used for the production of storage organs (Tekalign and Hammes, 2005). This is what happened in our study in treatment T1, where total dry matter was relatively high but the HI was lower than in the other treatments.

During the vegetative growth stage, the foliage is the main sink for assimilates; however the tubers gradually become the main sink after tuberization is initiated. Chen and Settetr (2012) found that the strong leaf and stem sink capacity formed before tuber initiation will compete with developing tubers for photosynthates. Furthermore, if there is a build-up of assimilates in source leaves and this, in turn, can decrease the rate of photosynthesis. Dry matter accumulated in tubers comes from the carbohydrates photosynthesized during tuber development and retransferred from leaves and stems. So both the redistribution of dry matter between above and underground components and the maintenance of the photosynthetic organs can influence tuber yield. Alva

(2004) also demonstrated that excessive N applications contribute to excess vine growth at the expense of the tuber production. Rosen (1995) demonstrated that reducing N application at planting, delaying N application to emergence/hilling and applying N at post hilling can increase potato yield. In a 2-year study conducted by Wilson et al. (2009), total and marketable tuber yields obtained with the application of slow release urea were similar to those with split applications of soluble N, even though weather conditions were hotter and drier than average. Shoji et al. (2001) also found that 134 kg N/ha with controlled release fertilizer was able to generate the same potato yield as commercial farmers' traditional practice that applied 269 kg N/ha. This may be because of a better synchrony between N supply and plant demand. These results coincide with our study in treatment T3, where total tuber yield and HI were higher than other treatments but the dry matter accumulated in vine was relatively low.

Sensitivity analysis of a predictive model for N management in potato indicated that N inflow in the soil-plant system was the key trait affecting potato tuber yield (Lia et al., 2006). In China, most native potato producer applied all N fertilizer at planting (T1), a few producer will dress N at squaring stage (T2) for later-maturing varieties and much earlier for early-maturing variety. The excessive N supplement during vegetative growth stage lead to a vigorous vine and most producers firmly believe that it is high yield appearance though they only got the yield no more than 20 t/ha. Some international corporations and some scientific research groups can get a high yield more than 30 t/ha at the cost of irrigating frequently and large dose of fertilizer, 300 kg N/ha, for example.

The incorrect N dressing time and ratio will also result in unreasonable N supplement during some special periods though the total N application level is reasonable. The excessive vine of T1 still occupied one quarter dry matter of the whole plant and many tubers were kept in small size even at harvest. In order to get an ideal yield, the plant should be keeping green during tuber bulking stage to produce the carbohydrate and be kept from green at harvest to promote the carbohydrate distributed to tubers. The insufficient N during the vegetative growth stage led to a poor productivity while the large amount of dressing N induced the N-hungry plant overgrowth and more than one quarter assimilates still kept in vine even at harvest (T4). The significantly lower source/sink ratios of T2 and T3 (Table 2) indicated that they had a stronger sink for partitioning of assimilates than the other two treatments. However, T4 indicated that HI and tuber yield did not improve with in-season N dressing if N was not applied at planting (Fig. 4).

Tuber specific gravity is one of the most widely accepted measures of potato internal quality, and because of its close relationship to tuber starch content, total solids and mealiness, it is commonly used by the potato processing industry for assessing acceptability (Marwaha and Kumar, 1987). Some farmers refuse to dress N in season to avoid decreasing dry matter content and specific gravity, especially for seed tuber production. Several studies had also suggested that N applications in season can result in delayed maturity and decreased specific gravity levels, especially if more than 34 kg N/ha was applied within four to five weeks of vine kill (MacLean, 1984). The period from tuber initiation to mid-tuber bulking was a critical stage for potato plant because the vegetative growth and reproductive growth advanced together. Adequate N supplement during this stage was necessary to satisfy the demand of N and boost the development of tubers. It was not found in the present study that the dressing N would depress the specific gravity and dry matter content of tubers.

References

Alva A K. 2004. Effects of pre-plant and in-season nitrogen management practices on tuber yield and quality of two potato cultivars. Journal of Vegetable Crop Production, 10(2): 43-60

Arduini I, Masoni A, Ercoli L, et al. 2006. Grain yield, and dry matter and nitrogen accumulation and remobilization in durum wheat as affected by variet and seeding rate. Eur. J. Agron., 25:309-318

Arriaga F J, Lowery B, Kelling K A. 2009. Surfactant impact on nitrogen utilization and leaching in potatoes. Am. J. Potato Res., 86: 383-390

Bao S D. 2005. Soil, Plant and Fertilizer Analysis. 3rd Ed. Beijing: Chinese agriculture Publication Company

Basu P S, Sharma A, Garg I D. 1999. Tuber sink modifies photosynthetic response in potato under water stress. Environmental and Experimental Botany, 42(1): 25-39

Borkhard B, Skiot M, Mikkelsen R, et al. 2005. Expression of a fungal endo-α-1,5-L-arabinanase during stolon differentiation in potato inhibits tuber formation and formation and results in accumulation of starch and tuber-specific transcripts in the stem. Plant Science, 169:872-881

Cerny J, Balk J, Kulhnek M, et al. 2010. Mineral and organic fertilization efficiency in long-term stationary experiments. Plant Soil Environ., 56:28-36

Chen C T, Setter T L. 2012. Response of potato dry matter assimilation and partitioning to elevated CO_2 at various stages of tuber initiation and growth. Environmental and Experimental Botany, 80: 27-34

Christos D. 2009. Dry matter, nitrogen and phosphorus accumulation, partitioning and remobilization as affected by N and P fertilization and source-sink relation. Europ. J. Agronomy, 30:129-139

Darwish T M, Hajhasan S, Haidar A, et al. 2006. Nitrogen and water use efficiency of fertigated processing potato. Agricultural Water Management, 85: 95-104

Delgado J A, Dillon M A, Richard T S, et al. 2007. A decade of advances in cover crops: cover crops with limited irrigation can increase yields, crop quality, and nutrient and water use efficiencies while protecting the environment. Journal of Soil and Water Conservation, 62:110-117

Dordas C. 2009. Dry matter, nitrogen and phosphorus accumulation, partitioning and remobilization as affected by N and P fertilization and source-sink relations. Europ. J. Agronomy, 30: 129-139

El-Sirafy Z M, Abbady K A, El-Ghamry A M, et al. 2008. Potato yield quality, quantity and protability as affected by soil and foliar potassium application. Res. J. Agric. Biol. Sci., 4: 912-922

Ferreira T C, Goncalves D A. 2007. Crop-yield/water-use production functions of potatoes (*Solanumtuberosum*,L.) grown under differential nitrogen and irrigation treatments in a hot, dry climate. Agricultural Water Management, 90: 45-55

Goffart J P, Olivier M, Frankinet M. 2008. Potato crop nitrogen status assessment to improve N fertilization management and efficiency: past-present-future. Potato research, 51: 355-383

Haase T, Schler C, Hess J. 2007. The effect of different N and K sources on tuber nutrient uptake, total and graded yield of potatoes (*Solanumtuberosum* L.) for processing. Eur. J. Agron., 26:187-197

Hamouz K, Lachman J, Dvork P, et al. 2005. The effect of ecological growing on the potatoes yield and quality. Plant Soil Environ., 51:397-402

Iritani W M. 1978. Seed productivity: stem numbers and tuber set. Proceeding Annual Washington State Potato Conference, 17:1-4

Ierna A, Pandino G, Lombardo S, et al. 2011. Tuber yield, water and fertilizer productivity in early potato as affected by a combination of irrigation and fertilization. Agricultural Water Management, 101: 35-41

Jacobs B C, Clarke J. 1993. Accumulation and partitioning of dry matter and nitrogen in traditional and improved cultivars of taro (*Colocasia esculenta* L. Schott) under varying nitrogen supply. Field Crops Research, 31(3-4): 317-328

Kulhnek M, Balk J C J, Schweitzer K, et al. 2008. Evaluation of phosphorus quantity/intensity parameters in soil with different systems of organic fertilising. Plant Soil Environ., 54:389-394

Li H, Parent L E, Karam A, et al. 2003. Efficiency of soil and fertilizer nitrogen in a humid, cool, and acid sod-potato system. Plant Soil, 251:23-36

Li H, Parent L E, Tremblay C, et al. 1999. Potato response to crop sequence and nitrogen fertilization following sod breakup in a Gleyed Humo-Ferric Podzol. Can. J. Plant Sci., 79:439-446

Lia H, Parentc L E, Karamc A. 2006. Simulation modeling of soil and plant nitrogen use in a potato cropping system in the humid and cool environment. Agriculture, Ecosystems & Environment, 115(1-4):248-260

MacLean A A. 1984. Time of application of fertilizer nitrogen for potatoes in Atlantic Canada. Am. Potato. J., 61:23-29

Marwaha R S, Kumar R. 1987. Relationship between specific gravity and dry matter content of potato tubers. Int. J. Trop. Agric., 5:227- 230.

Muchow R C. 1988. Effect of nitrogen supply on the comparative productivity of maize and sorghum in a semi-arid tropical environment. I. Leaf growth and leaf nitrogen. Field Crops. Res., 18:1-16

Peres L E P, Carvvalho R F, Zsogon A, et al. 2005. Grafting of tomato mutants onto potato rootstocks: an approach to study leaf-derived signaling on tuberization. Plant Science, 160: 680-688

Prystupa P, Savin R, Slafer G A. 2004. Grain number and its relationship with dry matter, N and P in the spikes at heading in response to NP fertilization in barley. Field Crops. Res., 90:245-254

Poljak M, Horvat T, Majic A, et al. 2008. Nitrogen management for potatoes using rapid test methods. Cereal Res. Commun., 36:1795-1798

Rosen C J. 1995. Nitrogen management studies in Minnesota. Proceedings of Wisconsin Annual Potato Meeting, 8:15-23

Shari M, Zebarth B J, Hajabbasi M A, et al. 2005. Dry matter and nitrogen accumulation and root morphological characteristics of two clonal selections of Russet Norkotah potato as affected by nitrogen fertilization. J. Plant Nutr., 2: 2243-2253

Shoji S, Delgado J A, Mosier A, et al. 2001. Use of controlled release fertilizers and nitrification inhibitors to increase nitrogen use efficiency and to conserve air and water quality. Journal of Communication in Soil Science and Plant Analyses ,32:1051-1070.

Shrestha R K, Cooperb L R, MacGuidwin A E. 2010. Strategies to reduce nitrate leaching into groundwater in potato grown in sandy soils: case study from North Central USA. Am. J. Pot. Res., 87: 229-244

Sincik M, Turan Z M, Gksoy A T. 2008. Responses of potato (*Solanumtuberosum* L.) to green manure cover crop and nitrogen fertilization rates. Am. J. Potato Res., 85:150-158

Srek P, Hejcman M, Kunzov E. 2010. Multivariate analysis of relationship between potato (*Solanumtuberosum* L.) yield, amount of applied elements, their concentrations in tubers and uptake in a long-term fertilizer experiment. Field Crops Research, 118:183-193

Tadesse M, Lommen W J M, Struik P C. 2001. Effects of nitrogen pre-treatment of transplants from in vitro produced potato plantlets on transplant growth and yield in the field. Netherlands Journal of Agricultural Science, 49: 67-79

Tekalign T, Hammes P S. 2005. Growth and productivity of potato as influenced by cultivar and reproductive growth. I. Stomatal conductance, rate of transpiration, net photosynthesis, and dry matter production and allocation. Scientia Horticulturae, 105(13):13-27

Vos J. 1999. Split nitrogen application in potato: Effects on accumulation of nitrogen and dry matter in the crop and on the soil nitrogen budget. Journal of Agriculture Science, 133: 263-374

Vos J. 2009. Nitrogen responses and nitrogen management in potato. Potato Res., 52: 305-317

Westermann D T. 2005. Nutritional requirements of potatoes. Am. J. Potato Res., 82:301-307

Wilson M L, Rosen C J, Moncrief J F. 2009. Potato response to a polymer-coated urea on an irrigated, coarse-textured soil. Agronomy Journal, 101:897-905.

Zebarth B J, Arsenault W J, Sanderson J B. 2006. Effect of seedpiece spacing and nitrogen fertilization on tuber yield, yield components, and nitrogen use efficiency parameters of two potato. Amer. J. of Potato Res. Cultivars, 83:289-296

Zebarth B J, Rosen C J. 2007. Research perspective on nitrogen BMP development for potato. American Journal of Potato Research, 84:3-18.

Zelalem A, Tekalign T, Nigussie D. 2009. Response of potato (*Solamum Tuberosum* L.) to different rates of nitrogen and phosphorus fertilization on vertisols ant Debre Berhan, in the central highlands of Ethiopia. African Journal of Plant Science, 3(2):16-24

4.3 Fertilization and catch crop strategies for improving tomato production in North China

4.3.1 Introduction

The total area of land devoted to vegetable production in China has exceeded 18.4 million ha over the last two decades and accounted for 11.6% of the total agricultural cropping area in 2009. Tomato is the most widely grown vegetable, and is typically produced in poly-tunnel greenhouses occupying more than 60000 ha (Chinese Ministry of Agriculture, 2010). Conventional practices use frequent and excessive fertilization (especially nitrogen (N)) and irrigation to ensure maximum yields. For example, Chen et al. (2004) reported that greenhouse tomatoes in Beijing received more than 1000 kg N/ha per growing season from manure and fertilizer applications, while He (2006) reported that between 1994 and 2004 the mean application rate per growing season was 2227 kg N/ha for greenhouse tomatoes in Shouguang, Shandong Province. Such practices result in excessive root zone nutrient loadings, especially nitrate (NO_3^-) that can be easily lost due to the shallow rooting systems of some vegetable crops (Thorup-Kristensen, 2006; Verma et al., 2007). Consequently, increased soil NO_3^- concentrations and subsequent leaching have degraded surface and ground water quality (Zhang et al., 2013).

Furthermore, during the wet summer season (July and August) in North China, the polyethylene film is typically removed from the greenhouses, leaving the soil fallow, and thus allowing any mobile soil nutrients such as NO_3^- to be easily lost through leaching. The practice of "reduced fertilization" considers the balance between crop N uptake and soil supply. For tomato crops, fertilizer N inputs can be reduced by 70% with a subsequent reduction in soil N loss of 54% (Ren et al., 2010), while for cucumber (*Cucumis sativus* L.) crops, reducing N fertilizer by 53% has reduced soil N losses by 45% (Guo et al., 2008). Although reduced fertilization strategies consider the N supply during the crop growing season, to date, they have not considered residual soil N or the N released by mineralization during fallow periods to be vulnerable to leaching.

Catch crops are quick-growing plants sown during fallow periods between main crops to reduce NO_3^- leaching (Sainju and Singh, 1997). Nitrate leaching has been reported to decrease by 29%-94% in systems with non-leguminous catch crops (mean 70% reduction reported by

Tonitto et al., 2006) and by – 6%-48% with leguminous catch crops, when compared with bare fallow systems (Sainju and Singh, 1997). The potential impact of catch crops for reducing NO_3^- leaching is sufficient that some countries e.g. Denmark (Munkholm and Hansen, 2012), have now adopted them as key elements in their national N management strategies. Plant species with deep rooting systems, rapid biomass accumulation and high nutrient uptake capacity (especially C4 plant species) are particularly efficient as summer catch crops (Snapp et al., 2005; Wang et al., 2005). Ju et al. (2007) also demonstrated the high risk of NO_3^- leaching in the summer season in China and suggested that deep-rooted species such as maize can be used as a catch crop to intercept soil NO_3^- deep in the soil profile and consequently reduce NO_3^- leaching. Catch crops can also deliver other benefits such as increased soil organic matter content, resulting in carbon sequestration services and improved soil quality, as well as providing additional productivity from land which would otherwise be left fallow. This can take the form of bio-energy crops, animal fodder (Munkholm and Hansen, 2012), green manures (Sorensen and Thorup-Kristensen, 2011), and despite the short growing season, even food for human consumption (e.g., baby sweet corn in the case of maize).

While the potential benefits of catch crops are becoming more widely recognized, research is still required into their application in different systems and optimization under different fertilizer regimes, cropping systems and soil types. The combined effects of reduced fertilization and growing of catch crops have not been studied for tomato cropping systems in North China. We hypothesize that a combination of catch crops and reduced fertilization strategies can reduce the potential for nutrient leaching while maintaining productivity of greenhouse tomato systems in this area. To test this, tomato plants were grown in greenhouses for four growing seasons using three different cropping systems; conventional fertilization, reduced fertilization or reduced fertilization with the inclusion of a maize catch crop during fallow periods. Tomato and maize yield, nutrient uptake, soil mineral N and estimated macro-nutrients (N, P and K) quantities were determined.

4.3.2 Materials and Methods

1. *Experimental design*

Continuous double-crop tomato experiments were carried out in greenhouse poly-tunnel systems over four growing periods (winter-spring (WS) from February to June and autumn-winter (AW) from September to January) between 2008 and 2009 in Changping (40°10′56.36″N, 116°15′52.52″E), Beijing, China. After the June harvest, the greenhouse covering was removed, leaving the soil open to the atmosphere during the fallow period (July-August). The soil was fluvo-aquic containing 760, 230, and 10 g/kg, sand, silt and clay, respectively. Its chemical properties were: pH (H_2O) 7.2; total N 2.2 g/kg; Olsen-P 97 mg/kg; NH_4OAc extractable K 210 mg/kg; organic matter 21.2 g/kg (Jiang, 2009).

Conventional fertilization (CF), reduced fertilization (RF) and reduced fertilization with maize as a summer catch crop (RF+C) were evaluated using a randomized block design with three replicates of 6 m × 4 m plots. Four-week-old tomato (Xianke 1) seedlings were transplanted into the greenhouses in February (WS season), and September (AW season) each year. Fruit harvest commenced in May and November and ended in June and January, respectively. For the catch crop treatment, three-leaf maize (Tianzi 22) seedlings were transplanted at a spacing of 60 cm × 30 cm at the end of June and harvested in September.

Chemical NPK fertilizer was surface applied during each growing period (Table 1). The amount of chemical fertilizer for CF treatment is according to the conventional practice in this area, and that for RF and RF+C treatments based on the determination of soil nutrient status. A blended fertilizer was used in 2008 (N∶P∶K ratio 2.38∶1∶3.38), while in 2009, urea (46% N), calcium monophosphate(12% P_2O_5) and potassium sulfate (50% K_2O) were applied separately for each treatment. A uniform application of chicken manure (16.7, 20, and 30 t/ha, respectively) was also applied to each treatment at the start of each growing season. Table 2 shows the amount of N, P and K applied via the manure.

Table 1 Chemical fertilizer application rates for conventional fertilization (CF), reduced fertilization (RF) and reduced fertilization with catch crop (RF+C) treatments from 2008 to 2009

Season	Date	N (kg/ha)			P_2O_5 (kg/ha)			K_2O (kg/ha)		
		CF	RF	RF+C	CF	RF	RF+C	CF	RF	RF+C
2008WS[①]	04-23	98	85	85	31	30	30	140	95	95
	05-06	50	40	40	30	25	25	90	75	75
	05-18	50	40	40	30	24	24	90	70	70
	06-12	90	70	70	30	20	20	90	95	95
	Total	288	235	235	121	99	99	410	335	335
2008AW[②]	10-17	124	90	90	93	38	38	130	110	110
	11-10	70	42	42	90	25	25	90	90	90
	11-26	70	42	42	90	25	25	90	85	85
	12-22	120	70	70	90	20	20	90	80	80
	Total	384	244	244	363	108	108	400	365	365
2009WS	03-28	90	60	60	90	0	0	90	60	60
	04-13	60	25	25	60	0	0	60	25	25
	04-25	60	25	25	60	0	0	60	25	25
	05-19	60	50	50	60	0	0	60	50	50
	Total	270	160	160	270	0	0	270	160	160
2009AW	10-18	90	50	50	90	0	0	90	80	80
	11-18	60	25	25	60	0	0	60	40	40
	12-01	60	20	20	60	0	0	60	40	40
	12-25	60	35	35	60	0	0	60	80	80
	Total	270	130	130	270	0	0	270	240	240

Notes: ①WS: winter-spring season; ②AW: autumn-winter season.

Total weekly water application via drip irrigation was 159 mm and 116 mm in the WS and AW seasons in 2008, and 185 mm and 151 mm in the WS and AW seasons in 2009, respectively. The greenhouse was covered with polyethylene film except between 9 August to the 6 September each year. Precipitation during those periods was 55.5 and 74.2 mm for 2008 and 2009, respectively. Except for an initial 35 mm of irrigation at planting, no additional fertilizer or irrigation were applied to the maize crop.

2. Plant and soil sampling and NPK measurement

Immediately before planting and after harvest of tomatoes and maize plants, three soil cores were collected from each plot. Three rows of plants were selected in each plot, and a sample collected 10 cm from the base of the stem (but along the line of the row) of a randomly selected plant in that row (excluding those at the end). Samples were taken to a depth of 180 cm in 30 cm increments (total of six per sampling point). Mineral soil N (NH_4^+-N + NO_3^--N), referred to as N_{min}, was determined using a continuous flow analysis technique (TRACS 2000 system, Bran and Luebbe, Norderstedt, Germany) after shaking 12 g of fresh soil with 100 ml 1 moL/dm^3 KCl for 1 h at 25℃ and filtering.

Mature tomato fruits were harvested every 2-3 days to assess fresh weight production. At the end of the harvest period, plant and fruit samples were oven-dried at 70℃ for 48 h, weighed and digested in a mixture of concentrated H_2SO_4 and H_2O_2. Digests were analyzed for total N content by the Kjeldahl method (Nelson and Somers, 1973), total P by the molybdo-vanadophosphate method (Soon and Kalra, 1995) and total K by flame photometer.

At harvest, maize root samples were collected in soil cores sampled from each plot at two randomly selected locations situated equi-distant between rows of maize. Samples were collected with an Eijkelkamp root auger (8 cm diameter×15 cm depth) to a depth of 105 cm in 15 cm increments. Soil was washed from the roots of each sample with water, and the root length was measured using a scanner (Epson 1680, Indonesia) and WinRhizo software (Regent Instruments Inc., Quebec, QC, Canada). Root-length density (RLD) for each treatment was calculated from the root length and the soil core volume.

Nutrient surplus in the soil was calculated according to the method described by Ren et al. (2010):

$$N_{surplus} = N_{min\ initial} + N_{manure} + N_{fer} - N_{crop} - N_{min\ harvest} \qquad (1)$$

$$P(K)_{surplus} = P(K)_{manure} + P(K)_{fer} - P(K)_{crop} \qquad (2)$$

Where, N (P\K)$_{surplus}$ is N (P\K) surplus; $N_{min\ initial\backslash harvest}$ is soil N_{min} to 30 cm depth before transplanting\at harvest; N (P\K)$_{manure}$ is total N (P\K) from chicken manure; N (P\K)$_{fer}$ is N (P\K) from chemical fertilizer; N (P\K)$_{crop}$ is total N (P\K) uptake by tomato and maize.

3. Statistical analysis

In general, soil and tomato plant parameters for each treatment were analyzed using means for the four growing seasons, while maize yield, nutrient uptake and root data were analyzed using means for the two cropping seasons. A one-way ANOVA was conducted using the SAS statistical software. Significant differences among means were determined by Least Significance Difference tests at the 0.05 level of probability.

4.3.3 Results

1. Yield and NPK uptake

Mean tomato fruit yield was significantly lower in the AW than WS season, probably as a result of lower air temperatures during fruiting (mean values were 25℃ and 29℃ for AW and WS seasons, respectively). Reduced fertilization and/or catch crop establishment did not reduce tomato fruit yield during any of the growing seasons (Fig. 1). No significant difference was observed in mean tomato fruit yield between the CF (101±6 t/ha), RF (103±10 t/ha) and RF+C (106±9 t/ha) treatments over the four growing seasons despite a 37% reduction in chemical fertilizer N input in the RF and RF+C treatments (Fig. 1). Mean fresh yields of maize were 13.4±0.3 and 7.5±0.4 t/ha in 2008 and 2009, respectively.

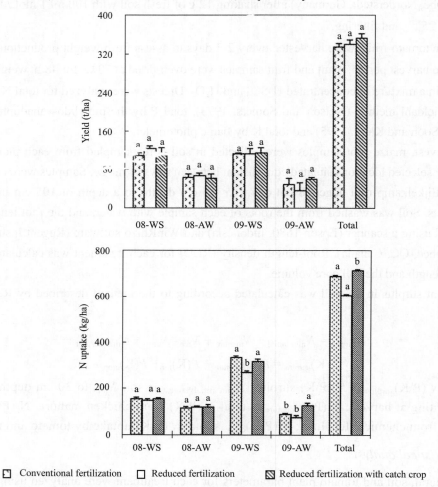

Fig. 1 Tomato fruit yield with different N and catch crop treatments in the double greenhouse tomato cropping system in 2008 and 2009 in Beijing, China

WS represents winter-spring season, AW represents autumn-winter season, bars represent SD of means with three replicates per treatment, the same letter in the same growing season denotes no significant differences among different treatments by LSD ($P<0.05$)

Tomato N uptake in the WS season in 2009 (305±8.8 kg N/ha) was significantly higher than that in 2008 (153±8.8 kg N/ha). There was no significant difference in mean tomato N uptake between the CF (135±6.4 kg N/ha), RF (133±6.5 kg N/ha) and RF+C (137± 6.8 kg N/ha) treatments in 2008. However, in 2009, mean tomato N uptake in the RF+C (221±10.2 kg N/ha) treatment was significantly higher than that of the RF (170±8.5 kg N/ha) treatment (Fig. 1). The mean P and K uptake by tomato fruits and plants at harvest was 37-39 kg P/ha, and 193-211 kg K/ha in the CF, RF and RF+C treatments, respectively. Mean N, P and K uptake by maize at harvest was 150±6, 39±3 and 208±9 kg/ha, respectively (Table 2).

Table 2 Apparent N, P and K surplus for conventional fertilization (CF), reduced fertilization (RF) and reduced fertilization with catch crop (RF+C) treatments in a greenhouse tomato cropping system from 2008 to 2009

	N (kg/ha)						P (kg/ha)						K (kg/ha)					
	CF		RF		RF+C		CF		RF		RF+C		CF		RF		RF+C	
	2008	2009	2008	2009	2008	2009	2008	2009	2008	2009	2008	2009	2008	2009	2008	2009	2008	2009
Nutrient input[1]	1468	1424	1275	1103	1275	1103	762	1020	485	480	485	480	1341	1354	1231	1214	1231	1214
N_{min}[2] at transplanting	378	246	378	175	378	175	-	-	-	-	-	-	-	-	-	-	-	-
Manure	418	638	418	638	418	638	278	480	278	480	278	480	531	814	531	814	531	814
Chemical fertilizer	672	540	479	290	479	290	484	540	207	0	207	0	810	540	700	400	700	400
Nutrient output[3]	576	934	486	784	663	996	89	65	83	64	130	96	529	269	521	249	724	533
N_{min} at harvest	306	513	220	444	216	426	-	-	-	-	-	-	-	-	-	-	-	-
Maize nutrient uptake	0	0	0	0	173	128	0	0	0	0	44	33	0	0	0	0	224	191
Tomato nutrient uptake	270	421	266	340	274	442	89	65	83	64	86	63	529	269	521	249	500	342
Nutrient surplus[4]	892	490	893	490	789	319	673	955	402	416	355	384	812	1085	710	965	507	681

Notes: ①Nutrient input is equal to nutrient from manure and chemical fertilizer (plus N_{min} at transplanting for N). ②0-30 cm soil N_{min}. ③Nutrient output is equal to nutrient uptake by tomato and maize (plus N_{min} at harvest for N). ④Nutrient surplus is equal to nutrient input minus nutrient output.

2. Root distribution of maize

In both years, maize roots penetrated to a depth of at least 105 cm with root length below 30 cm accounting for 15%-22% of the total. Over 83% of the total root length measured to a depth of 75 cm was found within the top 30 cm of the soil profile in both 2008 and 2009. The mean dry weight of maize roots was 1103 kg/ha at harvest (Table 3).

Table 3 Profile distribution of maize roots in the soil for RF+C (reduced fertilization with catch crop) treatment in the summer fallow period in 2008 and 2009

Soil depth (cm)	Root dry weight (kg/ha)				Root length density (cm/cm^3)			
	2008		2009		2008		2009	
	9 Aug	7 Sep	9 Aug	7 Sep	9 Aug	7 Sep	9 Aug	7 Sep
0-15	443.4	842.3	250.2	1050.1	1.1	0.9	0.6	0.7
15-30	64.3	80.9	49.9	65.1	0.3	0.2	0.3	0.3
30-45	16.5	40.0	10.0	24.0	0.2	0.1	0.1	0.1
45-60	N[1]	34.6	3.8	21.5	N	0.1	0.1	0.1
60-75	N	20.8	N	6.9	N	0.1	N	0.1
75-90	N	7.2	N	4.9	N	N	N	0.1
90-105	N	2.3	N	4.8	N	N	N	N

Note: ①N means not detected.

3. Soil mineral nitrogen distribution in the soil profile

Reducing fertilization and growing of a catch crop decreased N_{min} throughout the 180 cm soil profile. Average N_{min} values were 97 and 74 kg/ha for the CF and RF treatments respectively at tomato plant harvest. After the fallow period, the N_{min} concentration was similar in the CF and RF treatments with values of 109 and 100 kg/ha respectively, but it was reduced to 44 kg/ha in the RF+C treatment. The N_{min} reduction for both RF and RF+C treatments occurred within the 0-30 cm layer. Summed over the entire 180 cm profile, the RF treatment decreased N_{min} by 98 to 207 kg/ha compared to the CF treatment, at tomato harvest (Fig. 2:

Jun 2008 and 2009, Jan 2009 and 2010). Meanwhile, soil N_{min} concentrations of the RF+C treatment were reduced by 267 and 415 kg/ha at maize harvest compared with the RF treatment in 2008 and 2009 respectively (Fig. 2: Sep 2008 and 2009). Soil N_{min} in the RF+C treatment was 47%-65% lower than that in the RF treatment (Fig. 2). Generally, the greatest concentrations of N_{min} were in the top 30 cm for all treatments.

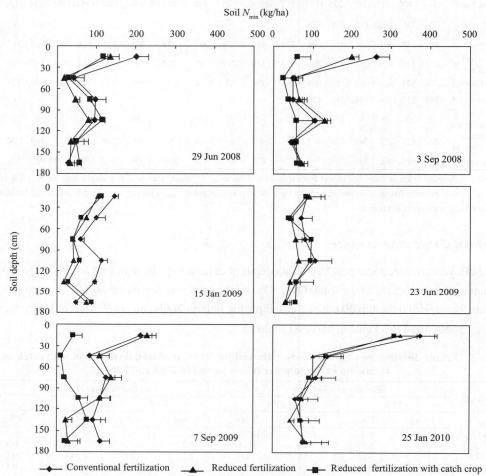

Fig. 2 N_{min} distribution in the 0-180 cm soil profile at harvesting in different treatments in the double greenhouse tomato cropping system in 2008 and 2009

Bars represent SD of means with three replicates per treatment

4. NPK surplus

Mean annual N input for the CF treatment was 1340 kg N/ha compared to an output of 606 kg N/ha, thus creating a mean annual surplus of 734 kg N/ha. The RF treatment decreased the N surplus by 15% in 2008 and 27% in 2009, but the mean surplus was still 593 kg N/ha (Table 2). The RF+C treatment reduced the N surplus an additional 23% in 2008 and 65% in 2009.

The mean annual P/K surplus in the RF treatment was 405 kg P/ha and 111 kg K/ha less than in the CF treatment, mainly due to the lower (mean 409 kg P/ha and 125 kg K/ha respectively) chemical fertilizer P/K inputs. The annual mean P/K surplus in the RF+C treatment was even lower than for the RF treatment because of additional P/K uptake by the maize crop (Table 2).

4.3.4 Discussion

1. Effect of reduced fertilization and catch crop on soil N_{min}

Many studies have shown that N leaching is a major N loss pathway for greenhouse vegetable cropping systems (Ramos et al., 2002; Cao et al., 2005; Yu et al., 2006; Zotarelli et al., 2007). This study showed that early August is an important time for N leaching because the greenhouses' plastic film is removed during the warm season. The RF treatment reduced N_{min} by 98 to 207 kg/ha compared to the CF treatment at harvest (Fig. 2), thus reducing the potential for N leaching during the subsequent wet period. The study also showed that to accurately assess the availability of plant nutrients derived from manure and to ensure optimal fertilization, it is necessary to account for the nutrients remaining in soil from previous manure applications. Residual N that will be released during subsequent cropping cycles will increase the risk of excess N application if it is not accounted for when calculating N fertilizer application rates. This emphasizes that long-term effects of organic fertilizers due to their slow N release must be considered to ensure optimum fertilizer use (Webb et al., 2013).

Constantin et al. (2010) reported that catch crops were the most efficient technique for reducing N leaching, reporting mean N_{min} reductions of between 36% and 62%. Wyland et al. (1996) showed a 65%-70% reduction in NO_3^- leaching from the catch-cropped plots compared with the fallow control during winter, because plant roots in the surface soil removed NO_3^- and water that would have otherwise been lost from the profile. The benefits of catch cropping were also confirmed by several experiments which showed that soil N depletion by catch crops was highly correlated with rooting depth (Thorup-Kristensen, 2001; Kristensen and Thorup-Kristensen, 2004). Sapkota et al. (2012) showed in both field experiments and model simulations that fodder radish (*Raphanus sativus* L.) developed the deepest root system and depleted N from deeper soil layers than chicory (*Cichorium intybus* L.) and ryegrass (*Lolium perenne* L.). In our study, a large proportion of maize roots penetrated to a depth of 105 cm, depleting N and water that would have otherwise have been susceptible to loss from the soil profile. Catch crops in vegetable cropping systems are also reported to increase the storage of N in organic compounds in the soil (Thorup-Kristensen et al., 2003; Sainju and Singh, 2008). In our study, the reduced soil N_{min} concentrations (368-531 kg N/ha) by the RF+C treatment, compared with the RF treatment, were higher than the quantity of N taken up by maize (128-173 kg N/ha). We postulate this discrepancy is the result of maize facilitating immobilization of N into soil organic matter.

2. Effect of reduced fertilization and catch crop on NPK surplus

The measurement of nutrient surplus in soils provides information on the sustainability of land use systems and potential environmental impacts (Oborn et al., 2003; Janssen and

Willigen, 2006). Schroder et al. (2010) and Webb et al. (2013) showed that the N surplus (i.e. the difference between total N input and harvested N) was a better predictor of NO_3^- leaching from grasslands on sandy soils than total N input from mineral fertilizer and slurry. Nutrient surplus needs to be small and positive in order to avoid unnecessary nutrient accumulation and reduce the risk of N losses to the environment, while maintaining optimal or acceptable crop yields. Vos and Putten (2000) reported that some catch crops (radish, rapeseed (*Brassica napus* L.) and rye (*Secale cereale* L.)) did not affect nutrient (NPK) surplus because they had very limited growth. However, in our experiment, due to its rapid growth and high nutrient uptake, the maize catch crop significantly reduced the nutrient surplus. The low N surplus obtained was mainly due to low N inputs and high N exports via the crops. The imbalance in NPK between manure inputs and crop removals means that regular manure use will elevate P and K concentrations in the soil if crops are not supplied with alternative N sources. Accumulation of P in soils has been shown to result in leaching and increased risk of P loss to surface and ground waters (Heckrath et al., 1995; Webb et al., 2013). Yanai et al. (1996) reported catch crops reduced K concentrations in soil water either directly by active uptake, or indirectly by increasing evapotranspiration rates and reducing the flow of leachate. The positive K surplus measured in our study suggests that the soil K pool was accumulating, which may result in K leaching. However, Margrethe and Jørgen (2008) reported that the growing of catch crops reduced K leaching by 28%. These results show that while catch crops can be effective for reducing NO_3^- leaching, their effectiveness for other nutrients may be different, and careful consideration needs to be given to the balance of nutrients in fertilizers such as manures to prevent potentially problematic accumulations of nutrients such as P and K. In some systems it has been reported that certain types of catch crops can reduce the productivity of the primary crop, due to competition for soil nutrients (Uchino et al., 2009; Munkholm and Hansen, 2012). However this is reported only for crops planted simultaneously, rather than sequentially, as in our study. Indeed, in our sequential catch crop system we have shown that not only is the productivity of the primary crop not reduced, even with reduced fertilizer applications, but that additional productivity can be attained with a catch crop which can potentially provide biomass for bioenergy, fodder or human consumption.

References

Cao L K, Chen G J, Lu Y T. 2005. Nitrogen leaching in vegetable fields in the suburbs of Shanghai. Pedosphere, 15: 641-645

Chen Q, Zhang X S, Zhang H Y, et al. 2004. Evaluation of current fertilizer practice and soil fertility in vegetable production in the Beijing region. Nutr. Cycl. Agroecosys., 69: 51-58

China Ministry of Agriculture. 2010. China Agriculture Statistical Data. Beijing: China Agriculture Press: 101 (in Chinese)

Constantin J, Mary B, Laurent F, et al. 2010. Effects of catch crops, no till and reduced nitrogen fertilization on nitrogen leaching and balance in three long-term experiments. Agr. Ecosyst. Environ., 135: 268-278

Guo R Y, Li X L, Christie P, et al. 2008. Influence of root zone nitrogen management and a summer catch crop on cucumber yield and soil mineral nitrogen dynamics in intensive production systems. Plant Soil, 313: 55-70

He F F. 2006. Studies on optimizing nitrogen management and environmental implications in greenhouse tomato cropping system.Beijing: China Agricultural University Ph D Thesis (in Chinese with English summary)

Heckrath G, Brookes P C, Poulton P R, et al. 1995. Phosphorus leaching from soils containing different phosphorus concentrations in the Broadbalk experiment. J. Environ. Qual., 24: 4-10

Janssen B H, Willigen P. 2006. Ideal and saturated soil fertility as bench marks innutrient management. II. Interpretation of chemical soil tests in relation to ideal and saturated soil fertility. Agr. Ecosyst. Environ., 116: 147-155

Jiang C G. 2009. Effects of water & fertilizer input and rotating waxy corn on the nutrient utilization in year-roung greenhouse tomato system. Beijing: China Agricultural University Master Thesis (in Chinese with English summary)

Ju X T, Gao Q, Christie P, et al. 2007. Interception of residual nitrate from a calcareous alluvial soil profile on the North China Plain by deep-rooted crops: a ^{15}N tracer study. Environ. Pollut., 146: 534-542

Kristensen H L, Thorup-Kristensen K. 2004. Uptake of ^{15}N labeled nitrate by root systems of sweet corn, carrot and white cabbage from 0.2-2.5 meters depth. Plant Soil, 265: 93-100

Margrethe A, Jørgen E. 2008. Residual effect and leaching of N and K in cropping systems with clover and ryegrass catch crops on a coarse sand. Agr. Ecosyst. Environ., 123: 99-108

Munkholm L J, Hansen E M. 2012. Catch crop biomass production, nitrogen uptake and root development under different tillage systems. Soil Use Manage., 28: 517-529

Nelson D W, Somers L E. 1973. Determination of total nitrogen in plant material. Agron. J., 65: 109-112

Oborn I, Edwards A C, Witter E, et al. 2003. Element balances as a tool for sustainable nutrient management: a critical appraisal of their merits and limitations within an agronomic and environmental context. Eur. J. Agron., 20: 211-225

Ramos C, Agut A, Lidón A L. 2002. Nitrate leaching in important crops of the Valencian Community region (Spain). Environ. Pollu., 118: 215-223

Ren T, Christie P, Wang J G, et al. 2010. Root zone soil nitrogen management to maintain high tomato yields and minimum nitrogen losses to the environment. Sci. Hortic., 125: 25-33

Sainju U M, Singh B P. 1997. Cover crops for sustainable agricultural systems: influence on soil properties, water quality, and crop yields. HortScience, 32: 21-28

Sainju U M, Singh B P. 2008. Nitrogen storage with cover crops and nitrogen fertilization in tilled and non-tilled soils. Agron. J., 100: 619-627

Sapkota T B, Askegaard M, Lægdsmand M, et al. 2012. Effects of catch crop type and root depth on nitrogen leaching and yield of spring barley. Field Crop. Res., 125: 129-138

Schroder J J, Assinck F B T, Uenk D, et al. 2010. Nitrate loss from grassland on sandy soils, as affected by soil properties and the rate and nature of the N-input. Grass Forage Sci., 65: 49-57

Snapp S S, Swinton S M, Labarta R, et al. 2005. Evaluating cover crops for benefits, costs and performance within cropping system niches. Agron. J., 97: 1-11

Soon Y K, Kalra Y P. 1995. A comparison of plant tissue digestion methods for nitrogen and phosphorus analyses. Can. J. Soil Sci., 75: 243-245

Sorensen J N, Thorup-Kristensen K. 2011. Plant-based fertilizers for organic vegetable production. J. Plant Nutr. Soil Sci., 127: 321-332

Thorup-Kristensen K. 2006. Root growth and nitrogen uptake of carrot, early cabbage, onion and lettuce following a range of green manures. Soil Use Manage., 22: 29-38

Thorup-Kristensen K, Magid J, Jensen L S. 2003. Catch crops and green manures as biological tools in nitrogen management in temperate zones. Adv. Agron., 79: 227-302

Thorup-Kristensen K. 2001. Are differences in root growth of nitrogen catch crops important for their ability to reduce soil nitrate-N content, and how can this be measured? Plant Soil, 230: 185-195

Tonitto C, David M B, Drinkwater L E. 2006. Replacing bare fallows with cover crops in fertilizer-intensive cropping systems: a meta-analysis of crop yield and N dynamics. Agr. Ecosyst. Environ., 112: 58-72

Uchino H, Iwama K, Jitsuyama Y, et al. 2009. Yield losses of soybean and maize by competition with interseeded cover crops and weeds in organic-based cropping systems. Field Crop. Res., 113: 342-351

Verma P, George K V, Singh H V, et al. 2007. Modeling cadmium accumulation in radish, carrot, spinach and cabbage. Appl. Math. Model., 31: 1652-1661

Vos J, Putten D. 2000. Nutrient cycling in a cropping system with potato, spring wheat, sugar beet, oats and nitrogen catch crops. I . Input and offtake of nitrogen, phosphorus and potassium. Nutr. Cycl. Agroecosys., 56: 87-97

Wang Q, Li Y, Klassen W. 2005. Influence of summer cover crops on conservation of soil water and nutrients in a subtropical area. J. Soil Water Conserv., 60(1): 58-63

Webb J, Sorensen P, Velthof G, et al. 2013. An assessment of the variation of manure nitrogen efficiency throughout Europe and an appraisal of means to increase manure-N efficiency. Adv. Agron., 119: 371-442

Wyland L J, Jackson L E, Chaney W E, et al. 1996. Winter cover crops in a vegetable cropping system: impacts on nitrate leaching, soil water, crop yield, pests and management costs. Agr. Ecosyst. Environ., 59: 1-17

Yanai J, Linehan D J, Robinson D, et al. 1996. Effects of inorganic nitrogen application on the dynamics of the soil solution composition in the root zone of maize. Plant Soil, 180: 1-9

Yu H R, Li Z Z, Gong Y S, et al. 2006. Water drainage and nitrate leaching under traditional and improved management of vegetable-cropping systems in the North China Plain. J. Plant Nutr. Soil Sci., 169: 47-51

Zotarelli L, Scholberg J M, Dukes M D, et al. 2007. Monitoring of nitrate leaching in sandy soils: comparison of three methods. J. Environ. Qual., 36: 953-962

Zhang F S, Chen X P, Vitousek P. 2013. An experiment for the world. Nature, 497: 33-35

4.4 Weed community composition after 26 years of fertilization of late rice

4.4.1 Introduction

To assess the influence of long-term fertilization on weed communities of early and late rice crops, the weed species composition was investigated in experimental plots initiated in 1981 at the Key Field Experimental Monitoring Station of the Reddish Paddy Soil Eco-environment in Wangcheng, China. The treatments were: ① a control (CK), no fertilizer; ② NP, no K; ③ NK, no P; ④ PK, no N; ⑤ NPK; ⑥ NPKCa, N, P, and K plus lime; ⑦ NPS, N and P plus additional rice straw return; ⑧ NPKS, N, P, and K plus additional rice straw; ⑨ NKM, N and K plus swine manure. The results showed that weed flora composition and density were influenced by the different fertilization treatments. Multivariate analyses indicated that changes in the weed community composition were primarily due to soil-available N, followed by light intensity on the field surface and soil-available P. More weed species and total weed density were observed in the control and PK plots than in plots in which N, P, and K were applied together. Omission of N application had a greater effect on the weed community than the omission of P or K applications. Nutrients derived from synthetic fertilizers and organic manure or the additional application of lime had no obvious effect on the weed community of late rice crops.

Agricultural production is the result of the growth, development, and yield of individual crops plants in the field. The weeds endemic to most crops may well be as old as agriculture itself. Weeds are notorious yield reducers and as such are, in many situations, economically more important than insects, fungi, or other pest organisms (Savary et al., 1997). From agriculture's very beginnings, farmers have been aware that the presence of those unsown species interferes with the growth of the crops they were intending to produce. This recognition led to the coevolution of agroecosystems and weed management (Ghersa et al., 1994). Competition between the undesired plants and the crops was to be avoided if reasonable yields were to be achieved. The yield loss due to weeds was almost always caused by an assemblage of different weed species that differed substantially in their competitive abilities (Weaver and Ivany, 1998; Milberg and Hallgren, 2004). Thus, in order to rid their fields of weed, save labor, and increase crops yields, increasing amounts of herbicides have been applied by farmers.

Rice is one of the most important crops in China. Its seeded area reached 2.93×10^7 hm^2 in 2006 and annual production amounted to almost 1.83×10^8 t in 2006, accounting for 36.7% of total foodstuffs (Editorial Committee of China Agriculture Yearbook, 2007). In Hunan Province, rice is the crop with the largest seeded area and the highest yield. It contributes 13.7% of the total rice yield in China. Herbicides have been used widely in rice paddies but they may contaminate nearby aquatic systems through a variety of mechanisms, including drift, surface runoff and leaching. This can result in non-point-source pollution and have adverse effects on non-target aquatic plants that play a major role in the aquatic ecosystem (Ueji and Inao, 2001). Furthermore, herbicides have negative effects on the microbial properties of soil. Specifically, soil microbial carbon (C) and nitrogen (N) biomass as well as basal soil respiration rates decreased significantly in soils treated with herbicides (Yao and Zhang, 2008).

Many agronomic aspects of land management influence the composition and density of a community of weeds (Buhler et al., 1997), such as tillage practices (Fryer and Evans, 1970; Cardina et al., 1991; Mulugeta and Stoltenberg, 1997) and fertilization (Theaker et al., 1995; O'Donovan et al., 1997; Yin et al., 2005). Previous studies have demonstrated that changes in weed community composition brought about by fertilization were mainly due to shifts of soil C, N (Blackshaw et al., 2002; Davis, 2007), and P (Santos et al., 2004). The rate of change and the type of N also influence weed community structure; plant communities treated with liquid urea differ from those receiving either sulfate or nitrogen fertilizers (Cathcart et al., 2004).

Long-term field experiments are important with a view to evaluating changes in weed community composition and are likely to provide insight into the effects of prolonged fertilization over time (Norris, 2007). Nonetheless, little research on the effects of long-term fertilization with respect to weed community composition in the paddy soils of early and late rice has been conducted, and many questions remain unanswered. Therefore, the objective of this study was to elucidate the effects of various nutrient inputs on the weed community composition in late rice, in order to reduce the rate of herbicide application. Our conclusions are based on the results of a fertilization experiment carried out over a period of 26 years.

4.4.2 Materials and Methods

A long-term field experiment was initiated in 1981 at the Key Field Experimental Monitoring Station of the Reddish Paddy Soil Eco-environment in Wangcheng, China. The station is located in the village of Huangjin (28° 37′N, 112° 80′ E, 100 m altitude), Wangcheng County, Hunan Province, in the central region of the Xiangjiang River (a branch of Dongting Lake). The climate of this area is subtropical monsoonal, with an average annual precipitation of approximately 1370 mm and an annual mean temperature of 17℃. The soil, derived from Quaternary red clay, is clay loamy, with 35.5 g organic matter/kg, 2.05 g total N/kg, 10.2 mg available P/kg, 62.3 mg exchangeable K/kg, and a pH of 5.3, as determined in 1981 at the beginning of the experiment. Nine treatments with three replicates in completely randomized blocks were established, i.e. 27 plots (6.7 m × 9.5 m for each plot). Plots were separated by concrete ridges (width 30 cm), and the treatment blocks by irrigation furrows (width 50 cm) and two ridges. The treatments were: ① a control (CK), no fertilizer; ② NP, no K; ③ NK, no P; ④ PK, no N; ⑤ NPK, applied as urea, superphosphate, and potassium chloride, respectively; ⑥ NPKCa, N, P, and K fertilizer at the same rate as in the NPK treatment plus lime applied at a rate of 1000 kg/hm^2; ⑦ NPS, total K applied at the same rate as in the NPK treatment but derived from rice straw plus chemical N and P fertilizers as in the NPK treatment; ⑧ NPKS, N, P, and K fertilizer applied at the same rate as in the NPK treatment plus additional rice straw applied at a rate of 2620 kg/hm^2; ⑨ NKM, total P applied at the same rate as in the NPK treatment but derived from swine manure plus chemical N and K fertilizers as in the NPK treatment. The application rates of N, P, and K were 150, 90, and 120 kg/hm^2 for early rice and 180, 90, and 120 kg/hm^2 for late rice. Swine manure was applied at 1.5 t/hm^2 and rice straw at 2.62 t/hm^2 to both early and late rice. Rice straw was cut into 5 to 8 cm pieces and swine manure was composted for 4 weeks. According to the experimental design, urea was split between a basal application (70%) and a top dressing (30%) at early tilling. P, K and swine manure were applied as basal fertilizers. Rice straw was ploughed into the soil when the field was furrowed by a cow. Herbicides were not used over the entire course of the experiment, and all weeds were removed by hand twice per year in early rice and late rice from 1981 to 2007.

In this experiment, early and late rice cropping was followed by a fallow period in winter. The varieties of early rice and late rice planted in 2007 were Xiangzaoxian 31 and Weiyou 46, and the plant spacing was 13 cm × 20 cm for early rice and 20 cm × 20 cm for late rice. Early rice was transplanted on April 25 and late rice on July 13. The plants were harvested on July 7 and October 11, respectively.

Weed presence was recorded in three 1 m^2 quadrats distributed randomly in each treatment plot during the tasseling stage of late rice, on September 1st, 2007, i.e. 48 days before the late rice was harvested. Five late-rice plants and all weeds present in the quadrats were clipped, sorted according to species, counted, and oven-dried at 70℃ for 48 h before weighing. Soil samples were taken in each plot to analyze soil-available N, P, exchangeable K, and pH as described by Page et al. (1982). At the same time, light transmittance within the canopy was calculated from ten measures of light intensity per plot. For this purpose, a digital light meter (TES-1330, TES Electrical Electronic Corp. China) was placed on the soil surface and on top of the crop canopy.

Data were analyzed with the program DPS 6.5. Data on light transmittance, soil-available N, P, exchangeable K, pH, and the dry biomass of individual late rice plants were compared using the program LSD. Weed community composition was determined using principal coordinates analysis (PCoA) (Zhao et al., 2007). For the parametric analysis, weed densities were transformed using an arcsine square root function, and only species that occurred frequently enough to have normal error distributions were analyzed (Swanton et al., 1999). The degree of association between weeds and nutrient input systems was assessed by a vector diagram, with the direction of the vector indicating the type of association between a weed species and any treatment, and the strength of the association being proportional to the vector length.

4.4.3 Results and Discussion

1. Soil nutrients, pH, light intensity, and late-rice growth

The results demonstrated that different fertilization treatments applied over the long-term resulted in significant differences in available N, P, exchangeable K, pH, light transmittance, and dry aboveground biomass per quadrat in late rice (Table 1). The amounts of soil N, P, and K resulting from the different fertilizers differed, yielding a range of soil inorganic N levels. In the control plots in which light transmittance was highest, the amounts of N, P, and K were very low and the growth of late rice was the worst. In PK-, NK-, or NP-treated plots, late rice suffered from N, P, and K deficiencies, respectively, but the reduction in growth under NP was not as serious as under PK and NK, although light transmittance in these latter plots was higher than in NP plots. Late-rice biomass was high in those plots in which N, P, and K had been applied together, regardless whether they were of organic or chemical origin. The additional application of rice straw at the rate of 2620 kg/hm^2 did not cause an obvious increase in soil-available N, P, and exchangeable K in the NPKS treatment, whereas the application of lime, at a rate of 1000 kg/hm^2, increase soil pH in NPKCa plots.

Table 1 Soil-available N, P, exchangeable K, pH, light transmittance, and dry aboveground biomass per late-rice plant under different treatments*

Treatment	Available N (mg/kg)	Available P (mg/kg)	Exchangeable K (mg/kg)	pH	Light transmittance (%)	Dry aboveground biomass of late rice (g/m^2)
CK	4.8a	3.7a	42.7a	5.40b	42.3e	362.5a
PK	4.0a	31.8c	140.0d	5.02a	27.8d	442.5b
NK	34.2c	2.7a	116.7c	5.27ab	15.8c	592.5c
NP	21.3b	24.8b	36.7a	5.00a	7.0b	695.0d
NKM	18.3b	4.9a	66.7b	5.10a	1.1a	877.5e
NPS	16.9b	26.1bc	58.7b	4.99a	2.1a	867.5e
NPK	18.2b	19.3b	64.0b	4.99a	1.3a	902.5e
NPKCa	17.7b	20.0b	63.3b	5.96c	1.9a	887.5e
NPKS	18.8b	22.2b	70.0b	4.90a	1.5a	905.0e

Note: *Significant differences indicated by different letters within columns ($P < 0.05$).

2. *Weed community*

Fourteen weed species were identified during the growth of late rice (Table 2). Of these, *Rotala indica*, *Lindernia procumbens*, and *Eleocharis tetraguetra* were the most abundant. Weed species and their densities in the community varied with the treatment type. More weed species and higher total weed densities were observed in the control and PK plots. There were 11 weed species in these two plots, with total weed densities of 555.3 and 652.7 plants/m^2, respectively. Nevertheless, fewer weed species and a lower total density occurred in those plots in which N, P, and K were applied together, especially in combination with rice straw or lime. Only two weed species were detected in NPK plots and three in NPKCa plots, respectively. There were more weed species and a higher total density in N-deficient treatments than in those in which P or K was omitted. Managing crop fertilization may be an important practice to protect crop yield and reduce weed populations, since the impact of different fertilization protocols differs relative to the individual weed species (Carlson and Hill, 1985). Under the different fertilization treatments of this study, soil-available N, P, exchangeable K, and light intensity varied significantly. These differences had a strong impact on the weed community and on the densities of the various weed species. Balanced applications of N, P, and K resulted in rice that grew well, thus decreasing the light intensity needed for weed growth.

Table 2 Weed species density under different treatments

Weed species	Density (plant/m^2)									Total
	CK	PK	NK	NP	NKM	NPS	NPK	NPKCa	NPKS	
Monochoria vaginalis (Burm.f.) Presl ex Kunth	0.3	2.7	2.7	11.3	12.7	5	7	0.3	0.7	42.7d
Sagittaria pygmaea Mig.	10.7	2.3	41.3		1.7	2		1.7		59.7 d
Cyperus difformis L.	33	17.7	0.7							51.4 d
Echinochloa crusgalli (L.) Beauv.	1.7	3.3				1.7				6.7c
Polygonum hydropiper L.	1	11.7		1	0.3				2	16c
Rotala indica (Willd.) Koehne	91	318.7	4	8.7	2	23.7	0.3	0.7		449.1f
Lindernia procumbens (Krock.) Philcox	2	283.7		4	3	0.7	0.3	21.3	20.3	335.3 e
Leptochloa chinensis (L.) Nees	1.3		0.7							2 b
Eleocharis tetraguetra Nees	364	1.3	0.3							365.6 e
Eleocharis congesta Subsp. Japonica	48.3									48.3 d
Lindernia antipoda (L.) Alston	1.7	0.7		0.3						2.7b
Commelina communis L.		0.3					0.3			0.6 a
Paspalum paspaloides (Michx.) Scribn.				0.7						0.7a
Alternanthera philoxeroides (Mart.) Griseb.	0.3	10.3		0.7						11.3d
Total density	555.3	652.7	49.7	26	20.3	33	8	24	23	1392

Based on the spatial representation of the weed species across all 27 plots, the weed species present in the different plots were further analyzed by PCoA plots of axis 1 and 2 (Fig. 1). The principal axis 1 and 2 obtained in the analysis accounted for 55.8% and 14.1% of the total variation, respectively (cumulative value = 69.9%). The 27 plots were divided into three groups, with the three CK plots comprising the first group, the three PK plots the second group, and the remaining plots the third group. No obvious distinctions were observed in the weed-community composition of plots treated with a combined application of N, P, and K, regardless of their organic or chemical origin or the additional application of lime or rice straw.

Logarithmic correlation analyses showed a significant relationship between the axis 1 and both soil-available N and light transmittance (Fig. 2 and Fig. 3). The relationship coefficient between axis 1 and soil-available N was 0.8377, while that between axis 1 and light transmittance was 0.6802. This result suggested that soil-available N, rather than light intensity, is the most important factor determining weed-community composition. This finding may be relevant to the observation that the competition for soil-available N between late rice and weeds was more serious than for early rice. This result was in agreement with other reports, in which N was usually considered the most important factor (Freyman et al., 1989; Moss et al., 2004). The value of axis 1 increased with increasing light transmittance and decreased with increasing available N. There was also a significant logarithmic correlation between axis 2 and soil-available P (Fig. 4) that increased with increasing available P.

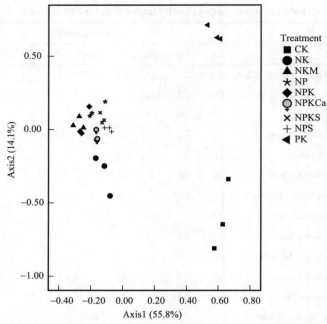

Fig. 1　Ordination plot (PCoA, principal coordinate analysis) of weed communities under different treatments

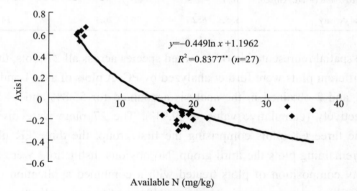

Fig. 2　Relationship between available N and axis 1 of the PCoA

$y = -0.449\ln x + 1.1962$
$R^2 = 0.8377^{**}$ ($n=27$)

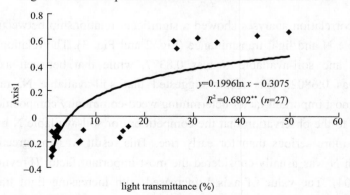

Fig. 3　Relationship between light transmittance and Axis 1 of the PCoA

$y = 0.1996\ln x - 0.3075$
$R^2 = 0.6802^{**}$ ($n=27$)

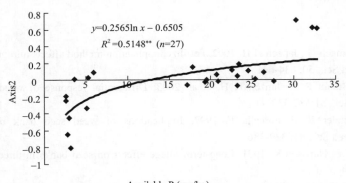

Fig. 4 Relationship between available P and axis 2 of the PCoA

3. *Implications of fertilizers management for weed control*

Fewer studies exist on the effect of long-term fertilization on weed community composition in paddy soil. The results of our study illustrated that the unfertilized control plots and PK-treated plots resulted in soil available N deficiencies. And more agricultural weeds species and higher densities were found in these plots during late rice growth. The more weed species and higher densities resulted in the competition for soil nutrients and light. This also implied that a greater herbicide dose should be required to kill them in the no N-treated plots. The balanced applications of N, P, and K reduced weed species and densities during late rice growth. The additional application of rice straw provided the largest effect on reducing weed species. Results of our study suggested that some weed species growing worst under the balanced fertilzation plots may alter the relative abundance of weed species and species density. Fewer weed species and lower densities in paddy field would help reduce the competition for soil nutrients and light between rice and weed, which should improve efficiency of rice production. Significantly negative effect of soil available N level on weed density was found in this study during the late rice-growing season, which was in contrast with the previous reports (Blackshaw et al., 2003; Cathcart et al., 2004). It probably was due to that high soil N content might have favored rice growth and repressed weed growth in the rice system, especially conbined application of chemical fertilizers with additional application of rice straw or swine manure.

Therefore, nutrient management may be an important factor to change weed communities and density. Fertilizer banding as a weed-control strategy should be recommended in agricultural weed management programs. Our results indicated differences in weed sensitivity to long-term different fertilization in paddy soil resulting from soil available N, light intensity and soil available N content when grown under natural conditions. The balanced fertilization of N, P and K, especially combined application of chemical fertilizer with rice straw or swine manure, can be integrated into rice production with associated weed suppression, increase in yield, and reduction in herbicide inputs. Therefore, in order to maintain soil fertility and promote agricultural byproduct utilization, combined application of chemical fertilizer with rice straw or organic manure should be introduced to farmers as an effective agricultural practice in cultivated paddy soil.

References

Blackshaw R E, Semach G, Janzen H H. 2002. Fertilizer application method affects nitrogen uptake in weeds and wheat. Weed Sci., 50: 634-641

Blackshaw R E, Brandt R N, Janzen H H, et al. 2003. Differential response of weed species to added nitrogen.Weed Sci., 51:532-539

Buhler D D, Hartzler R G, Forcella F. 1997. Implications of weed seed bank dynamics to weed management.Weed Sci., 45: 329-336

Cardina J, Regnier E, Harrison K. 1991. Long-term tillage effects on seed banks in three Ohio soils. Weed Sci., 39: 186-194

Carlson H L, Hill J E. 1985. Wild oats competition with spring wheat: plant density effects. Weed Sci., 33: 176-181

Cathcart R J, Chandler K, Swanton C J. 2004. Fertilizer nitrogen rate and the response of weeds to herbicides. Weed Sci., 52:291-296

Clements D R, Weise S F, Swanton C J. 1994. Integrated weed management and weed species diversity. Phytoprotection, 75:1-18

Davis A S. 2007. Nitrogen fertilizer and crop residue effects on seed mortality and germination of eight annual weed species. Weed Sci., 55: 123-128.

Editorial Committee of China Agriculture Yearbook. 2007. China Agriculture Yearbook. Beijing: Chinese Agriculture Press

Freyman S, Kowalenko C G, Hall J W. 1989. Effect of nitrogen, phosphorus and potassium on weed emergence and subsequent weed communities in south coastal British Columbia. Can. J. Plant Sci., 69: 1001-1010

Fryer J D, Evans S A, 1970. Weed control Handbook. Principles. England: The Newdigate Press LCT: 220-221

Ghersa C M, Roush M L, Radosevich S R, et al. 1994. Coevolution of agroecosystems and weed management. BioScience, 44: 85-94

Milberg P, Hallgren E. 2004. Yield loss due to weeds in cereals and its large-scale variability in Sweden. Field Crops Res., 86: 199-209

Moss S R, Storkey J, Cussans J W, et al. 2004. The Broadbalk long-term experiment at Rothamsted: what has it told us about weeds? Weed Sci., 52: 864-873

Mulugeta D, Stoltenberg D E. 1997. Weed and seedbank management within tegrated methods as influenced by tillage. Weed Sci., 45: 706-715

Norris R F. 2007. Weed fecundity: current status and future needs. Crop Prot., 26: 182-188

O'Donovan J T, Mandrew D W, Thomas A G. 1997. Tillage and nitrogen influence weed population dynamics in barley. Weed Technol., 11:502-509

Page A L, Miller R H, Keeney D R. 1982. Methods of Soil Analysis, Part 2. Chemical and Microbiological Properties. Agronomy 9. 2nd Ed. Madison, Wisconsin, USA: American Society of Agronomy, Inc.

Santos B M, Dusky J, Stall W M, et al. 2004. Effects of phosphorus fertilization on the area of influence of common lambsquarters (*Chenopodium album* L.) in lettuce. Weed Technol., 18:1013-1017

Savary S, Srivastava R. K, Singh H M, et al. 1997. A characterisation of rice pests and quantification of yield losses in the rice-wheat system of India. Crop Protect., 16: 387-398

Swanton C J, Shrestha A, Roy R C, et al. 1999. Effect of tillage systems, N, and cover crop on the composition of weed flora. Weed Sci., 47: 454-461

Theaker A J, Boatman N D, Froud-Williams R J. 1995. The effect of nitrogen fertilizer on the growth of Bromus sterilis in field boundary vegetation. Agric. Ecosyst. Environ., 53: 185-192

Ueji M, Inao K. 2001. Rice field herbicides and their effects on the environment and ecosystems. Weed Biology and Management, 1: 71-79.

Weaver S E, Ivany J A. 1998. Economic thresholds for wild radish, wild oat, hemp-nettle and corn spurry in spring barley. Can. J. Plant Sci., 78: 357-361

Yao B, Zhang C N. 2008. Effect of three herbicides on microbial biomass C, N and respiration in paddy soil. Ecology and Environment, 17: 580-583

Yin L C, Cai Z C, Zhong W H. 2005. Changes in weed composition of winter wheat crops due to long-term fertilization. Agric. Ecosyst. Environ., 107: 181-186

Zhao C F, Chen G J, Wang Y H. 2007. Genetic variation of hippophae rhamnoides populations at different altitudes in the Wolong Nature Reserve Based on RAPDs. Chin. J. Appl. Environ. Biol., 13:753-758

4.5 Risk assessment of potentially toxic element pollution in soils and rice (*Oryza sativa*) in a typical area of the Yangtze River Delta

4.5.1 Introduction

The increasing trend of PTE concentrations in the environment has caused great concern worldwide. PTEs in the environment may arise from natural or anthropogenic routes and their concentrations are elevated due to solid-waste disposal, atmospheric deposition, fertilizer and pesticide use, and the application of sewage sludge and wastewater irrigation on land (Cui et al., 2004, 2005; Wilson and Pyatt, 2007; Zheng et al., 2007a; Khan et al., 2008). The spatial distribution of PTEs in the environment can be related to the sources of the elements.

Soil contamination by PTEs has become one of the most important environmental problems in many developed and developing countries in recent years (Rodriguez et al., 2007). With the economic development of society, many PTEs from industrial or agricultural sources have entered crop tissues due to crop absorption from contaminated soils, wastewater irrigation, and polluted air (Huang et al., 2007; Chary et al., 2008). Excessive PTEs not only restrain the growth of crops, but also affect their quality and safety throughout the food chain. Furthermore, the chronic low-level intake of PTEs can pose a detrimental effect on human health which may not be easily reversed through medical treatment (Huang et al., 2007). As a staple food crop in China, rice uses 24% of all agricultural land to produce about 40% of the overall yield (Hu et al., 2002), suggesting that the concentrations of PTEs in rice may be a critical problem for food safety in China.

The Yangtze River Delta is a rapidly developing region with a high population density, where PTE is one of the most important environmental issues. Heavy metal pollution has been reported for the Yangzhong district and the HJH (Hangzhou-Jiaxing-Huzhou) water-network plain in this region (Huang et al., 2007; Liu et al., 2006). However, research on this topic is lacking in the high-risk area of Changshu City, one of the most rapidly developing areas in the Yangtze River Delta.

The primary objectives of this research were to analyze PTE concentrations and their spatial distributions in soils and rice samples from the farmland surrounding industrial areas in Changshu City, a typical area in the Yangtze River Delta, and to evaluate health effects from PTE due to daily consumption of rice. The results of our study may provide some insight into

PTE accumulation in the agricultural ecosystem and serve as a basis for comparison to other regions both in China and worldwide.

4.5.2 Materials and Methods

1. *The study area*

Changshu is an important city in southern Jiangsu Province in East China, with a total area of 1264 km² and a population of 1.05 million at the end of 2005. The agricultural area is 58531 ha, consisting of paddy fields (44359 ha) and dry-lands (14172 ha). The region is located at 31°33′-31°50′ N and 120°33′-121°03′ E (Fig. 1) at an elevation of 3 to 7 m above sea level and the topography is flat. The area has a warm and humid subtropical climate with an annual temperature of 15.4℃, rainfall of 1054 mm, and average sunshine of 2130 h. The soils are mainly paddy soils formed on calcareous deposits of the Yangtze River and the farming systems are highly mechanized. The area is therefore important for food production. However, many industrial operations have started since the 1980s and have threatened the safety of agricultural production. For example, high concentrations of heavy metals were detected in rice from a typical E-waste (electronic and electrical waste) recycling area by Fu et al. (2008).

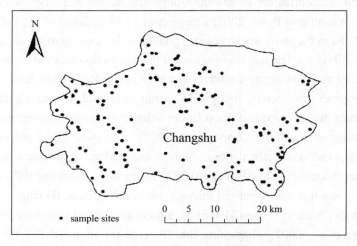

Fig. 1 Distribution of sampling sites in Changshu City

2. *Sampling and sample preparation*

The sampling sites were selected on the basis of terrain and landform as well as the layout and structure of industries in Changshu City, and all sampling sites were located in paddy fields in high-risk areas including fields near industries such as printing and dyeing, engineering, electroplating, thermoelectricity, electronics, metallurgy, accumulators, chemicals, pesticides, and livestock and poultry breeding farms. These industries usually produce a variety of wastes and pollutants that enter the fields in their vicinity. A total of 155 surface soil samples from 155 sampling sites and corresponding 155 rice grain samples were collected from the study area (Fig. 1).

At each site a monolith was dug for the collection of surface soil and a composite of three subsamples was taken from one paddy field. All soil samples (3 kg each) were taken to a depth of 20 cm from each field and air-dried, and stones and coarse plant roots or residues were removed. Soil samples were thoroughly mixed, crushed, passed through a 2-mm mesh sieve and 0.149-mm mesh sieve, and then stored in polyethylene bags at ambient temperature prior to chemical analysis.

Rice samples (2 kg each) were also collected in triplicate from the corresponding sites. Replicate samples were washed with deionized water and air-dried to constant weight. The rice hulls were separated from the white rice grains using a mill. The white rice and rice hull samples were ground and stored in polyethylene bags at ambient temperature for further analysis.

3. *Analytical procedures*

Soil pH was measured in a 1 : 2.5 soil-to-water suspension using a pH meter (HI98182, Hanna Instruments, Italy) with an accuracy of 0.001 pH unit. The pH values fell within the range 4.83-8.19 and the arithmetic mean was 6.33. Both available and total concentrations of Pb, Zn, Cu, Cd and Cr as well as total Hg and As were determined in the soil samples. For determination of total Hg and As, soil sub-samples (1 g) were digested with a mixed acid (1 : 3 : 4 HNO_3 : HCl : H_2O) (Shi et al., 2005). The total concentrations of the other PTEs in the soil samples (1 g) were determined by digestion with counter-aqua regia (3 : 1 HNO_3 : HCl) (Burt et al., 2003). The available heavy metals (Pb, Zn, Cu, Cd and Cr) in the soil samples were extracted using the DTPA method as described by Lindsay and Norvell (1978) in which 10 g of air-dried soil (<2 mm) was shaken for 2 h with 20 mL of a solution containing 0.005 mol/L diethylenetriaminepentaacetic acid (DTPA), 0.01 mol/L $CaCl_2$, and 0.1 mol/L triethanolamine (TEA) at pH 7.30.

To determine Hg and other PTEs absorbed by rice, white grain samples (1 g) were digested in HNO_3 and H_2O_2 (Han et al., 2004, 2006) and in 3 : 1 HNO_3 : $HClO_4$ (Zu et al., 2004; Li et al., 2006), respectively. The concentrations of the PTEs (except for Hg and As) in the solutions (obtained by digestion or DTPA extraction) were determined by inductively coupled plasma-atomic emission spectrometry (ICP-AES) (IRIS Advantage, Thermo Jarell Ash Co., USA) (Demirak et al., 2006; Li et al., 2006). The detection limits for Cr, Cu, Ni, Pb, Cd, and Zn in the extracts were 1.0, 1.5, 1.0, 3.0, 0.3 and 0.2 µg/L, respectively. For the solutions obtained by digestion or DTPA extraction, when ICP-AES was insufficiently sensitive for the measurement, the PTE concentrations were determined by graphite furnace atomic absorption spectrometry (GF-AAS) (Varian SpectrAA 110/220, Varian, Inc., USA) (Zheng et al., 2007a). The concentrations of Hg and As in all the digest solutions were determined by hydride generation atomic fluorescence spectrometry (HG-AFS) (AFS-930, Beijing Titan Instruments Co., Ltd., China) (Fu et al., 2008), in which the detection limits for Hg and As were below 0.001 and 0.01 µg/L, respectively.

All acids used in this study were ultra-pure grade and other reagents were analytical grade. Deionized water was obtained from a Millipore machine (18.3 MΩ · cm resistivity). All glassware was soaked in nitric acid solution for 24 h and rinsed with deionized water. Three replicates were used for the analysis of both soil and rice grain samples. Reagent blanks, a standard reference soil sample (GBW07403) and standard plant samples (GBW07602 and GBW10010) were employed in the analysis to ensure accuracy and precision.

4. Bio-accumulation factor

The BAF (bio-accumulation factor, the ratio of the concentration of an element in the grain to that in the corresponding soil) was calculated for each rice sample to quantify the bio-accumulation effect of rice in the uptake of PTEs from the soils (Liu et al., 2005). The BAF was computed as:

$$\mathrm{BAF} = \frac{C_r}{C_s}$$

Where, C_r and C_s represent the PTE concentrations in extracts of rice grain and soils, respectively, on a dry weight basis.

5. Human risk assessment of PTEs through rice consumption

(1) *Estimated daily intake (EDI)*

Rice is the staple food for daily consumption in the region according to dietary custom and the dietary intake survey by Zhong et al. (2006). Exposure via rice is therefore very important to the health of the local population. According to some reports (Liu et al., 2005; Fu et al., 2008) the daily intake of PTEs depends on both the concentration of the element in food and the daily food consumption. Furthermore, human body weight can also influence tolerance of the pollutants (Fu et al., 2008). Considering these factors, Zheng et al. (2007a) proposed that the estimated daily intake (μg/(kg·d) B_w) of PTEs via rice consumption can be calculated as:

$$\mathrm{EDI} = \frac{C \times C_{\mathrm{on}} \times \mathrm{EF} \times \mathrm{ED}}{B_w \times \mathrm{AT}}$$

Where, C (μg/g) is the concentration of PTEs in the contaminated rice; C_{on} (g/(person · d)) is the daily average consumption of rice in the region; B_w (kg/person) represents body weight; EF is exposure frequency (365 days/a); ED is exposure duration (70 years, equivalent to the average lifespan); and AT is average time (365 days/a number of exposure years, assuming 70 years in this study). The average daily rice intakes of adults and children were considered to be 389.2 and 198.4 g/(person · d), respectively (Zheng et al., 2007a), and average adult and child body weights were considered to be 55.9 and 32.7 kg, respectively, as used in many previous studies (Ge, 1992; Wang et al., 2005; Zheng et al., 2007a,2007b; Khan et al., 2008).

(2) *Target hazard quotient (THQ)*

According to standard EPA methods, the risk of non-carcinogenic effects is expressed as the ratio of the dose resulting from exposure to site media compared to a dose that is believed to be without risk of effects, even in sensitive individuals (Zheng et al., 2007a). This ratio is referred to as the target hazard quotient (THQ). The THQ for the locals through consumption of contaminated rice can therefore be assessed based on the food chain and the reference oral dose (RfD) for each PTE. The applied RfD for Cr, Cu, Pb, Zn, Cd, Hg, and As were 1500, 40, 3.5, 300, 1.0, 0.7, and 50 μg/(kg · d), respectively. Oral reference doses were obtained from the Integrated Risk Information System (US-EPA, IRIS, 2008), with the exception of Pb, Hg, and As for which we used the formula RfD = PTWI / 7, where PTWI is the provisional tolerable weekly intake (μg/(kg · d)) by the Joint FAO/WHO Expert Committee on Food Additives (JECFA) (FAO/WHO, 1997; UNEP/FAO/WHO, 1992). The THQ is determined by

$$THQ = \frac{EDI}{RfD}$$

If the value of THQ is less than one it is assumed be safe for risk of non-carcinogenic effects. If it exceeds one it is believed that there is a chance of non-carcinogenic effects, with an increasing probability as the value of THQ increases (Zheng et al., 2007a).

(3) *Hazard index (HI)*

Some previous studies (Harrison and Chirgawi, 1989; Zheng et al., 2007a) reported that exposure to two or more pollutants may result in additive and/or interactive effects. THQs can therefore be summed across constituents to generate a hazard index (HI) for a specific receptor/pathway (e.g., diet) combination. The HI is a measure of the potential risk of adverse health effects from a mixture of chemical constituents in rice. The HI through daily average consumption of rice for a human being was calculated as follows:

$$HI = \sum_{n=1}^{i} THQ_i$$

6. *Data analysis*

All data were analyzed using Microsoft Excel and the SPSS 13.0 for Windows statistical package. Arithmetical mean and range were used to assess the contamination levels of PTEs in soils and rice. Correlation analysis was used to determine the relationship between PTE concentrations in soils and rice. Independent variables were determined for a linear regression model at 95% ($P<0.05$) and 99% ($P<0.01$) confidence levels with regression coefficients designated as r.

4.5.3 Results and Discussion

1. *PTE accumulation in soils*

Total PTE concentrations in the soils are presented in Table 1. Zinc had the highest total concentration (90.1 mg/kg) and Cd had the lowest (0.168 mg/kg). Total Cr and Pb were around 50 mg/kg, total Cu about 30 mg/kg, As 8.5 mg/kg, and Hg 0.555 mg/kg. Compared to surface soils in Yangzhong district (Huang et al., 2007), the concentrations of Hg and Pb in the soils of Changshu City were higher, although other elements studied in Changshu showed lower concentrations than in Yangzhong district. This indicates that the Hg and Pb contamination of Changshu City may be more serious than in Yangzhong district. The contamination may be derived from industrial factories in the vicinity of the fields.

DTPA-extractable metal concentrations in soil samples were in the order Cu > Pb > Zn > Cr > Cd (Table 1). In general, the DTPA-extractable forms accounted for only a small portion of the total metals present. In addition, the order of DTPA-extractable forms of these metals was different from the order of total soil concentrations, indicating different mobilities and bioavailabilities of the metals in the surface soil. Correlation analysis demonstrates significant relationships between DTPA-extractable forms and the total concentrations of Cu (0.548, $P < 0.001$), Pb (0.704, $P < 0.001$), Zn (0.311, $P < 0.001$), and Cd (0.822, $P < 0.001$). These results are in agreement with many published reports, suggesting that the mobility and bioavailability of these heavy metals are influenced by the total concentrations.

Table 1 Concentrations (mg/kg) of total and available PTEs in soils

			Cr	Cu	Pb	Zn	Cd	Hg	As
Total	Changshu ($n = 155$)	Mean	53.4	30.5	44.5	90.1	0.168	0.555	8.60
		Std. Deviation	18.3	18.0	22.9	46.4	0.181	1.231	1.93
		Range	18.0-112	3.1-89.3	13.9-178	20.3-355	0.020-1.36	0.109-14.8	4.46-16.10
	Yangzhong ($n=76$)	Mean	77.2	33.9	35.7	98.1	0.3	0.2	10.2
		Range	65.2-91	23-52.1	28.3-71	77.2-133	0.22-0.71	0.08-0.35	6-16
	Background value[①]		90	35	35	100	0.2	0.2	15
	MAC		250	50	80	200	0.30	0.30	30
DTPA extractable	Changshu ($n = 155$)	Mean	0.110	7.31	4.44	1.50	0.109	-	-
		Std. Deviation	0.051	3.44	5.01	1.78	0.089	-	-
		Range	0.003-0.277	2.19-25.0	1.14-54.4	0.15-12.3	0.024-0.746	-	-

Note: ①National Environmental Protection Agency of China (1995).

According to the Environmental Quality Standard for Soils (National Environmental Protection Agency of China, 1995), our arithmetical mean concentration of soil Hg is approximately three times the threshold value of the nationwide natural background, and Pb is also slightly above its threshold value. However, the other PTE concentrations in the soils were below their threshold values. These results suggest that Hg and Pb in the study area might be affected by anthropogenic activities. The extent of soil contamination can also be evaluated by comparing the maximum allowable concentrations (MAC) (National Environmental Protection Agency of China, 2006) of the metals in agricultural land. As soil pH partially governs the speciation and bioavailability of heavy metals, MAC values are adjusted according to soil pH (Fu et al., 2008). The pH values of soil samples in Changshu City ranged from 4.83 to 8.19 with an arithmetic mean of 6.33, which corresponds to Grade I of agricultural land environmental quality standards for edible agricultural products (HJ332-2006). As shown in Table 1, the arithmetic mean of Hg exceeded its MAC level by a factor of 1.9 but the mean values of the other PTEs were below their MAC levels. Moreover, Hg concentrations in 83 soil samples were above the MAC, accounting for 53.5% of all soil samples compared with the corresponding values for Cu of 21 soil samples (13.5%), Cd of 11 soil samples (7.1%), Pb of 10 soil samples (6.5%), and Zn of 5 soil samples (3.2%). This indicates that many different point-pollution sources might exist in Changshu City. These sources may be from dispersed industries. We believe that the soils in the industrial area of Changshu City are primarily contaminated by Hg, followed by Cu, Cd, Pb, and Zn, whereas Cr and As may not be regarded as pollutants. This suggests that there is a discrete contamination source of the five metal pollutants in the vicinity of the fields from which the samples were taken.

2. PTE accumulation in rice grain

The PTE concentrations in rice grain samples from the industrial area of Changshu City decreased in the order Zn > Cu > Cr > As > Pb > Cd > Hg (Table 2). The mean concentrations of Cr and As in rice were higher than those from Taizhou and in commercial rice in China (Fu et al., 2008). The concentration of Cu in the rice samples in the present study was also slightly higher than in commercial rice but was lower than that from Taizhou. Concentrations of Pb, Cd, and Hg were all below those from Taizhou and commercial rice. Importantly, the maximum concentrations of Cr (0.742 mg/kg), Cu (7.77 mg/kg), and As (0.587 mg/kg) were approximately two-fold greater than those from Taizhou and commercial rice, but the concentrations of Pb, Cd and Hg fell within the ranges determined by Fu et al. (2008) Table 2 also shows detailed results for PTE concentrations in the rice grain samples compared with previous studies. The results indicate that the concentrations of Cr, As, Pb, Cd, and Hg in rice from the industrial area of Changshu may have been affected by industrial activities. However, the high concentrations of Pb, Cd, and Hg in rice grain from Taizhou may result from E-waste utilization in Taizhou, which may have introduced large amounts of Pb and Cd, but had low recycling efficiencies (Puckett and Smith, 2002). The erosion of these metals from E-waste recycling activities may be discharged into the peripheral environment, which would therefore reflect contamination by these elements (Fu et al., 2008).

Table 2 Comparison of the concentrations (mg/kg dry matter) of PTEs in rice grain in this study with data available from some previous studies

		Cr	Cu	Pb	Zn	Cd	Hg	As
Changshu (n=155)	Mean	0.292	3.84	0.171	19.1	0.019	0.0145	0.199
	Std. Deviation	0.140	1.07	0.126	3.26	0.021	0.0103	0.114
	Range	0.024-0.742	0.87-7.77	LOD[①]-0.957	9.10-28.3	LOD[①]-0.201	LOD[①]-0.0602	LOD[①]-0.587
Taizhou[②] (n=13)	Mean	0.107	4.26	2.042	-	0.224	0.022	0.155
	Range	0.006-0.279	3.04-5.18	0.256-2.602	-	0.012-0.661	0.016-0.068	0.095-0.308
Commercial rice in China[①] (n=4)	Mean	0.199	3.33	0.356	-	0.035	0.029	0.070
	Range	0.062-0.424	2.81-4.48	0.167-0.745	-	0.004-0.070	0.024-0.036	0.047-0.122
Maximum levels of contaminants in foods[③]		1.0	10	0.2	50	0.2	0.02	0.7

Notes: ① Below the limit of detection. ② Fu et al. (2008); others obtained in this study. ③ Maximum levels of contaminants in foods (GB 13106–1991; GB 2762–2005; GB 15199–1994; GB 4810–1994).

Table 2 lists the maximum allowable concentrations (MAC) of Cr, Cu, Pb, Zn, Cd Hg, and As in foods recommended by the Chinese National Standard Agency. The mean concentrations of these elements in the rice grain were below their maximum allowable levels. However, there remained 46 rice samples (29.7%) containing Pb in excess of its maximum level in foods. For Hg and Cd the number of samples was respectively 32 (20.6%) and 1 (0.7%). Of all the contaminated rice grain samples, the highest concentrations of Pb, Hg, and Cd were 0.957, 0.060, and 0.201 mg/kg, respectively.

Mercury and As are considered two of the most important toxic elements found in the environment, because of its potential risk to ecosystems and human health. More importantly, they have potentially the most severe health effects from chronic low level exposure. This study focused on these two elements in the rice grain samples. Concentrations of Hg in rice grain ranged from 0 to 0.060 mg/kg with a mean of 0.015 mg/kg. A considerable amount of rice grain samples (20.6%) were above the MAC, although the mean was below its MAC (0.020 mg/kg). Mercury is a highly toxic element because of its accumulative and persistent character in the environment and biota (Qiu et al., 2006). The Hg in rice grain of the study area should therefore be the really remarkable PTE.

Food surveys had revealed that rice accumulates the highest amount of As of all cereals (Marin et al., 1993; Tao and Bolger, 1998), largely because of the high bioavailability of As under reduced soil conditions (Marin et al., 1993). Arsenic concentrations in rice grains varied from 0 to 0.578 mg/kg with a mean of 0.199 mg/kg. The concentrations of all samples were below the MAC (0.7 mg/kg) but were beyond the global "normal" range (0.082-0.202 mg/kg) in rice grains reported by Zavala and Duxbarg (2008). It is well-known that the speciation of As plays an important role in determining As toxicity to humans (Zavala et al., 2008), inorganic species (As_i) were thought to be the most toxic in As speciation (Petrick et al., 2000). There is no EU, US or WHO limits for either total As or As_i in food until recently (Francesconi, 2007). China is virtually the first and only country to have MAC for As of food contamination, and Chinese standards for As in rice are probably the most stringent in the world (Zhu et al., 2008a, 2008b). Zavala et al. (2008) reported that rice from Asia and Europe was the inorganic As type. Williams et al. (2005) also showed that southeast Asian rice where As_i was the dominate species. According to MAC (0.150 mg/kg) for As_i of China, there were probably 94 rice samples (60.6%) in the study area containing As_i in excess of its MAC. Furthermore, it seems that the Chinese statutory level may be too high for protection of human health (Zhu et al., 2008a, 2008b). As a result, the potential risk of As in rice grain of the study area to human beings should be more serious than we had thought. Rice hull is a byproduct of polishing whole grain rice, it can be used as a human dietary supplement or an animal feed (Sun et al., 2008). The As in rice hulls should therefore be concentrated. From our analysis, concentrations of As in rice hulls ranged from 0.017 to 3.033 mg/kg with a mean of 0.706 ± 0.502 mg/kg. Total As levels were much higher in hull samples than in white rice obtained from the same whole grain rice, this agrees well with some previous studies (Ren et al., 2006; Rahman et al., 2007). Meharg et al. (2008) and Zhu et al. (2008b) showed that whole grain rice had a higher proportion of total As content than white rice, due to the localization of As_i in the bran layer. Consequently, As in rice hulls should receive additional consideration. Moreover, there was no significant correlation between the rice grain and rice hull ($r = 0.015$, $P = 0.848$), this is in agreement with the result of Sun et al. (2008).

3. *Spatial distribution of PTEs in soils and rice grain*

The spatial distribution of PTEs in soils of Changshu demonstrates their distinct geographical distribution patterns (Fig. 2). The spatial distribution maps for Cr, Cu, Pb, Zn, and Cd show similar geographical trends, with low concentrations in the north and high concentrations in the south. The spatial patterns of Hg and As show similar geographical trends, being high in the southwestern area. In general, the spatial distributions of PTEs in Changshu are consistent with the distribution of industries. In Changshu the majority of the industrial factories are located at the southern area. Consequently, wastewater, solid wastes, and waste gases from these factories may have acted as sources of soil pollution. Moreover, many major roads are in the southern area and these may also contribute to soil pollution. Our observations suggest that the high concentrations of Hg and As in the southwest may be derived from some industries and pesticide use. The correlations among the PTEs were also analyzed (Table 3). The relationships among Cr, Cu, Pb, Zn, and Cd were all highly significant, implying that they had the same pollution sources. The relationship between Hg and As, however, was not statistically significant. The correlations among seven of the elements is exactly in accordance with the similarities in their spatial distributions.

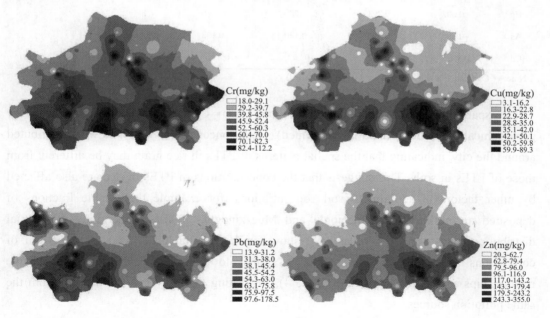

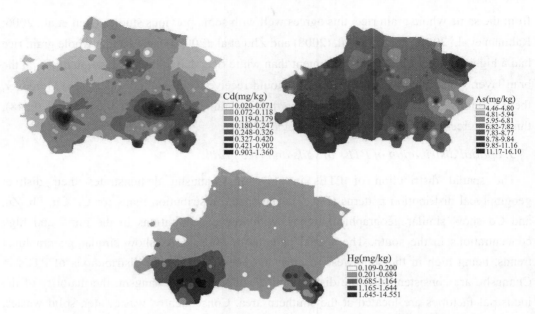

Fig. 2　Spatial distribution of PTE contents in surface soils in Changshu City

Table 3　Pearson correlations of total PTEs in soils ($n = 155$)

	Cr	Cu	Pb	Zn	Hg	As	Cd
Cr	1						
Cu	0.683**	1					
Pb	0.649**	0.510**	1				
Zn	0.800**	0.717**	0.566**	1			
Hg	−0.002	0.293**	0.088	0.089	1		
As	−0.080	0.009	0.130	−0.152	0.086	1	
Cd	0.350**	0.362**	0.273**	0.431**	0.072	−0.068	1

Note: ** Correlation is significant at the 0.01 level (2-tailed).

The spatial patterns of PTEs in rice grain from Changshu were irregular in their geographical distribution (data not presented). High concentrations of PTEs were distributed around the city, indicating that the spatial patterns of PTEs in rice grain may be different from those of PTEs in soils. This suggests that the concentrations of PTEs in rice are also affected by other factors in addition to soil concentrations, for example the soluble fractions of deposited air pollutants (Cizmecioglu and Muezzinoglu, 2008). Furthermore, the spatial distribution maps for Cu, Zn, and Cd show similar geographical trends. The pattern of correlations among these three metals is similar to their spatial distributions and the relationships are highly significant (Table 4), suggesting that they may be derived from the same pollution sources.

Table 4 Pearson correlations of PTEs in rice grain (n = 155)

	Cr	Cu	Pb	Zn	Hg	As	Cd
Cr	1						
Cu	0.022	1					
Pb	0.119	0.141	1				
Zn	0.125	0.671**	0.221**	1			
Hg	0.005	−0.024	0.032	−0.010	1		
As	0.197*	0.017	0.157	0.165*	−0.164	1	
Cd	−0.038	0.430**	0.148	0.333**	−0.109	0.056	1

Notes: * Correlation is significant at the 0.05 level (2-tailed); ** correlation is significant at the 0.01 level (2-tailed).

4. PTE transfer from soils to rice

The BAF values of rice in Changshu were in the order Zn (0.258) > Cu (0.196) > Cd (0.178) > Hg (0.047) > As (0.025) > Cr (0.006) > Pb (0.005), indicating that the accumulation capacity of these PTEs is different. Moreover, a significant negative correlation (r = −0.206, P < 0.01) was detected between Zn BAF values and soil pH. Similarly, a significant positive correlation (r = 0.212, P < 2580.01) was found between Cd DTPA-extractable concentrations in soils and rice grain samples.

5. Daily PTE intakes via rice consumption and the risk to human health

(1) Dietary PTE intake

Although there are many pathways of human exposure to PTEs, rice consumption has been identified as one of the major pathways. Therefore, as rice consumption rises in Changshu City, human exposure to PTE contamination increases. Table 5 shows the dietary intake of PTEs through rice consumption for adults and children in Changshu on the assumption that the local population consumes mostly local rice. The mean dietary intakes of Cr, Cu, Pb, Zn, Cd, Hg, and As through rice are estimated to be 2.0, 26.7, 1.2, 132.8, 0.1, 0.1, and 1.4 μg/(kg·d) for adults, and 1.8, 23.3, 1.0, 115.7, 0.1, 0.1, and 1.2 μg/(kg·d) for children. The mean estimated dietary intakes (MEDIs) of PTEs for adults are higher than those for children and this is in agreement with a previous study (Zheng et al., 2007a). The MEDIs of PTEs are all below the RfD for adults and children. For Cu, 6 out of 155 samples (3.9%) exceed the RfD for adults but for Pb and Cd, there are only 3 samples (1.9%) and 1 sample (0.6%) respectively which exceed the RfD for adults. The maximum daily intake of Cu, Pb, and Cd for adults from rice is 54.1, 6.7 and 1.4 μg/(kg·d) (calculated from the maximum concentrations of Cu, Pb and Cd in rice) in Changshu, which are 0.4-fold, 0.9-fold and 0.4-fold higher than the RfD of these metals. Only 3 samples (1.9%) of Cu are beyond the RfD for children, and, for Pb and Cd, there are 3 samples (1.9%) and 1 sample (0.6%). The maximum daily intakes of Cu, Pb and Cd from rice for children are 47.1, 5.8, and 1.2 μg/(kg·d) (calculated from the maximum concentrations of Cu, Pb, and Cd in rice) in Changshu, which are 0.2-fold, 0.7-fold, and 0.2-fold higher than the RfD of these metals. It must be noted that some individuals may consume more than twice the average amount of rice (Fu et al., 2008), and their exposure to these PTEs will be correspondingly higher.

Table 5 Estimated daily intake by human beings and potential health risk of PTEs due to rice consumption

Individuals	Element	RfD[①]	MEDI[②]	MinI[③]	MaxI[④]	% E-RfD[⑤]	THQ	HI
Adults	Cr	1500	2.0	0.2	5.2	0	0.001	
	Cu	40	26.7	6.1	54.1	3.9	0.688	
	Pb	3.5	1.2	0	6.7	1.9	0.343	
	Zn	300	132.8	63.3	196.7	0	0.443	1.726
	Cd	1	0.1	0	1.4	0.6	0.100	
	Hg	0.7	0.1	0	0.4	0	0.143	
	As	50	1.4	0	4.1	0	0.028	
Children	Cr	1500	1.8	0.1	4.5	0	0.001	
	Cu	40	23.3	5.3	47.1	1.9	0.583	
	Pb	3.5	1.0	0	5.8	1.9	0.286	
	Zn	300	115.7	55.2	171.4	0	0.386	1.523
	Cd	1	0.1	0	1.2	0.6	0.100	
	Hg	0.7	0.1	0	0.4	0	0.143	
	As	50	1.2	0	3.6	0	0.024	

Notes: ① Reference oral dose (μg/(kg·d)). ② Arithmetical mean estimated daily intake (μg/(kg·d)). ③ Estimated minimum daily intake (μg/(kg·d)). ④ Estimated maximum daily intake (μg/(kg·d)). ⑤ The percentage of PTE in rice grain samples which exceed the RfD.

(2) *Potential health risk of individual PTEs*

The THQs of individual PTEs through rice consumption for the residents (adults and children) in Changshu are listed in Table 5. There are no values for an individual PTE beyond 1 through rice consumption. For adults, the THQ of PTEs from rice consumption is in decreasing order Cu > Zn > Pb > Hg > Cd > As > Cr. The THQ of PTEs for children is similar to that for adults. Our results suggest that Cu ingestion has the highest potential health risk of adverse effects for adults and children and Cr ingestion has the minimum risk for adults and children. In view of the largest and most comprehensive review of chronic As exposure and associated cancer risks to date, US drinking water maximum contaminant level fell from 0.050 mg/L to the WTO endorsed level of 0.010 mg/L (Zhu et al., 2008b). Thus the potential health risk of As for adults and children would be increased.

(3) *Potential health risk of combined PTEs*

The Hazard Index (HI) for a specific receptor/pathway combination can be calculated (Zheng et al., 2007a). The HI values through consumption of rice for adults and children in Changshu are 1.726 and 1.523, respectively (Table 5). This indicates that adults and children may experience adverse health effects. The HI value for adults through rice consumption is slightly higher than that for children, suggesting that the risk of non-carcinogenic effects for adults is slightly higher than that for children. The relative contributions of Cr, Cu, Pb, Zn, Cd, Hg, and As to the HI are 0.1%, 38.7%, 19.9, % 25.7%, 5.8%, 8.3%. and 1.6% for adults, and 0.1%, 38.3%, 18.8%, 25.3%, 6.6%, 9.4%, and 1.6% for children, respectively. Therefore, Cu, Pb and Zn are key components contributing to the potential health risk of non-carcinogenic effects for adults and children, with Cd, Hg, and As being secondary and Cr the least important.

Although the daily intake metals or toxic elements through rice is an important pathway for the dietary exposure of local people to PTEs through food, many studies have reported that human beings are also significantly exposed to metals through other foods such as wheat, vegetables, fruit, fish, meat, eggs, water, and milk (Wang et al., 2005; Zheng et al., 2007a; Chary et al., 2008; Sipter et al., 2008). However, little attention has been paid to inhalation and dermal exposure (Granero and Domingo, 2002; Hough et al., 2004; Nadal et al., 2004; Grasmück and Sholz, 2005; Hellström et al., 2007; Sipter et al., 2008). In the report of Sipter et al. (2008) two exposure pathways were assumed in their study on land use of high-risk areas and the features of metals: ingestion of soil and ingestion of foods. In fact, the potential health risk of metals is far beyond our calculations. Furthermore, our calculations did not consider special groups, such as the elderly, pregnant women, children, and medical patients who are presumably more vulnerable to pollutants with non-carcinogenic or even carcinogenic effects.

References

Burt R, Wilson M A, Mays M D S. 2003. Major and trace elements of selected pedons in the USA. Journal of Environmental Quality, 32: 2109-2121

Chary NS, Kamala CT, Raj D S S. 2008. Assessing risk of heavy metals from consuming food grown on sewage irrigated soils and food chain transfer. Ecotoxicology and Environmental Safety, 69: 513-524

Cizmecioglu S C, Muezzinoglu A. 2008. Solubility of deposited airborne heavy metals. Atmospheric Research, 89: 396-404

Cui Y J, Zhu Y G, Zhai R H, et al. 2004. Transfer of metals from soil to vegetables in an area near a smelter in Nanning, China. Environment International, 30: 785-791

Cui Y J, Zhu Y G, Zhai R H, et al. 2005. Exposure to metal mixtures and human health impacts in a contaminated area in Nanning, China. Environment International, 31: 784-790

Demirak A, Yilmaz F, Tuna A L, et al. 2006. Heavy metals in water, sediment and tissues of Leuciscus cephalus from a stream in southwestern Turkey. Chemosphere, 63: 1451-1458

FAO/WHO. 1997. Food Consumption and Exposure Assessment of Chemicals. Report of FAO/WHO Consultation. WHO, Geneva: 17-25

Francesconi K A. 2007. Toxic metal species and food regulations-making a healthy choice. The Analyst., 132: 17-20

Fu J J, Zhou Q F, Liu J M, et al. 2008. High levels of heavy metals in rice (*Oryza sativa* L.) from a typical E-waste recycling area in southeast China and its potential risk to human health. Chemosphere, 71: 1269-1275

Ge K Y. 1992. The Status of Nutrient and Meal of Chinese in the 1990s. Beijing: Beijing People's Hygiene Press: 415-434 (in Chinese)

Granero S, Domingo J L. 2002. Levels of metals in soils of Alcalá de Henares, Spain: human health risk. Environment International, 28: 159-164

Grasmück D, Sholz R W. 2005. Risk perception of heavy metal soil contamination by high-exposed and low-exposed inhabitants: the role of knowledge and emotional concerns. Risk Analysis, 25: 611-622

Han F X, Sridhar B B M, Monts D L, et al. 2004. Phytoavailability and toxicity of trivatent and hexavalent chromium to Brassica juncea L.Czern. New Phytologist, 162: 489-499

Han F X, Su Y, Monts D L, et al. 2006. Binding, distribution, and plant uptake of mercury in a soil from Oak Ridge, Tennessee, USA. Science of the Total Environment, 368: 753-768

Harrison R M, Chirgawi M B. 1989. The assessment of air and soil as contributors of some trace metals to vegetable plants. III. Experiment with field-grown plants. Science of the Total Environment, 83: 47-63

Hellström L, Persson B, Brudin L, et al. 2007. Cadmium exposure pathways in a population living near a battery plant. Science of the Total Environment, 373: 447-455

Hough R L, Breward N, Young S D, et al. 2004. Assessing potential risk of heavy metal exposure from consumption of home-produced vegetables by urban populations. Environmental Health Perspectives, 112: 215-221

Hu P S, Zhai H Q, Wan J M. 2002. New characteristics of rice production and quality improvement in China. Review of China Agricultural Science and Technology, 4: 33-39 (in Chinese)

Huang S S, Liao Q L, Hua M, et al. 2007. Survey of heavy metal pollution and assessment of agricultural soil in Yangzhong district, Jiangsu Province, China. Chemosphere, 67: 2148-2155

Khan S, Cao Q, Zheng Y M, et al. 2008. Health risks of heavy metals in contaminated soil and food crops irrigated with wastewater in Beijing, China. Environmental Pollution, 152: 686-692

Li M S, Luo Y P, Su Z Y. 2006. Heavy metal concentrations in soils and plant accumulation in a restored manganese mineland in Guangxi, South China. Environmental Pollution, 147: 168-175

Lindsay W L, Norvell W A. 1978. Development of a DTPA soil test for Zinc, Iron, Manganese, and Copper. Soil Science Society of America Journal, 42: 421-428

Liu H Y, Probst A, Liao B H. 2005. Metal contamination of soils and crops affected by the Chenzhou lead/zinc mine spill (Hunan, China). Science of the Total Environment, 339: 153-166

Liu X M, Wu J J, Xu J M. 2006. Characterizing the risk assessment of heavy metals and sampling uncertainty analysis in paddy field by geostatistics and GIS. Environmental Pollution, 141: 257-264

Marin A R, Masscheleyn P H, Patrick W H. 1993. Soil redox-pH stability of arsenic species and its influence on arsenic uptake by rice. Plant Soil, 152: 245-253

Meharg A A, Lombi E, Williams P N, et al. 2008. Speciation and localization of arsenic in white and brown rice grains. Environmental Science & Technology, 42: 1051-1057

Nadal M, Schuhmacher M, Domingo J L. 2004. Metal pollution of soils and vegetation in an area with petrochemical industry. Science of the Total Environment, 321: 59-69

National Environmental Protection Agency of China. 1995. Environmental Quality Standard for Soils. State Environmental Protection Administration, China. GB 15618—1995

National Environmental Protection Agency of China. 2006. Farmland environmental quality evaluation standards for edible agriculture products. State Environmental Protection Administration, China. HJ 332-2006

Petrick J S, Ayala-Fierro F, Cullen W R, et al. 2000. Monomethylarsonous acid (MMA^{III}) is more toxic than arsenite in changing human hepatocytes. Toxicology and Applied Pharmacology, 163: 203-207

Puckett J, Smith T. 2002. Exporting harm: the high-tech trashing of Asia. The Basel Action Network (BNN) Silicon Valley Toxics Coalition (SVTC)

Qiu G L, Feng X B, Wang S F, et al. 2006. Environmental contamination of mercury from Hg-mining areas in Wuchuan, northeastern Guizhou, China. Environmental Pollution, 142: 549-558

Rahman M A, Hasegawa H, Rahman M M, et al. 2007. Accumulation of arsenic in tissues of rice plant (Oryza sativa L.) and its distribution in fractions of rice grain. Chemosphere, 69: 942-948

Ren X L, Liu Q L, Wu D X, et al. 2006. Variations in concentration and distribution of health-related elements affected by environmental and genotypic differences in rice grains. Rice Science, 13: 170-178.

Rodriguez L, Rincón J, Asencio, I, et al. 2007. Capability of selected crop plants for shoot mercury accumulation from polluted soils: phytoremediation perspectives. International Journal of Phytoremediation, 9: 1-13

Shi A B, Liang L N, Jiang G B, et al. 2005. The speciation and bioavailablity of mercury in sediments of Haihe River, China. Environment International, 31: 357-365

Sipter E, Rózsa E, Gruiz K, et al. 2008. Site-specific risk assessment in contaminated vegetable gardens. Chemosphere, 71: 1301-1307

Sun G X, Williams P N, Carey A M, et al. 2008. Inorganic arsenic in rice bran and its products are an order of magnitude higher than in bulk grain. Environmental Science & Technology, 42: 7542-7546

Tao S S H, Bolger P M. 1998. Dietary arsenic intakes in the United States: FDA total diet study, September 1991-December 1996. Food Additives & Contaminants, 16: 465-472

UNEP/FAO/WHO. 1992. Assessment of dietary intake of chemical contaminants. WHO/HPP/FOS/92.6, UNEP/GEMS/92.F2, United Nations Environmental Program, Nairobi

US-EPA, IRIS. 2008. United States, Environmental Protection Agency, Integrated Risk Information System. http://cfpub.epa.gov/ncea/iris/index.cfm?fuseaction=iris.showSubstanceList

Wang X L, Sato T, Xing B S, et al. 2005. Health risks of heavy metals to the general public in Tianjin, China via consumption of vegetables and fish. Science of the Total Environment, 350: 28-37

Williams P N, Price A H, Raab A, et al. 2005. Variation in arsenic speciation and concentration in paddy rice elated to dietary exposure. Environmental Science & Technology, 39: 5531-5540

Wilson B, Pyatt F B. 2007. Heavy metal dispersion, persistence, and bioaccumulation around an ancient copper mine situated in Anglesey, UK. Ecotoxicology and Environmental Safety, 66: 224-231

Zavala Y J, Duxbury J M. 2008. Arsenic in rice: I. Estimating normal levels of total arsenic in rice grain. Environmental Science & Technology, 42: 3856-3860

Zavala Y J, Gerads R, Gürleyük H, et al. 2008. Arsenic in rice: II. Arsenic speciation in USA grain and implications for human health. Environmental Science & Technology, 42: 3861-3866

Zheng N, Wang Q C, Zhang X W, et al. 2007a. Population health risk due to dietary intake of heavy metals in the industrial area of Huludao City, China. Science of the Total Environment, 387: 96-104

Zheng N, Wang Q C, Zheng D M. 2007b. Health risk of Hg, Pb, Cd, Zn, and Cu to the inhabitants around Huludao Zinc Plant in China via consumption of vegetables. Science of the Total Environment, 383: 81-89

Zhong J M, Yu M, Liu L Q, et al. 2006. Study on the dietary nutrition intake level in Zhejiang Province. Disease Surveillance, 21: 670-672 (in Chinese)

Zhu Y G, Sun G X, Lei M, et al. 2008a. High percentage inorganic arsenic content of mining impacted and nonimpacted Chinese rice. Environmental Science & Technology, 42: 5008-5013

Zhu Y G, Williams P N, Meharg A A. 2008b. Exposure to inorganic arsenic from rice: a global health issue? Environmental Pollution, 154: 169-171

Zu Y Q, Li Y, Christian S, et al. 2004. Accumulation of Pb, Cd, Cu and Zn in plants and hyperaccumulator choice in Lanping lead-zinc mine area, China. Environment International, 30: 567-576

4.6 Intraspecific variation in potassium uptake and utilization among sweet potato (*Ipomoea batatas* L.) genotypes

4.6.1 Introduction

Sweet potato (*Ipomoea batatas* L.) is an important versatile crop offering various products and diverse uses, and it is ranked seventh in annual production at approximately 9 Mt with a cultivated area of 110 Mha (Vincent, 2008; FAO, 2009). China, the leading producer of sweet potato, had an annual production of 75.6 million tons in 2011, which was 76.1% of the world production (FAO, 2011), and the sweet potato ranks fourth in Chinese food crops. Potassium (K) is associated with the activation of enzymes, negative charge neutralization, and osmo-regulation (Römheld and Kirkby, 2010). It has been estimated that to produce 1000 kg sweet potato dry matter requires approximate 10 kg K (Geoger et al., 2002). The arable land with a K-deficiency in China exceeded 22.7 million ha, accounting for 23% of the farmland (Wu et al., 2011). Due to intensive cropping, the increasing tuberous root yield indicates a negative K balance in arable land. Therefore, the improvement of K efficiency in crops is an attractive issue both for reducing costs in agricultural production and for increasing sweet potato production.

Nutrient use efficiency can be divided into two interrelated groups of plant factors, namely ①uptake efficiency and ② utilization efficiency (Sattelmacher et al., 1994; Swiader et al., 1994). K uptake efficiency is the nutrient uptake relative to the K supply, and genotypes may differ in root morphological parameters, such as root length, root density or the frequency of root hairs (Wang and Chen, 2012), or in root physiological parameters, such as the capacity for high affinity K uptake or the capacity to alter K availability in the rhizosphere (Brouder and Cassman, 1990; Yang et al., 2011); therefore, K content was assumed to represent an efficiency for K uptake (Damon and Rengel, 2007). K utilization efficiency (KIUE) is the relative ability of plants to produce maximal amounts of dry matter or yield for each unit of accumulated K. Genotypic differences in KIUE are mainly attributed to ① differences in partitioning and redistribution of K at cellular and whole plant levels, ② the substitution of K by other ions, and ③ the partitioning of resources into the economic product. Yang et al. (2004) reported that high KIUE rice cultivars under a K deficiency had a two-fold higher concentration of K in the lower leaves and a 30% higher concentration in the upper leaves when compared with K-inefficient rice genotypes at the booting stage.

Significant intraspecific variation for KIUE has been reported for many crops, including rice (Yang et al., 2003, 2004), wheat (Zhang et al., 1999; Damon and Rengel, 2007), soybean (Sale and Campell, 1987) and tomato (Chen and Gabelman, 1995). As for sweet potato, Bourke (1985) reported that the K level influenced tuber yield via an increase in the proportion of dry matter diverted to the tubers and a rise in tuber number per plant. George et al. (2002) found that a K application rate of up to 300 kg/ha of K_2SO_4 produced the greatest increase in tube root yield, and the increase was mainly due to the increase in the root/top ratio. Additionally, he added the influence of K fertilizer on the quality of sweet potato and screened four genotypes that could be used in breeding programs to improve KIUE while taking into account both yield and quality. The development of crop cultivars that are more efficient in nutrient use or better adapted to nutritionally marginal environments requires effective and reliable techniques for the re-evaluation of the nutrient use efficiency of regional varieties and national cultivars. Therefore, the aims of the study were ① to screen the genotypes representing widely diverse geographic areas that are more efficient in K uptake and KIUE for further work which will identify relevant genes and/or markers for use in sweet potato breeding programs, ② to determine the extent of genotypic variation in the KIUE of sweet potato, including identifying sweet potato genotypes contrasting in KIUE, and ③ to determine some plant traits associated with KIUE.

4.6.2 Materials and Methods

1. *Experiment* 1: *preliminary screening*

The experiments were conducted in Jiangyan City, Jiangsu Province (32°28′8.8″ N, 120°05′52.24″ E, 992 mm annual rainfall) in 2012-2013. The soil was deep (>100 cm) sandy loam, orthic aquisols derived from the ancient Yellow River. The available K content (extracted by 1 mol NH_4OAc and determined by flame photometry) in the soil was 49 mg/kg and 55 mg/kg soil in experiment 1 and experiment 2, respectively. In 2012, the preliminary experiment was carried out to evaluate sweet potato genotypes for their KIUE to screen genotypes for further research. In total, 108 sweet potato genotypes or breeding lines with different morphologies and yield components were selected. Vine cuttings of each genotype, ~30 cm long, were planted 0.3 m apart within three ridges 6 m in length and 0.5 m in width. Nitrogen and phosphorus fertilizers in the form of urea (N 120 kg/ha) and superphosphate (P_2O_5 90 kg/ha), respectively, were applied before the ridges were built. Two potassium levels (0 and 300 kg/ha as K_2SO_4) were applied. Fertilizer was applied to the soil in a band and in the ridges, which were built prior to planting, at a height of 30 cm.

2. *Experiment 2: advanced screening*

In 2013, based on the preliminary study of the KIUE of 108 sweet potato varieties screened in 2012, seven genotypes, Xu22, Nan88, Wan5, Wan6, Xu085, Xu083 and Ji22, with similar growth conditions but different KIUEs were planted in the same soil as in the preliminary study. A completely randomized block design with three replications was used. The plot area was 15 m^2 and plant density was 50000 plant/ha. Four levels of K_2SO_4, 0, 150, 300, and 450 kg/ha were supplied. Nitrogen and phosphate fertilizers were applied in the same manner as in the preliminary study.

Ten vine cuttings from each plot were sampled at intervals of 30, 70, and 100 days after plantation (DAP) to determine the dry matter yield, and the K concentration and accumulation. In the two experiments, harvesting was done 130 days after plantation. At maturity, the remaining plants from each genotype were harvested to determine the dry matter weight, and the recorded tuberous root numbers and dry matter of tuberous roots were used to calculate the yield. Additionally, 10 plants from each genotype were randomly selected to determine the K concentration and accumulation in the two experiments.

3. *Chemical analysis and parameter calculation*

In the two experiments, leaf, petiole and stem samples were oven dried at 90℃ for 30 min and then air-oven dried at 60℃ until a constant weight. Tuberous roots of sweet potato were washed thoroughly with tap water and then with deionized water three times. They were then surface dried in the laboratory, cut into thin slices and air-oven dried at 60℃. The dried samples were ground with a cyclone mill. Before determining K concentration by flame photometry, the samples were moist-ashed with H_2SO_4-H_2O_2.

The following parameters were calculated based on dry matter and the K concentrations of various organs:

(1) K accumulation value (KAV): dry matter weight × K concentration;

(2) KIUE: biomass or tuberous root (KIUE-T) yield produced per unit of K (Yang et al., 2003);

(3) Relative tuberous root yield: tuberous root yield at deficient K supply/ tuberous root yield

with adequate K supply (KER-T);

(4) Relative biomass dry matter weight: biomass of dry matter at deficient K supply/ tuberous root yield with adequate K supply (KER-B);

(5) K harvest index (KHI): KAV in tuberous root/ KAV in all organs.

4.6.3 Results and Analysis

1. *Preliminary screening of different sweet potato genotypes*

Large differences were observed in tuberous root yield and dry matter weight, as well as K concentration and accumulation at harvest stage among the sweet potato genotypes (Table 1). The maximum tuberous root yield and the highest dry matter weight were 3.74 times and 3.17 times, respectively, higher than that of the minimum under a K deficiency. Similarly, the maximum K concentration at a deficient K supply was 2.40 times higher than that of the minimum value; the greatest coefficient of variation (CV) occurred with KAV, and the ratio of maximum to minimum values was 7.13, showing that the difference in KAV among different genotypes at the deficient K supply was greater than that of the tuberous root yield, dry matter weight and K concentration.

Table 1   Variation among genotypes in shoot and root K concentration, shoot and root K accumulation, KIUE and KHI in field conditions of 2012

Parameter	Mean	Maxi.	Mini.	CV (%)	F-test	Relative value (K_0/K_2)	Significance between (K_0/K_2)
Tuberous root yield for deficient K (g/plant)	107.0	186.9	49.9	28.5	**	0.79	0.01
Dry matter weight for deficient K (g/plant)	147.0	221.9	69.9	23.4	**	0.80	0.05
K conc. of plant deficient K (g/kg)	6.76	10.43	4.34	21.2	**	0.63	0.01
K accumulation of plant deficient K (g/plant)	0.80	2.28	0.32	35.5	**	0.57	0.01
KIUE-T deficient K	104.3	160.5	49.5	19.5	**	1.62	0.05
KHI (K harvest index) of deficient K (%)	69.6	85.5	21.1	15.9	*	1.15	ns

Notes: F-test was used for assessing the variations of tuberous root yield, dry matter yield, K concentration, K accumulation, KIUE and KHI of different genotypes or breeding lines of sweet potato. Student's t-test was used for assessing the variations of different parameters between the deficient K (K_0) and adequate K application (K_2 300 kg/ha K_2SO_4). *, **, significant at $P=0.05$, $P=0.01$, respectively. ns: not significant.

There were significant differences in KAV, KIUE and KHI as revealed through the F-test (Table 1). In terms of KHI, the K harvest index with a deficient K supply ranged from 21.1 to 85.5. KIUE-T ranged from 49.5 to 160.5 with a mean of 104.3. The cultivars Wan5 and Xu28 had greater KIUE values of 157.7 and 160.5 g DW/g, respectively, while Wan6, Ji22 and Xu083 had comparatively smaller KIUE-Ts of 59.4, 69.9 and 71.4 g DW/g, respectively.

There were significant differences among all parameters, except KHI at the deficient K and adequate K supplies (t-test in Table 1). KER-T and KER-B were 0.79 and 0.80, respectively;

while the relative values of the K concentration to KAV at the deficient K supply and adequate K supply were 0.63 and 0.57, respectively, indicating the flexibility of sweet potato to adapt to low K stress under field conditions.

2. *The results of ANOVA: advanced screening*

Seven genotypes were selected from experiment 1 for advanced screening in experiment 2 based on their KIUE-Ts and tuberous root yields at the deficient and adequate K supply levels. There were highly significant effects of K and genotype on biomass, tuberous root yield and two yield components, tuberous root amount and weight per tuberous root (Table 2). While no significant K × genotype interaction appeared to affect tuberous root yield or the two yield components, the KAVs of the vines and tuberous roots of various sweet potato cultivars showed differences. There were also significant effects of biomass, KAV in tuberous root and KIUE with K supply levels, genotypes and their (K × genotype) interaction.

Table 2 Mean square of various sources of variation and their significant test for different characters in field experiment in 2013

	Block	Source of variation			
		Potassium	Genotype	G×K	Error
DF	2	3	6	18	54
Root dry matter yield/plant	136.54 ns	9924.63**	32742.23**	1370.05 ns	1775.4 ns
Root amount/plant	0.37 ns	8.44**	29.9**	1.71 ns	0.50 ns
Weight per tuberous root	11.4 ns	222.05**	4194.72**	133.63**	61.06 ns
Biomass/plant	1491.22 ns	14905.69**	37066.72**	2374.18**	2977.91 ns
K% in vine	0.07 ns	0.32**	0.08*	0.02 ns	0.02 ns
K% in root	0.01 ns	0.06**	0.1**	0.01 ns	0.01 ns
Kacc in vine	0.05 ns	0.77**	0.39**	0.07*	0.03 ns
Kacc in root	0.02 ns	0.61**	0.64**	0.04**	0.01 ns
KIUE-T	64.4 ns	3414.5**	2572.77**	1298.3**	20.2 ns

Notes: Kacc: K accumulation. *, **, significant at $P=0.05$, $P=0.01$, respectively. ns: not significant.

3. *Tuberous root yields and yield components: field experiment 2*

There were highly significant effects of K and genotypes on sweet potato tuberous root yield (Table 3). The K application increased the yield of sweet potato to a certain extent up to 300 kg K_2SO_4/ha (K_2), and then the increased rate was reduced. From Table 3, the difference in yield between K_2 and K_3 (450 kg K_2SO_4/ha) was not significant. This indicated that diminishing returns occurred after the K application rate reached 300 kg K_2SO_4/ha. With the K fertilizer application rate increasing, the yield components, including tuberous root number per plant and weight per tuberous root, increased. The degree of the increase of the two components among the different K levels was different, in term of weight of tuberous roots, and there was a significant difference between genotypes with and without K supplies. While the treatments with the different K supplies showed no significant differences, the tuberous root number varied significantly between genotypes at K_0 and K_2, and K_1 and K_3, but not at K_1

and K_2, or K_2 and K_3. K application rates were closely correlated with tuberous root number per plant and, to a lesser degree, with weight per tuberous root (Fig. 1).

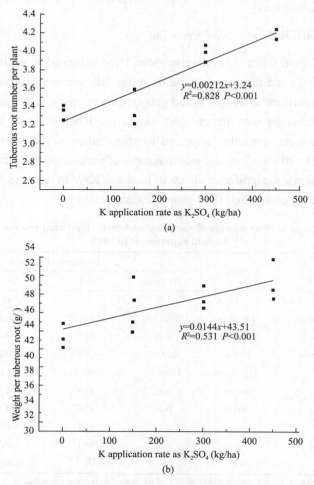

Fig. 1 Correlations between K fertilizer application rate and tuberous root number or weight per tube root

There were differences in the reaction to K nutrition among the seven genotypes. On an average of four K levels, Nan88 had the largest yield, followed by Xu28 and Wan5, but there was little difference between them (Table 3). The genotype order was the same among those varieties under K deficient conditions and also showed no differences (Table 5). Ji22 had the lowest yield under the different K supplies, followed by Wan6 and Xu083. The effect of genotype on tuberous root number per plant and weight per tuberous root was inconsistent and varied among genotypes.

Table 3 Tuberous yields and yield components under different K levels for seven genotypes in 2013

	Yield (g/plant)	Weight per tuberous root (g)	Tuberous root number per plant
K_0	130.1	42.5	3.34
K_1	153.6	47.3	3.42
K_2	170.3	47.5	3.98

	Yield (g/plant)	Weight per tuberous root (g)	Tuberous root number per plant
K_3	178.8	49.6	4.20
$LSD_{0.05}$	16.6	4.35	0.63
Genotype			
Wan5	181.3	72.4	3.12
Xu28	194.8	47.3	4.15
Nan88	199.5	74.1	2.79
Xu082	173.7	32.3	5.46
Wan6	124.4	40.1	3.07
Xu083	145.5	29.5	5.11
Ji22	78.6	34.9	2.04
$LSD_{0.05}$	24.0	5.04	0.57

4. Genotypic differences in K concentration and accumulation

For the K concentration in roots and different parts of shoots at harvest, the effects of K supply and genotypes were all significant (Table 2). On average, for all treatments, petioles had the highest K concentrations ranging from 6.29 g/kg to 12.43 g/kg at the four K levels, and from 6.83 g/kg to 11.30 g/kg in the seven genotypes. Leaves had the second highest K concentration. The stems had the lowest K concentration compared with other shoot parts but their concentrations were higher than those of tuberous roots (Table 4).

Table 4 K concentration and accumulation in different parts of shoot and tuberous root at maturity in field experiment in 2013

		K concentration (g/kg)					K accumulation (g/plant)					
		Leaf	Petiole	Stem	Shoot	Root	Leaf	Petiole	Stem	Shoot	Root	Total
K level	K_0	5.58	6.29	4.28	5.46	3.97	0.33	0.16	0.31	0.79	0.57	1.37
	K_1	6.84	7.63	4.80	6.10	4.46	0.35	0.18	0.36	0.90	0.79	1.69
	K_2	8.67	8.72	5.21	7.07	5.08	0.46	0.26	0.47	1.18	0.96	2.15
	K_3	8.88	12.43	5.86	8.27	5.58	0.56	0.33	0.52	1.40	1.17	2.57
	$LSD_{0.05}$	1.74	2.24	0.83	1.03	0.81	0.07	0.04	0.05	0.13	0.16	0.30
Genotype	Wan5	6.32	6.83	4.34	5.80	4.12	0.31	0.17	0.27	0.75	0.85	1.60
	Xu28	6.40	7.64	4.37	5.66	4.34	0.39	0.20	0.38	0.96	0.94	1.90
	Nan88	7.44	9.69	4.78	6.89	4.46	0.38	0.22	0.39	0.99	1.14	2.13
	Xu082	8.23	8.27	5.06	6.60	4.84	0.47	0.25	0.52	1.24	1.03	2.26
	Wan6	10.54	11.30	6.36	9.18	6.15	0.56	0.33	0.57	1.46	0.94	2.39
	Xu083	5.86	7.32	4.50	5.71	4.39	0.22	0.17	0.27	0.66	0.79	1.46
	Ji22	8.19	8.75	5.86	7.22	5.08	0.35	0.20	0.47	0.75	0.44	1.20
	$LSD_{0.05}$	0.91	1.78	1.63	0.96	1.13	0.06	0.06	0.09	0.14	0.20	0.36

There were significant differences in KAV between shoot and root at the four K levels. Although the shoot had a higher K concentration response to K fertilizer application than roots, roots had greater dry matter weight than shoots, resulting in a higher KAV in shoots than roots relative to the K application on average in the seven genotypes. The cultivars responded differently to K supply levels. Wan5, Nan88 and Xu083 had higher KAVs in roots than in shoots, and Xu28, Wan6 and Ji22 showed the opposite trend.

5. *K uptake efficiency and utilization efficiency*

Wan5 and Xu28 were selected for advanced screening as K-efficient genotypes for their high KIUE and good growth under a deficient K supply. Genotype Nan88 and Xu082 had moderate KIUEs, and were selected for further research as K-efficient genotypes for their excellent growth under K deficient and adequate K supplies. Furthermore, Nan88 and Xu082 had higher KAVs at harvest in 2012 (K uptake efficient genotype). Xu083, Wan6 and Ji22 were selected as K-inefficient for their low KIUEs and poor growth under deficient K supply levels.

Fig. 2 shows plant KAVs at the different growth stages under K a deficient supply in the field experiment of 2013. All sweet potato genotypes showed increases in KAVs under K deficient supply levels and the peak occurred about 3 months after planting (DAP100), and then reduced at maturity (DAP130). The variety Ji22 slightly decreased from 930.0 mg KAV per plant at DAP100 to 903.6 mg KAV per plant at DAP130. The other genotypes moderately declined at DAP130, ranging from a 12.4% to 29.3% decrease compared with DAP110.

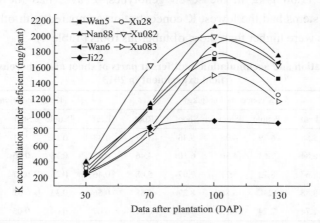

Fig. 2　K accumulation in different sweet potato genotypes under K deficient at different growth stages

DAP100 is the tuberous root booming stage when plants absorb more K than at the other growth stages and have more variation among the different genotypes than at maturity or the other growth stages. Nan88 had the highest KAV both at DAP100 and DAP130, Ji22 had the lowest KAV. Xu083 had the next to lowest KAV at both DAP100 and DAP130, with Xu082 and Wan6 having slightly higher KAVs at DAP100, but Wan6 had a higher KAV than Xu082 by DAP130. Despite Xu28 having a higher KAV at DAP100 than Wan5, there was an opposite trend by DAP130. Therefore, there was no consistent KAV between the harvest stage and booming period among the different sweet potato genotypes.

There were significant differences in KAV among the seven genotypes at the seeding stage (DAP30). Nan88 and Xu082 had significantly higher KAVs than Wan5, Wan6, Xu083 and Ji22.

Compared with Ji22, Nan88 and Xu082 had about one and half fold higher K uptake rates. Xu28 had a moderate KAV, which was not significantly different from those of the other genotypes, with KAV characteristics that conformed at DAP90 to those at DAP30. Based on the above results, genotypes Nan88 and Xu082 ranked as K uptake efficient, Wan5, Wan6 and Xu28 as K uptake moderate-efficient, and Ji22 and Xu083 as K uptake inefficient.

KIUE-B and KIUE-T values under K deficient conditions in the seven genotypes and the KER (the K efficiency ratio = tuber yield at deficient K supply/tuber yield at adequate K supply) are presented in Table 4. Both the KIUE-B and KIUE-T values of the seven genotypes ranked in similar orders in 2012 and 2013, with highly significant differences existing among genotypes. While KERs of different genotypes showed great fluctuations between the 2012 and 2013 seasons, they showed less significant differences among different genotypes than that of KIUE. In terms of KIUE, Wan5 and Xu28 had larger KIUE-T values than the others, followed by Nan88 and Xu082, while Wan6, Xu083, and Ji22 had smaller KIUE-T values than the others. Meanwhile, it was interesting to note that the variation in KIUE-B between 2012 and 2013 was higher than that of KIUE-T. Ji22 showed the largest variation over the different years, and its KIUE in 2012 and 2013 were 90.8 and 232.6, respectively, while Wan5 and Xu28 remained relatively unchanged in 2012 and 2013. According to the yield and yield potential results (Table 3 and Table 4, respectively), the varieties Wan5 and Xu28 adapted to K poor soils. Ji22, Wan6 and Xu083 responded poorly under deficient and adequate K supplies, while Nan88 and Xu082 had moderate KIUE-T values and high tuberous root dry matter yields in K deficient soils. However, they performed well in tuberous yield and had great yield increase potentials under an adequate K supply, suggesting that these genotypes had a good growth response to K fertilizers.

6. Correlation coefficients between KIUE and other characteristics

It could be seen that dry matter yield, root dry matter yield and root dry matter yield accumulation after the expanding stage (DAP60) did not correlate with KIUE-T and KIUE in the 2012 and 2013 seasons (Table 5). However, the root dry matter yield accumulation after the expanding stage correlated well with KIUE-T in the 2013 season ($r^2 = 0.415^{**}$), and it appeared that the KIUE was influenced by root dry matter yield accumulation after the swelling stage, and that the K-efficient genotypes had more root dry matter yield accumulation after the expanding stage than the inefficient genotypes at the low K supply levels. KER-T was significantly correlated with KIUE-T ($r^2 = 0.455^{**}$, 0.584^{**}) and KIUE ($r^2 = 0.353^{**}$, 0.385^{**}) in both seasons, but relative shoot dry matter yield did not correlate with KIUE-T and KIUE.

There were significantly negative correlations between the K concentration of root and KIUE-T, and the K concentration of whole plant and KIUE-T. K shoot concentration showed significantly negative correlations with KIUE in 2012 and KIUE-T in 2013, while K shoot concentrations did not correlate with KIUE-T in 2012 or KIUE in 2013(Table 6).

Table 5 Yield and KER and KIUE based on tuberous root and biomass for different genotypes

Geno.	2012 season					2013 season				
	Tuberous root yield K_0 (g/plant)	KER (K_0/K_2)	KIUE-B K_0	KIUE-T K_0	KIUE-T K_2	Tuberous root yield K_0 (g/plant)	KER (K_0/K_2)	KIUE-B K_0	KIUE-T K_0	KIUE-T K_2
Wan5	155.0	0.87	227.8	157.7	113.8	160.7	0.78	213.1	150.2	94.1
Xu28	158.5	0.81	203.0	160.5	95.7	171.0	0.77	235.9	164.4	104.7
Nan88	168.4	0.89	160.3	122.5	80.7	176.0	0.70	252.5	111.3	82.3
Xu082	171.1	0.88	144.6	97.0	59.5	149.0	0.77	184.4	108.5	79.8
Wan6	93.2	0.62	109.9	59.4	80.9	103.3	0.80	157.5	72.8	69.1
Xu083	86.3	0.81	106.9	71.4	75.9	114.5	0.72	167.4	98.5	85.2
Ji22	75.3	0.76	90.8	69.9	49.0	46.2	0.58	232.6	74.4	67.6
$LSD_{0.05}$	44.0	0.26	44.7	43.6	37.3	33.5	0.19	55.9	24.8	29.9

Table 6 Correlation coefficients between the K use efficiency and some plant and K nutrition characters

Variables	2012		2013	
	KIUE-T	KIUE-B	KIUE-T	KIUE-B
Dry matter yield	0.061	0.115	0.006	0.010
Tuberous root matter yield	0.132 ns	0.187	0.14	0.01
Dry matter yield accumulation after swelling stage	-	-	0.0308	0.0018
Tuberous yield accumulation after swelling stage	-	-	0.415**	0.0326
Relative shoot dry matter yield	ns	ns	ns	ns
Relative tuberous root dry matter yield	0.455**	0.353**	0.584**	0.385**
K concen. of root	–0.690**	–0.670**	–0.286**	0.047
K concen. of shoot	–0.141ns	–0.212**	–0.317**	–0.137 ns
K concen. of plant	–0.686**	–0.985**	–0.481**	–0.558**
Relative K accumulation in shoot	–0.161	–0.268**	–0.141 ns	–0.605**
Relative K accumulation in root	–0.121	–0.089	0.438**	0.411**
KHI (KAV root/KAV plant)	0.145	0.089	0.394**	0.028

Notes: *, **, significant at $P=0.05$, $P=0.01$, respectively. ns: not significant.

4.6.4 Discussion

The present study focused on the K uptake and utilization characteristics of sweet potato genotypes in a wide genetic pool and, unlike most previous studies that examined K uptake at a single harvest, provided the first detailed examination of the changes in K uptake throughout the growth cycle. Moreover, comparing and identifying genetic variation in K utilization using KIUE and KER produced evidence of genetic variation among sweet potato genotypes. Both field experiments showed that genetic variation in nutrient uptake and use efficiency occurred between genotypes and breeding lines. Furthermore, K fertilizers obviously increased sweet potato yields, mainly by increasing the tuber number. K efficiency can be divided into two multi-factorial components: effectiveness in taking up nutrients from the soil and/or the efficiency of produce yield or biomass per unit K absorbed by the plant. There are many ways to define nutrient efficiency in crops, and studies on genotypic differences in nutrient uptake have been made by many researchers in different crops (Woodend and Glass, 1993; Rengel and Damon, 2008). For sweet potato, genotypic differences in KAV and distribution in different plant parts at maturity was reported (George et al., 2002). The nutrients taken up by sweet potato plants come from both indigenous soil sources and applied fertilizers, and there was a considerable amount of K as litter or roots that excludes returning to the soil (Xie et al., 2000). Therefore, the K uptake rate in the maturity stage is frequently lower than that at the peak growth of biomass. In the study presented here, the K efficient genotype Xu28 had the highest K translocation by the whole plant from the tuberous root booming stage to maturity with a decrease of 29.2% KAV, while the K inefficient genotype Ji22 had a 2.8% decrease in KAV. If studies on the KAV in sweet potato concentrated on a single growth stage only, it would not reveal the variation in plant K accumulation during different growth stages, and thus some valuable information may have been camouflaged or even distorted. In the present study, by investigating the K uptake rate of sweet potato varieties at seeding, expanding, booming and maturing stages, and according to yield, two K uptake-efficient genotypes were identified (Nan88 and Xu082), which had a great ability to absorb K during the whole growth period and produce high tuberous dry matter yields. Despite the slightly higher KAV observed at harvest, there were lower levels of K absorption in the seedling and booming stages by the moderately K uptake efficient cultivars Wan6, Wan5 and Xu28. Due to the poor K absorption at the seedling stage and lower KAV in the booming and maturity stages, Xu082 and Ji22 were defined as inefficient K acquisition genotypes.

Several studies have associated genotype with KER and evaluated the ability of plants to adapt to a K deficiency（Gerloff and Gabelman, 1983; Samal et al., 2010）. Genotypes with a small accumulation of certain mineral nutrients and a high photosynthetic rate are the most economical since they require the lowest concentration of mineral fertilizer to produce a high yield (Damon et al., 2007; Xia et al., 2011), and numerous researchers use KIUE to screen K use efficiency (Siddiqi et al., 1987; EI-Dessougi, et al., 2002; Wu et al., 2011). In the present study, KIUE was significantly different among sweet potato genotypes and more stable than the KER between different seasons. Therefore, according to the response of KIUE-T, KAV and tuberous root dry matter yield to K levels, the selected seven genotypes can be divided into highly efficient, moderately efficient and inefficient: ① Nan88 and Xu082 have higher yields and moderate yield responses to K levels, but moderate utilization efficiency, ② Wan5 and Xu28 have high utilization efficiencies and moderate uptake efficiencies, ③ Ji22 and Xu083 have lower yields, moderate yield responses and are K inefficient in uptake and utilization, and ④ Wan6 has a low K utilization efficiency, but with moderate KAV in the seeding and booming stages, and a moderate yield response to K levels, resulting in a moderate uptake efficiency and lower utilization efficiency.

The correlation of KIUE-T and root dry matter yield and dry matter yield were not significant in either 2012 or 2013, which did not coincide with the results of George et al. (2002). The reason may be related to variation among genotypes in KIUE and greater genetic differences in yield and biomass production. However, KIUE-T had a significant negative correlation with the K concentration in the whole plant in 2012 and 2013, and a positive correlation with net tuberous growth rates after expanding in 2013. The KHI were also significantly correlated with KIUE-T, indicating that the highly efficient cultivars have the ability to absorb K from soils to produce more tuberous dry matter than inefficient cultivars.

Römheld and Kirkby (2010) proposed that genotypic differences could influence the capacity to produce high economic yield per unit K. Genotypic differences in KIUE have been reported for species, including sweet potato. However, it is important to identify the intraspecific variation within a more widely diverse group of geographic and genetic genotypes or breeding lines. For sweet potato, George et al. (2002) reported that Nan88 had the highest KIUE at 131 g DW/g among 84 genotypes and breeding lines measured in a field experiment. In the present study, Nan88 had KIUE values of 122.5 and 113.1 g DW/g in 2012 and 2013, respectively, illustrating the great progress in sweet potato breeding and the potential to improve K efficiency. K levels had an obvious influence on the tuber yield and their components. The significant correlation between root dry matter yield, tuber number and mean tuber weight indicated that K fertilizers increased root dry matter via an increase of tuber number, and, to a lesser degree, mean tuber weight. This contradicted the results of Enyi (1972) using the lesser yam (*Dioscorea esculenta* L.) who found that K increased mean tuber dry matter weight, but agreed with the results of Bourke (1985), although those results lacked a significant correlation between mean tuber weight and tuber number per plant.

In summary, sweet potato plants are highly responsive to the soil K supply in terms of root yield, K uptake and use efficiency. The features of K accumulation in early growth and K redistribution in later growth suggest some valuable information may have been camouflaged or even distorted in studies of K accumulation that concentrated on maturity only. The KIUE of roots correlated with tuberous root increments from root expansion to harvest, while having no correlation with root dry matter yield, indicating high KIUE genotypes have a greater ability to propagate and expand.

References

Bourke R M. 1985. Influence of nitrogen and potassium fertilizer on growth of sweet potato (*Ipomoea batatas*) in Papua New Guinea. Field Crops Res., 12: 363-375

Brouder S M, Cassman K G. 1990. Root development of two cotton cultivars in relation to potassium uptake and plant growth in a vermiculitic soil. Field Crops Res., 23(3/4): 187-203

Chen J J, Gabelman W H. 1995. Isolation of tomato strains varying in potassium acquisition using a sand-zeolite culture system. Plant Soil, 176: 65-70

Damon P M, Osborne L D, Rengel Z. 2007. Canola genotypes differ in potassium efficiency during vegetative growth. Euphytica, 156: 387-397

Damon P M, Rengel Z. 2007. Wheat genotypes differ in potassium efficiency under glasshouse and field conditions. Aust. J. Agr. Res., 58: 816-825

EI-Dessougi H, Claassen N, Steingrobe B. 2002. Potassium efficiency mechanisms of wheat, barley, and sugar beet grown on a K fixing soil under controlled conditions. J. Plant Nutr. Soil Sci., 165: 732-737

Enyi B A C. 1972. Effects of staking, nitrogen and potassium on growth and development in lesser yams:Dioscorea esculenta. Ann. Appl. Biol., 72: 211-219

FAO. 2009. World information and early warning system on plant genetic resources. Food and Agriculture Organization of the United Nations, Rome

FAO. 2011. Food and agricultural commodities production. Food and Agriculture Organization of the United Nations, Rome

George M S, Lu G Q, Zhou W J. 2002. Genotypic variation for K uptake and utilization efficiency in sweet potato (*Ipomoea batatas* L.). Field Crops Res., 77: 7-15

Gerloff G C, Gabelman W H. 1983. Genetic basis of inorganic plant nutrition. In: Lauchli A, Bieleski R L. Encyclopedia of Plant Physiology. Berlin, Germany: Spring-Verlag: 453-480

Rengel Z, Damon P M. 2008. Crops and genotypes differ in efficiency of potassium uptake and use. Physiol. Plantarum, 133(4): 624-636

Römheld V, Kirkby E A. 2010. Research on potassium in agriculture: needs and prospects. Plant Soil, 335: 155-180

Sale P W G, Campell L C. 1987. Differential response to K deficiency among soybean cultivars. Plant Soil, 90: 231-237

Samal D, Kovar J L, Steingrobe B, et al. 2010. Potassium uptake efficiency and dynamics in the rhizosphere in maize (*Zea mays* L.), wheat (*Triticum aestivum* L.) and sugar beet (*Beta vulgaris* L.) evaluated with a mechanistic model. Plant Soil, 332: 105-121

Sattelmacher B, Horst W J, Becker H C. 1994. Factors that contribute to genetic variation for nutrient efficiency of crop plants. Zei. für. Pflanz. und. Bodenk., 157: 215-224

Siddiqi M Y, Glass A D M, Hsiao A I. 1987. Genotypic differences among wild oat lines in K uptake and growth in relation to K supply. Plant Soil, 9: 93-105

Swiader J M, Chyan Y, Freiji F G. 1994. Genotypic differences in nitrate uptake and utilization efficiency in pumpkin hybrids. J. Plant Nutr., 17: 1687-1699

Vincent L. 2008. Tropical Root and Tuber Crops: Cassava, Sweet Potato, Yams and Aroids. Oxford shire, UK: CABI Publishing: 97

Wang L, Chen F. 2012. Genotypic variation of potassium uptake and use efficiency in cotton(*Gossypium hirsutum* L.). J. Plant Nutr. Soil Sci., 175: 303-308

Woodend J J, Glass A D M. 1993. Genotype-environment interaction and correlation between vegetative and grain production measures of potassium use-efficiency in wheat (*T. aestivum* L.) grown under potassium stress. Plant Soil, 151: 39-44

Wu J T, Zhang X Z, Li T X, et al. 2011. Differences in the efficiency of potassium (K) uptake and use in barley varieties. Agricultural Sciences in China, 10(1): 101-108

Xia Y, Jiang C C, Chen F, et al. 2011. Differences in growth and potassium use efficiency of two cotton genotypes. Commun. Soil Sci. Plant, 42: 132-143

Xie J C, Zhou J M, Härdter R. 2000. Potassium in Chinese agriculture. Nanjing: Hehai University Press: 28

Yang F Q, Wang G W, Zhang Z Y, et al. 2011. Genotypic variations in potassium uptake and utilization in cotton. J. Plant Nutr., 34(1): 83-97

Yang X E, Liu J X, Wang W M, et al. 2003. Genotypic differences and some associated plant traits in potassium internal use efficiency of lowland rice (*Oryza sativa* L.). Nutr. Cycl. Agroecosys., 67: 273-282

Yang X E, Liu J X, Wang W M, et al. 2004. Potassium internal use efficiency relative to growth vigor, potassium distribution and carbohydrate allocation in rice genotypes. J . Plant Nutr., 27: 837-852

Zhang G P, Chen J X, Eshetu A T. 1999. Genotypic variation for potassium uptake and utilization efficiency in wheat. Nutr. Cycl. Agroecosys., 54: 41-48

4.7 Use of FTIR-PAS combined with chemometrics to quantify nutritional information in rapeseeds (*Brassica napus*)

4.7.1 Introduction

Rapeseed (*Brassica napus* L.) is one of the most important oilseed crops (Cardone et al., 2003). Nutritional information such as nitrogen, phosphorus and potassium content is an important concern for rapeseed breeders and farmers. Fast and accurate acquisition of nitrogen, phosphorus and potassium content in rapeseeds carries significant implication for quality improvement in the breeding industry and for fertilizer recommendations during field production. Traditional methods to obtain such nutrition include the Kjeldahl method and the Dumas combustion method (Simonne et al., 1997) for nitrogen, molybdenum blue colorimetric method (Martin and Doty, 1949) for phosphorus, and flame photometry (Fox, 1951) for potassium. Additionally, inductively coupled plasma atomic emission spectroscopy (ICP-AES) can be used for the determination of phosphorus and potassium. However, all of these methods necessitate seed grinding and, a series digestion procedures including dry ashing and wet ashing with different mixtures of reagents (Rodushkin et al., 1999). The analysis process is time-consuming, laborious and even dangerous. Also, these methods are destructive to the seed and unsuitable for the breeding selection of small samples of rapeseed varieties. Recently, Fourier transform infrared photoacoustic spectroscopy (FTIR-PAS) has been applied as a powerful alternative to traditional chemistry procedures for analysis of agricultural products (Sivakesvava and Irudayaraj, 2001). FTIR-PAS requires no sample pretreatment and can analyze nearly any type of sample in a rapid way. Thorough elaboration on PAS technique principles and its applications can be found in Ref. (Tam, 1986; Haisch, 2012).

Multivariate calibrations are commonly developed for spectral quantification of major constituents as well as microcomponent and other qualities in agricultural and food products (Martens and Naes, 1989). Partial least squares (PLS) has been extensively used as a standard multivariate calibration method (Wold et al., 2001). Pre-processing methods can be used before the PLS calibration to remove information unrelated to the chemical components in samples to improve predictive results. A recent method called direct orthogonal signal correction (DOSC) (Westerhuis et al., 2001) has been developed for this purpose. This technique, a modification of orthogonal signal correction (OSC) (Wold et al., 1998), can subtract spectral variances irrelevant to the component of interest in samples. DOSC corrected spectra are then utilized to establish a PLS model, i.e. DOSC-PLS model. The DOSC-PLS model usually involves fewer latent variables and has an improved predictive performance over OSC-PLS. There is much evidence to demonstrate the effectiveness of DOSC to refine calibration models (Pizarro et al., 2004; Zhu et al., 2008; Laghi et al., 2011).

PAS technique has been introduced as an analytical tool for more than four decades (Rosencwaig, 1975), but the application of PAS technique to analyze agro-food systems is rather limited (Irudayaraj et al., 2001). Systems studied include cheese (McQueen et al., 1995a, 1995b), pea seeds (Letzelter et al., 1995), meat (Yang and Irudayaraj, 2001), soybean seed (Caires et al., 2008), edible oils (de Albuquerque et al., 2013). Recently, PAS has been firstly used for analysis of rapeseeds (Lu et al., 2014). The study presented below is to further use PAS to quantify the nitrogen, phosphorus and potassium content in rapeseeds. The objectives of the present study were: ① to evaluate the feasibility of FTIR-PAS combined with PLS to determine nitrogen, phosphorus and potassium content in rapeseeds; ② to compare the predictive ability between the PLS model and its corresponding DOSC-PLS model.

4.7.2 Material and Methods

1. Samples

A total of 180 rapeseed samples were provided by Oil Crops Research Institute of the Chinese Academy of Agricultural Sciences. They were harvested from a field experiment (Table 1) in the Yingtan Ecology Experimental Station of the Chinese Academy of Sciences. Each sample represented one cultivar, thus ensuring that samples have a relatively broad variation in nutrition levels. All rapeseeds were air-dried to make the moisture content approximately equivalent among samples, and then stored in plastic bags at room temperature.

Table 1 Details of the field experiment

	Soil properties			Fertilization scheme			
pH	$AN^{①}$ (mg/kg)	$AP^{②}$ (mg/kg)	$AK^{③}$ (mg/kg)	Urea (kg/ha)	Superphosphate (kg/ha)	Potassium chloride (kg/ha)	Borax (kg/ha)
5.59	43	5.71	113.52	270	750	225	15

Notes: ①AN= available nitrogen; ② AP= available phosphorus; ③ AK= available potassium.

2. FTIR-PAS measurements

Air-dried intact samples were directly used for spectral measuring. Photoacoustic spectra were recorded for all samples using a Fourier transform infrared spectrometer (Nicolet 6700, USA) equipped with a photoacoustic cell (Model 300, MTEC, USA). After placing the sample (about 100 mg) in the cell holding cup (diameter 5 mm, height 3 mm) and purging the cell with dry helium (10 mL/min) for 10 s, the scans were conducted in the range of 500-4000 cm^{-1} with a resolution of 4 cm^{-1} and a mirror velocity of 0.32 cm/s. A carbon black reference was used for spectra-intensity normalization. Thirty-two successive scans were recorded, and the average for each sample was used for the chemometric analysis.

3. Chemical methods

After spectral measurement, seed samples were ground before they are analyzed by means of chemical methods which provided the reference data for the development of spectral calibration model. Ground samples were weighed using an electronic lab balance (Sartorius BS210S, Germany) with a readability of 0.0001 g, and were then digested with concentrated sulfuric acid and hydrogen peroxide in the digestion instrument (LabTech EHD36, USA), and afterwards, nitrogen content was determined using the Kjeldahl method with an automatic Kjeldahl apparatus (Buchi339, Switzerland), and phosphorus and potassium contents were determined using ICP-AES (Thermo element IRIS, USA).

4. Chemometric methods and software

The Savitzky-Golay (Savitzky and Golay, 1964) smoothing filter with 21 points and a polynomial of order 1 was applied as the spectral pretreatment based our previous experience, to de-noise the original photoacoustic spectra. The smoothed spectral matrix was divided into calibration and prediction subsets of 135 and 45 samples, respectively, according to the Kennard-Stone algorithm (Kennard and Stone, 1969). The detailed statistics of N, P and K content in each sample set are given in Table 2. Calibration models were subsequently established using PLS and DOSC-PLS algorithms, respectively. The brief description of the DOSC algorithm is given below.

Table 2 Statistics of nitrogen, phosphorus and potassium in rapeseeds

Properties	Sample set	Minimum	Maximum	Mean	Standard deviation	CV* (%)
N content (%)	Calibration	2.688	3.775	3.214	0.2040	6.35
	Prediction	2.871	3.564	3.231	0.1635	5.06
P content (%)	Calibration	0.634	0.896	0.746	0.0496	6.65
	Prediction	0.651	0.823	0.740	0.0462	6.24
K content (%)	Calibration	0.481	0.742	0.610	0.0565	9.26
	Prediction	0.498	0.701	0.619	0.0494	7.98

Note: *CV= coefficient of variation.

The first step is to divide Y (the parameter matrix needs to be predicted) into two parts, $\hat{Y}$ and F which are the projection of Y on X (the spectral matrix), and the residual part orthogonal to X, respectively:

$$Y = P_X Y + A_X Y = \hat{Y} + F \quad (1)$$

Where, P_X is defined as the projector onto the column space of X, A_X is the anti-projector with respect to X-space.

Next, X is decomposed into two orthogonal parts; $\hat{X}$ and E, which are related to $\hat{Y}$ and orthogonal to $\hat{Y}$, respectively:

$$X = P_Y Y + A_Y Y = \hat{X} + E \quad (2)$$

Then, principal component analysis (PCA) is applied to E to find a small number of principal components t corresponding to the DOSC components to be subtracted. t is expressed as linear combinations of X:

$$t = Xr \quad (3)$$

r is the weight matrix in the principal orthogonal directions, which can be calculated as below:

$$r = X^+ t \quad (4)$$

Where, X^+ is the Moore-Penrose inverse of X.

The DOSC-corrected spectra of the calibration data can be written as:

$$X_{DOSC} = X - t\,P' \tag{5}$$

Where, $t\,P'$ is the orthogonal part removed from the original spectral matrix X with the loading matrix P:

$$P = X'\,t(t't)^{-1} \tag{6}$$

With the weights r and the loadings P, the DOSC-corrected spectra for new data X_{new} can be directly obtained as below:

$$X_{DOSC,new} = X_{new} - X_{new}\,rP' \tag{7}$$

However, due to the constraint of complete orthogonality, a generalized inverse X^- is generally adopted instead of X^+ for the calculation of r to avoid overfilled of DOSC components. X^- is obtained from a solution of principal component regression (PCR) between X and t. In conducting DOSC, the number of DOSC components and a tolerance factor need to be tuned according to a given spectral dataset. More details on the DOSC algorithm referred to Westerhuis et al. (2001).

In PLS calibration, the optimal number of latent variables of a calibration model was decided based on the minimal root-mean-squared error of cross validation (RMSECV) through the leave-one-out cross validation procedure (Browne, 2000). In performing DOSC, two parameters, i.e., the number of DOSC components and a tolerance factor, needed to be tuned. The two parameters were also optimized based on the lowest RMSECV of the calibration set. RMSECV was calculated as follows:

$$\mathrm{RMSECV} = \sqrt{\frac{1}{I_c}\sum_{i=1}^{I_c}(\hat{y}_i - y_i)^2} \tag{8}$$

Where, $\hat{y}_i$ is the predicted value of the ith observation; y_i is the chemical value of the ith measurement; and I_c is the number of measurements in the calibration set. Leave-one-out cross validation was performed by first defining certain number of latent variables. Then, one sample was removed from the sample set, and, afterwards, the calibration model was built with the remaining samples and then predicted the removed sample. The procedure was repeated for all samples and finally the RMSECV was calculated.

Savitzky-Golay smoothing was implemented using the Unscrambler v9.8 (CAMO Software AS, Norway). A home-made Kennard-Stone procedure, the PLS algorithm from PLS_Toolbox 4.2 (Eigenvector Technology, USA) and the DOSC code downloaded from the website of "Biosystems Data Analysis Group of the Universiteit van Amsterdan" were realized in the MATLAB R2011b (The Math Works, Natick, USA).

5. Model evaluation standards

For the calibration set, calibration models were compared using RMSECV and the coefficient of determination r_{cal}^2 in the process of cross validation. For the validation set, the root-mean-square error of prediction (RMSEP) and r_{pre}^2, were used to evaluate the predictive performance of calibration models (Wu et al., 2010). Besides, an integrated criterion, RPD (ratio of prediction to deviation, which was defined as the ratio of the standard deviation (SD) of the prediction set to the RMSEP) was taken into account (Williams and Sobering, 1993). According to Viscarra et al. (2006), RPD should be greater than 2 for good quantitative prediction and 1.8 was the lowest acceptable value for prediction. In general, RPD should be as high as possible for a good model.

4.7.3 Results and Discussion

1. Spectral analysis

Well-defined spectral features could be observed from the average smoothed photoacoustic spectrum of all rapeseed samples (Fig. 1 (a)). The major peaks could be interpreted as follows (Fan, 2001): a broad peak around 3365 cm^{-1} corresponded to O—H stretching and N—H stretching. The peak at 2925 cm^{-1} was assigned to C—H stretching. The peak at 1635 cm^{-1} was associated with O—H deformation vibration overlapping with the amide Ⅰ C—O stretching. A shoulder peak around 1555 cm^{-1} corresponded to the amide Ⅱ (the out-of-phase combination of

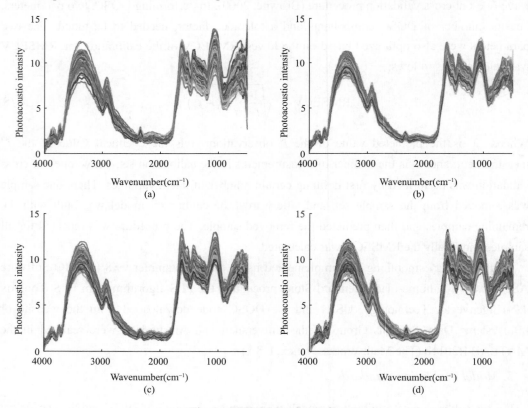

Fig. 1 FTIR-PAS spectra of rapeseed

(a) Smoothed spectra; (b) DOSC-corrected spectra for nitrogen content; (c) DOSC-corrected spectra for phosphorus content; (d) DOSC-corrected spectra for potassium content

the N—H deformation vibration and C—N stretching vibration). The peak around 1435 -1380 cm^{-1} corresponded to C—H deformation vibration. In the fingerprint range, 1200 - 500 cm^{-1}, a broad band around 1060 cm^{-1} was mainly assigned to the C—O and C—C stretching.

2. *Principal component score plots*

The smoothed spectra of all rapeseed samples were further processed by DOSC using the contents of nitrogen, phosphorus and potassium as the response variable, respectively. The DOSC-corrected spectra are shown in Fig. 1 (c), (b) and (d). According to Fig. 1, no significant differences were observed between smoothed spectra and the DOSC-corrected spectra. However, the spectral discrepancies in certain regions such as around 3000-3700 cm^{-1} and 500-1000 cm^{-1} in the DOSC-corrected spectra were to some degree narrowed down compared to the smoothed spectra, since the systematic variations irrelevant to response variables had been extracted out by DOSC.

To investigate the effect of DOSC, principal component analysis (PCA) was implemented on the spectra of all samples (Pizarro et al., 2004). Before PCA, every sample was numbered in an increasing order according to the value of the response variable considered. The score plots of the first two principal components in the smoothed spectral matrix and DOSC corrected spectral matrix are shown (Fig. 2). The first principal component (PC1) of the smoothed spectra accounted for much less spectral variance than that of the DOSC corrected spectra. More

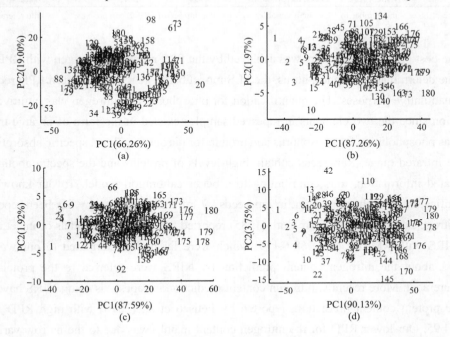

Fig. 2   PCA distribution based on FTR-PAS spectra

(a) score plot of the smoothed spectra; (b) score plot of DOSC corrected spectra for nitrogen; (c) score plot of DOSC corrected spectra for phosphorus; (d) score plot of DOSC corrected spectra for potassium. PC1 denotes the first principal component, and PC2 denotes the second principal component

importantly, no specific distribution pattern of samples could be recognized in the score plot of the smoothed spectra. Instead, all samples were almost exactly arranged along the PC1 axis in each score plot of the DOSC-corrected data, which demonstrated the effectiveness of DOSC. It was then logical to deduce that the DOSC-corrected spectra would lead to the improvement of a calibration model.

3. *PLS models and DOSC-PLS models*

Smoothed spectra of the calibration set were directly used to develop PLS models for nitrogen, phosphorus and potassium, respectively. For comparison, DOSC-corrected spectra were utilized to build the corresponding DOSC-PLS models. The prediction set was employed to test individual calibration models. The statistics including r_{cal}^2, RMSECV, r_{pre}^2, RMSEP and RPD of each PLS model are shown in Table 3.

Table 3 Rsults of PLS models and DOSC-PLS models

Properties	Models	Parameters ($m^①, n^②, t^③$)	Calibration set		Prediction set		
			r_{cal}^2	RMSECV	r_{pre}^2	RMSEP	RPD
N	PLS	$m=10$	0.829	0.0843	0.737	0.0839	1.95
	DOSC-PLS	$m=3, n=2, t=0.001$	0.882	0.0702	0.846	0.0642	2.54
P	PLS	$m=7$	0.725	0.0260	0.698	0.0254	1.81
	DOSC-PLS	$m=3, n=2, t=0.001$	0.817	0.0212	0.773	0.0220	2.10
K	PLS	$m=8,$	0.763	0.0275	0.658	0.0289	1.71
	DOSC-PLS	$m=2, n=1, t=0.001$	0.794	0.0224	0.746	0.0254	1.94

Notes: ① m = the number of PLS latent variables; ② n = the number of DOSC components; ③ t = the tolerance factor of DOSC.

The best prediction accuracy was achieved by the PLS model for nitrogen with RPD of 1.95. For the determination of phosphorus, the PLS model barely reached minimum RPD requirement for quantitative purposes. The quantification for phosphorus and nitrogen was intuitively out of question since the two elements in rapeseed samples existed in the form of organic molecules, such as phospholipids for phosphorus and protein for nitrogen, and had specific absorption bands in the infrared range. Rapeseeds contain high levels of proteins and the spectra abounded with N-related information, thus contributing to a better calibration model. To our knowledge, no quantification phosphorus content in rapeseeds by infrared spectroscopy has been reported. For the nitrogen determination, Starr et al. (1985) reported the calibration models of intact rapeseeds by NIRS with RMSEP of 0.168-0.179, which were far higher compared to our work. More reports about the nitrogen content prediction by NIRS were referred to the protein content measurement. More recently, nitrogen content prediction in rapeseeds using NIRS have referred to the protein content prediction, reported by Petisco et al. (2010) with high RPD values of 3.93-4.95. Our lower RPD for the nitrogen content mainly was due to the narrow variability in reference values.

However, the quantification for potassium was unsatisfactory due to the RPD less than 1.8. This result was not surprising considering that potassium had no direct spectral adsorption in the infrared range. However, some applications of near infrared spectroscopy (NIRS) for determination of potassium content in plant tissues had been reported. Clark et al. (1987) applied NIRS to determine potassium content in forages, resulting in the r_{pre}^2 of 0.71-0.81. Ciavarella et al. (1998) reported better calibration models accounting for 85%-96% of the potassium concentrations in the validation set. These results could be explained based on a basic assumption that potassium was associated with some organic compounds which had infrared absorption. However, the association of potassium in certain samples seemed unintuitive and the PLS model sometimes could not give a desirable result. A low RPD of 1.66 was obtained using NIRS calibration for determination of potassium in straw samples without milling (Huang et al., 2009). The RPD of 1.74 in our work might also indicate this scenario.

DOSC was applied to the smoothed spectral matrix and then DOSC-PLS calibration models were established for the prediction of the contents of nitrogen, phosphorus and potassium. After DOSC, an improvement in the performance of each PLS model was observed. The best result was still obtained from the PLS model for nitrogen. Its RPD substantially increased from 1.95 to 2.54, which indicated good quantitative predictions (Viscarra et al., 2006). The RPD of the PLS model for phosphorus also increased beyond 2. The PLS model for potassium was initially unacceptable; however, after DOSC the RPD increased to 1.94, which made the DOSC-PLS model suitable for the basic quantitative prediction. DOSC seemed to have intensified potassium-related information hidden in the spectra. These results verified the ability of DOSC to eliminate the spectral information unrelated to components of interest and highlight relevant information, thus leading to better calibration models. The results reported by Pizarro et al. (2004) and Zhu et al. (2008) also confirmed this point. After DOSC, the number of latent variables for each PLS model was significantly reduced. Only two or three components were sufficient to obtain desirable results without the presence of obvious over-fitting. DOSC was found to be a quite appropriate pre-processing method for PLS calibration.

The PLS and DOSC-PLS models have preliminarily shown the feasibility of FTIR-PAS for quantifying nutritional information in rapeseeds, although more rapeseed samples from a wide range of cultivation locations, plant varieties and sampling years should be incorporated to improve prediction accuracy. Besides, FTIR-PAS hold merits including rapid spectral measurement (about one minute per test), non-destructiveness and no preparation of samples, and minimal sample requirement (about 100 mg) especially compared to several grams generally required by NIRS. However, it should be noted that due to the high complexity and sensitivity of photoacoustic systems, application for rapeseed analysis outside the laboratory was still unfeasible. But this situation can be expected to be changed in the near future due to the advances of photoacoustic instrumentation. Recently, compact, potable and reliable PAS instruments have been devised and available (Rabasovic et al., 2009; Bozoki et al., 2011). In our view, PAS technique is currently at the leading edge and hold great promise for practical application for quality analysis of agricultural products including rapeseeds.

References

Bozoki Z, Pogany A, Szabo G. 2011. Photoacoustic instruments for practical applications: present, potential and future challenges. Appl. Spectrosc. Rev., 46: 1-37

Browne M W. 2000. Cross-validation methods. J. Math. Psycho., 44: 108-132

Caires A R L, Teixeira M R O, Suarez Y R, et al. 2008. Discrimination of transgenic and conventional soybean seeds by Fourier transform infrared photoacoustic spectroscopy. Appl. Spectrosc., 62: 1044-1047

Cardone M, Mazzoncini M, Menini S, et al. 2003. Brassica carinata as an alternative oil crop for the production of biodiesel in Italy: agronomic evaluation, fuel production by transesterification and characterization. Biomass Bioenerg., 25: 623-636

Ciavarella S, Batten G D, Blakeney A B. 1998. Measuring potassium in plant tissues using near infrared spectroscopy. J. Near Infrared Spectrosc., 6: 63-66

Clark D H, Mayland H F, Lamb R C. 1987. Mineral analysis of forages with near infrared reflectance spectroscopy. Agron. J., 79: 485-90

de Albuquerque J E, Santiago B C L, Campos J C C, et al. 2013. Photoacoustic spectroscopy as an approach to assess chemical modifications in edible oils. J. Braz. Chem. Soc., 24: 369-374

Fan K N. 2001. Introduction to spectroscopy. Beijing: High Education Press: 63

Fox C L. 1951. Stable internal standard flame photometer for potassium and sodium analyses. Anal. Chem., 23: 137-142

Haisch C. 2012. Photoacoustic spectroscopy for analytical measurements. Meas. Sci. Technol., 23: 012001

Huang C J, Han L J, Yang Z L, et al. 2009. Exploring the use of near infrared reflectance spectroscopy to predict minerals in straw. Fuel, 88: 163-168

Irudayaraj J, Sivakesava S, Kamath S, et al. 2001. Monitoring chemical changes in some foods using Fourier transform photoacoustic spectroscopy. J. Food Sci., 66: 1416-1421

Kennard R W, Stone L A. 1969. Computer aided design of experiments. Technometrics, 11: 137-148

Laghi L, Versari A, Parpinello G P, et al. 2011. FTIR spectroscopy and direct orthogonal signal correction preprocessing applied to selected phenolic compounds in red wines. Food Anal. Methods, 4: 619-625

Letzelter N S, Wilson R H, Jones A D, et al. 1995. Quantitative determination of the composition of individual pea seeds by Fourier transform infrared photoacoustic spectroscopy. J. Sci. Food Agric., 67: 239-245

Lu Y, Du C, Yu C, et al. 2014. Classification of rapeseed colors using Fourier transform mid-infrared photoacoustic spectroscopy. Anal. Methods, 6: 1412-1417

Martens H, Naes T. 1989. Multivariate Calibration. Chichester: John Wiley & Sons Ltd: 1

Martin J B, Doty D M. 1949. Determination of inorganic phosphate. Anal. Chem., 21: 965-967

McQueen D H, Wilson R H, Kinnunen A, et al. 1995b. Comparison of two infrared spectroscopic methods for cheese analysis. Talanta, 42: 2007-2015

McQueen D H, Wilson R H, Kinnunen A. 1995a. Near and mid-infrared photoacoustic analysis principal components of foodstuffs. Trends Anal. Chem., 14: 482-492

Petisco C, Garcia-Criado B, Vazquez-de-Aldana B R, et al. 2010. Measurement of quality parameters in intact seeds of Brassica species using visible and near-infrared spectroscopy. Ind. Crop Prod., 32: 139-146

Pizarro C, Esteban-Diez I, Nistal A J, et al. 2004. Influence of data pre-processing on the quantitative determination of the ash content and lipids in roasted coffee by near infrared spectroscopy. Anal. Chim. Acta, 509: 217-227

Rabasovic M D, Nikolic M G, Dramicanin M D, et al. 2009. Low-cost, portable photoacoustic setup for solid samples. Meas. Sci. Technol., 20: 095902

Rodushkin I, Ruth T, Huhtasaari A. 1999. Comparison of two digestion methods for elemental determinations in plant material by ICP techniques. Anal. Chim. Acta, 378: 191-200

Rosencwaig A. 1975. Photoacoustic spectroscopy: a new tool for investigation of solids. Anal. Chem., 47: 592-604

Savitzky A, Golay M J E. 1964. Smoothing and differentiation of data by simplified least squares procedures. Anal. Che., 36: 1627-1632

Simonne A H, Simonne E H, Eitenmiller R R. 1997. Could the Dumas method replace the Kjeldahl digestion for nitrogen and crude protein determinations in foods? J. Sci. Food Agri., 73: 39-45

Sivakesvava S, Irudayaraj J. 2001. Analysis of potato chips using FTIR photoacoustic spectroscopy. J. Sci. Food Agri., 80: 1805-1810

Starr B C, Suttle J, Morgan A G, et al. 1985. A comparison of sample preparation and calibration techniques for the estimation of nitrogen, oil and glucosinolate content of rapeseed by near infrared spectroscopy. J. Agric. Sci. Camb., 104: 317-323

Tam A C. 1986. Applications of photoacoustic sensing techniques. Rev. Mod. Phys., 58: 381-431

Viscarra R R A, McGlynn R N, McBratney A B. 2006. Determining the composition of mineral-organic mixes using UV-vis-NIR diffuse reflectance spectroscopy. Geoderma., 137: 70-82

Westerhuis J A, de Jong S, Smilde A K. 2001. Direct orthogonal signal correction. Chemom. Intell. Lab. Syst., 56: 13-25

Williams P C, Sobering D C. 1993. Comparison of commercial near infrared transmittance and reflectance instruments for analysis of whole grains and seeds. J. Near Infrared Spectrosc., 1: 25-32

Wold S, Antti H, Lindgren F, et al. 1998. Orthogonal signal correction of near-infrared spectra. Chemom. Intell. Lab. Syst., 44: 175-185

Wold S, Sjöström M, Eriksson L. 2001. PLS-regression: a basic tool of chemometrics. Chemom. Intell. Lab. Syst., 58: 109-130

Wu D, He Y, Nie P C, et al. 2010. Hybrid variable selection in visible and near-infrared spectral analysis for non-invasive quality determination of grape juice. Anal. Chim. Acta, 659: 229-237

Yang H, Irudayaraj J. 2001. Characterization of beef and pork using Fourier transform infrared photoacoustic spectroscopy. LWT-Food Sci. Technol., 34: 402-409

Zhu D Z, Ji B P, Meng C Y, et al. 2008. The application of direct orthogonal signal correction for linear and non-linear multivariate calibration. Chemom. Intell. Lab. Syst., 90: 108-115